13th International Conference on Vibration Measurements by Laser and Noncontact Techniques (AIVELA 2018)

Journal of Physics: Conference Series
Volume 1149

Ancona, Italy
20 - 22 June 2018

Editor:

Enrico Primo Tomasini

ISBN: 978-1-5108-8027-6
ISSN: 1742-6588

Printed from e-media with permission by:

Curran Associates, Inc.
57 Morehouse Lane
Red Hook, NY 12571

Some format issues inherent in the e-media version may also appear in this print version.

Additional copies of this publication are available from:

Curran Associates, Inc.
57 Morehouse Lane
Red Hook, NY 12571 USA
Phone: 845-758-0400
Fax: 845-758-2633
Email: curran@proceedings.com
Web: www.proceedings.com

Table of contents

Volume 1149

13th International Conference on Vibration Measurements by Laser and Noncontact Techniques

20–22 June 2018, Ancona, Italy

Accepted papers received: 13 November 2018
Published online: 24 December 2018

Preface

Papers

Modal Analysis, Structural/Material Identification

NDT, Fatigue Test and Quality Control

Optical Techniques for Microvibration Measurements

Metrology and Calibration

Biomedical Applications

Advances in Laser Interferometry

Full-field Techniques

IOP Conf. Series: Journal of Physics: Conf. Series **1149** (2018) 011001 doi:10.1088/1742-6596/1149/1/011001

PROCEEDINGS OF THE AIVELA INTERNATIONAL CONFERENCE ON VIBRATION MEASUREMENTS BY LASER AND NONCONTACT TECHNIQUES
ANCONA, 20- 22 JUNE 2018

Since October 1994 the **AIVELA Conference on Vibration Measurements by Laser Techniques** established itself as the international forum where research ideas and technological innovation on Laser Vibrometry were presented and exchanged. Ancona, an ancient and fascinating Italian town situated on the Adriatic coast, became the biennial destination of the international laser vibrometry community.
The event was initially focused on Vibration Measurements by Laser Techniques and it immediately met large consensus among the International Research Community. In 2010 the Scientific Committee decided to widen the topics of the Conference by also including other noncontact techniques and the name of the event was changed to **International Conference on Vibration Measurements by Laser and Noncontact Techniques.**

The Conference aims at creating an active and stimulating forum where current research results and technical advances can be exchanged and the development of new systems for laboratory use, field testing and industrial application can be promoted. For this reason, experts in vibration and acoustics, manufacturers and authorities in the field of optical and non- invasive instrumentation and industrial users of such measurement devices have come from all over the world to present their activities and innovative approaches to vibration measurements.

The 2018 edition of the event (the 13th edition) was organised by A.I.VE.LA. in collaboration with the **Department of Industrial Engineering and Mathematical Sciences of Università Politecnica delle Marche** under the support of **EAA, EOS and OSA** and will take place in Ancona on 20- 22 June 2018.
The Conference Programme featured three particularly interesting Invited Lectures, delivered respectively by Prof. David J. Ewins, Imperial College, London, and BLADE- Bristol University, UK, Prof. Roman Schnabel, Institute for Laser Physics, University of Hamburg and Dr Adam Styk, Warsaw University of Technology, Poland, followed by high scientific quality presentations dealing with different measurement techniques: Blade Tip Timing, Continuous Scanning LDV, Laser Vibrometry, Optical Sensors, Vision Based Techniques, to cite a few. As far as the Application Areas addressed are concerned, they go from Biomedical Applications to Metrology & Calibration, Modal Analysis, NDT, Quality and Process Control, Civil Structures Monitoring and Diagnostics, etc.

A **Short Course** was held the day before the Conference to provide participants with the fundamentals of Vibration Measurements through theoretical lectures in the morning held by Prof. Christian Rembe, Clausthal University of Technology, Germany, and Prof. Emanuele Zappa, Politecnico di Milano, Italy, and hands-on lab sessions in the afternoon organised by the Group of Mechanical and Thermal Measurements of Polytechnic University of Marches in collaboration with Polytec GmbH, GOM GmbH and Julight Srl.

Moreover, as for the past editions of the event, an **Exhibition** was organised in conjunction with the Conference, to offer leading manufacturers of laser vibrometer systems, non-contact instrumentation, systems for vibro-acoustic testing, etc., the opportunity to bring their products directly to the attention of a selected audience.

The detailed Programme, together with other interesting information on the event, may be found online at: http://www.aivela.org/13th_Conference/conference-programme.html

Organising the AIVELA event is always quite a challenge and my staff and I have been working at it with great passion and enthusiasm. But of course none of all this would have been possible without the valuable contribution of all the authors, speakers, lecturers, exhibitors, supporting organisations, sponsors, members of the Scientific Committee etc., whom I wish to thank for their valuable time and effort.

Trusting the AIVELA event will meet your interest, I hope to have the opportunity to meet you soon in Ancona.

EDITOR
Enrico Primo Tomasini
President of A.I.VE.LA.

IOP Conf. Series: Journal of Physics: Conf. Series **1149** (2018) 011002 doi:10.1088/1742-6596/1149/1/011002

Peer review statement

All papers published in this volume of *Journal of Physics: Conference Series* have been peer reviewed through processes administered by the proceedings Editors. Reviews were conducted by expert referees to the professional and scientific standards expected of a proceedings journal published by IOP Publishing.

IOP Conf. Series: Journal of Physics: Conf. Series **1149** (2018) 012001 doi:10.1088/1742-6596/1149/1/012001

Modal analysis of a beam structure excited by the ultrasound radiation force

A. Sabato*, S. Chen, P. Avitabile, and C. Niezrecki

Department of Mechanical Engineering, University of Massachusetts Lowell, 1 University Avenue, Lowell, MA 01854, USA

*Corresponding author e-mail: Alessandro_Sabato@uml.edu

Abstract. The use of conventional modal excitation techniques (e.g., impact hammers, piezoelectric actuators, and mechanical shakers) may be challenging if high-frequency dynamic measurements on lightweight structures need to be performed. The use of the high-frequency radiation force generated by focused ultrasonic transducers (FUTs) can help to resolve this problem. In this research, the acoustic pressure generated by a FUT is measured using an acoustic microphone and used for determining the force distribution in the focal plane of the FUT by employing a boundary element model (BEM). The results indicate that the radiation pressure field has a spot size of ~3 mm in diameter and the radiation force has a maximum intensity equal to 2.5 10^{-5} N. To validate the capability of the FUT as an excitation technique for modal analysis, a back-to-back comparison with a modal test performed using a traditional excitation system (i.e., shaker) was completed. Analysis of the results when the two excitation techniques are used, shows a good agreement between the data sets. Excellent correlation is also observed when the mode shapes are compared using the Modal Assurance Criterion (MAC). Overall, the good match of both resonant frequencies and mode shapes indicates that the focused ultrasound radiation force excitation technique is helpful in performing modal tests.

1. Introduction

The capability to perform accurate dynamic measurements is essential for a wide range of applications. Conventional modal testing excitation techniques, typically rely on impact hammers, piezoelectric actuators, or mechanical shakers excitation. However, the use of these devices may be challenging if accurate high-frequency dynamic measurements on small or lightweight structural parts have to be performed [1]. Moreover, the physical connection may lead to mass and stiffness loading effects; issues that distort structures' dynamic characteristics (e.g., natural frequencies, mode shapes, and damping) [2]. Therefore, providing mechanical excitation to small or lightweight structures without interfering with their dynamic characteristics to obtain information to estimate Frequency Response Functions (FRFs), natural frequencies, mode shapes, and damping, is still a challenge that needs to be addressed.

In recent years, non-contact excitation methods have been explored as potential approaches for exciting and detecting vibration on structures having a size ranging from the micro- to macro-scale. Researchers performed a considerable amount of work to understand the physics of the acoustic radiation pressure and force. In particular, the work made by Westervelt has to be recognized as one of the first and most influential on this topic [3]. For instance, direct acoustic excitation has been used as a non-contact method for operational modal analysis in the audio range up to ~40 kHz [4, 5]. Unfortunately, the sound generated by those transducers is not focused, and the resulting acoustic force is imparted over a distributed area, preventing accurate estimation of FRFs. By employing transducers emitting sounds at a higher frequency, the dimension of the area over which the excitation is applied can be reduced to a few millimeters, and the radiated sound may be used as an effective non-contact method for structures such as micro-cantilevers and small turbine blades.

The steady acoustic radiation force has been used in a variety of applications, including the measurement of sound intensity and power out of Ultrasonic Transducers (UTs) [6], acoustic manipulation of microparticles [7], and acoustic levitation [8]. In addition to that, the UT-generated acoustic pressure has been widely employed in non-destructive testing applications, and it is one of the most used techniques in the field of imaging for biomedical applications [9, 10]. Some efforts about the possibility of using ultrasound radiated force as structural excitation technique have also been made for measuring the frequencies and operating deflection shapes (ODSs) [11 – 13].

To perform experimental modal testing, the input force needs to be known or estimated to obtain the FRF (i.e., output/input response) of the system being tested and identify critical modal parameters. To date, the inability to quantify the acoustic radiation force prevents this approach from being used as a practical technique for experimental modal testing and motivates this research. In this study, the dynamic focused ultrasound pressure generated by focused ultrasonic transducers (FUT) is quantified both experimentally and analytically by using a boundary element method (BEM) model based on the Rayleigh Integral. Experiments are used for validating the performance of the BEM in the lower frequency range (i.e., 50 - 80 kHz). Once validated, the model is used for predicting the behavior of the FUT at higher frequencies (i.e., above 300 kHz) and the computed pressure fields are used to calculate the acoustic radiation force generated by the transducer. Finally, the capability of the FUT as an excitation technique for modal analysis is validated. The FUT is used for exciting a cantilever beam during a modal test, and results are compared with those recorded as the same structure is excited using a shaker.

2. Theoretical background

Commercially available microphones can measure acoustic pressure fields up to 100 kHz. Therefore, a model is the only possible way to determine the pressure field first, and the acoustic force radiated by a FUT operating at frequencies higher than 300 kHz.

The FUT used in this research is a single element amplitude modulated (AM) device that can be driven by a double sideband suppressed carrier (DSB-SC) AM signal to generate sinusoidal signals that can be used to generate a harmonically varying force [14]. In this configuration, as two ultrasound beams with frequencies f_1 and $f_2 = f_1 + \Delta f$ are superimposed, the constructive and destructive interferences between the beams produce a radiation force having a vibration frequency equal to the difference frequency $\Delta f = f_2 - f_1$. The total pressure field $p(\mathbf{r}, t)$, generated by the two frequency components in a point distant $\mathbf{r}$ from the FUT can be evaluated using Eq. (1):

$$p(\mathbf{r},t) = P(\mathbf{r})\cos[2\pi f_1 t + \varphi_1(\mathbf{r})] + P(\mathbf{r})\cos[2\pi f_2 t + \varphi_2(\mathbf{r})] \quad (1)$$

If the two waves have difference frequency Δf far less than both f_1 and f_2; p_1 and p_2 can be considered to be almost the same and equal to $P^2(\mathbf{r})$ with phases $\varphi_1(\mathbf{r})$ and $\varphi_2(\mathbf{r})$ respectively [3]. Therefore, under the assumption that the waves are planar and propagate with speed c in a fluid having density ρ, the ultrasound beams will have an energy density $E(\mathbf{r}, t)$ governed by Eq. (2):

$$E(\mathbf{r},t) = \frac{p^2(\mathrm{r},t)}{\rho c^2} = \frac{P^2(\mathrm{r})\cos[2\pi\Delta f t + \Delta\varphi(\mathrm{r})]}{\rho c^2} \quad (2)$$

The total radiation force in time-varying form $F_{\Delta f}(\mathbf{r},t)$ over the surface area of the structure impacted by the incident sound waves can be calculated using the Eq. (3) for message frequencies much smaller than the carrier frequencies (i.e., < 1%) [15]:

$$F_{\Delta f}(\boldsymbol{r},t) = \int_S E(\boldsymbol{r},t)\, \mathrm{d_r}(\boldsymbol{r})\, \mathrm{dS} = \mathrm{d_r}(\boldsymbol{r}) \int_S \frac{(p_1+p_2)^2}{\rho c^2}\, \mathrm{dS} =$$
$$= \mathrm{d_r}(\boldsymbol{r}) \int_S \frac{p^2(r)}{\rho c^2} \cos[2\pi \Delta f t + \Delta\varphi(\boldsymbol{r})]\, \mathrm{dS} \quad (3)$$

where $\boldsymbol{d_r}$ is the drag coefficient vector as defined in [3], *dS* the scattered area, and p_1 and p_2 the pressure of the two incident planar waves considered. It should be noted that the resulting acoustic radiation force is in proportion to $(p_1+p_2)^2$, which include a DC component, a summation component (f_1+f_2), a doubling of the frequency component ($2f_1$ or $2f_2$), and a difference frequency component (f_1-f_2). As a result of the interaction of a test object with the ultrasound radiation force, the structure will be vibrated at the difference frequency Δf. This quantity is relevant to this research, while the other three components will induce vibrations in the structure that are out of its range of interest or that have negligible magnitude. Therefore, by adjusting Δf, different structural excitation frequencies over the desired bandwidth are achievable as this technique is employed. It implies that as the transducer is operated with a modulation frequency much smaller than the carrier frequency, the generated wave can be considered a single-frequency plane wave and used as a steady-state source of excitation.

For a focal transducer, the sound pressure of a field point $p_i\,(x, y, z)$ produced by a source element S_i can be calculated using Eq. (4):

$$p_i\,(x,y,z) = \frac{j\omega\rho U\,(x',y',z')\, e^{-jkr}}{2\pi r}\, \mathrm{dS} \quad (4)$$

where r is the distance between the field point and the center of the source element, ω is the circular frequency of the vibration of the transducer surface, ρ is the fluid density, $U\,(x', y', z')$ is the amplitude of vibration velocity of the source element (which is going to be experimentally measured at its geometrical center), *dS* is the area of the *i-th* source element, and k is the wave number. In this study, with reference to Figure 1, and to take into account the curvature of the transducer, the geometrical terms of Eq. (4) can be calculated as:

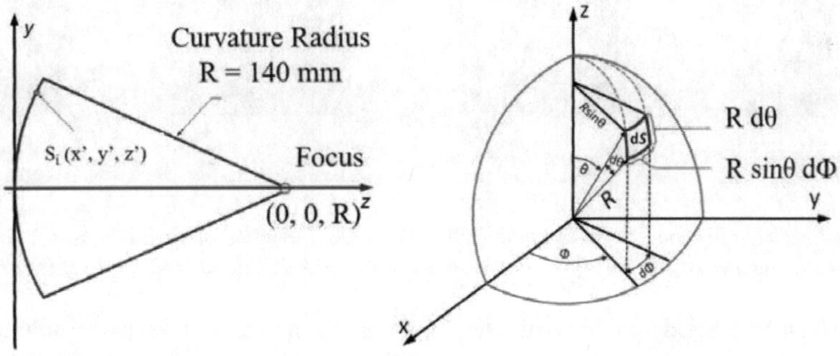

(a) (b)

Figure 1 – Diagram of the coordinate system and BEM used to simulate the acoustic field generated by spherically curved ultrasonic transducers. (a) Section of the curved transducer, (b) an individual source element dS that represents a small area of the transducer.

$$\mathrm{dS} = R\sin(\theta)\mathrm{d\Phi} \cdot R\mathrm{d\theta} \quad (5)$$

$$R = \sqrt{(x-x')^2 + (y-y')^2 + (z-z')^2} \quad (6)$$

where R is the curvature radius of the FUT and the terms x', y', z' in Eq. (6) can be calculated like $x' = R\sin(\vartheta)\cos(\Phi)$, $y' = R\sin(\vartheta)\sin(\Phi)$, and $z' = R - R\cos(\Phi)$ respectively. The pressure calculation method shown in Eq. (4) is derived for simulating pressure field of a plane piston. However, it can be considered accurate for slightly curved transducers (i.e., almost plane, which is the case analyzed in this study) as well. The overall pressure at a field point is the superposition of pressure contributed from all of the *N* source elements and can be formulated using Eq. (7):

$$P\,(x, y, z) = \sum_{i=1}^{N} p_i\,(x, y, z) \tag{7}$$

with N = 15000 in the case study here described. Eq. (7) is used to generate an acoustic radiation pressure profile for the FUT over numerous points representing an area of interest. The accuracy of the simulation is validated comparing the numerical results with those obtained as the radiated pressure is experimentally measured using an acoustic microphone in the lower frequency range and will help to understand the FUT's radiation characteristics (e.g., pressure, directivity, and spot size) in the higher frequency range.

3. Pressure field characterization

The BEM described in the previous paragraph has been validated through a set of laboratory experiments. In this research, a noncontact circular FUT (model NCG500-D50-P150 from the Ultran Group), with an actual focal length of ~140 mm, has been used. It has a nominal diameter of 50 mm and a nominal operating frequency of 500 kHz [16].

By using a scanning laser Doppler vibrometer (SLDV) (model PSV-400 manufactured by Polytec, Inc.), the velocity profiles *U (x', y', z')* of the transducer surface were measured at discrete frequencies (i.e., 50, 60, 70, and 80 kHz) and entered in Eq. (7) to obtain the pressure distribution of a field point. As shown in Figure 2, a pressure field of 100 × 100 × 450 mm with a spatial resolution of 5 mm was simulated in this research.

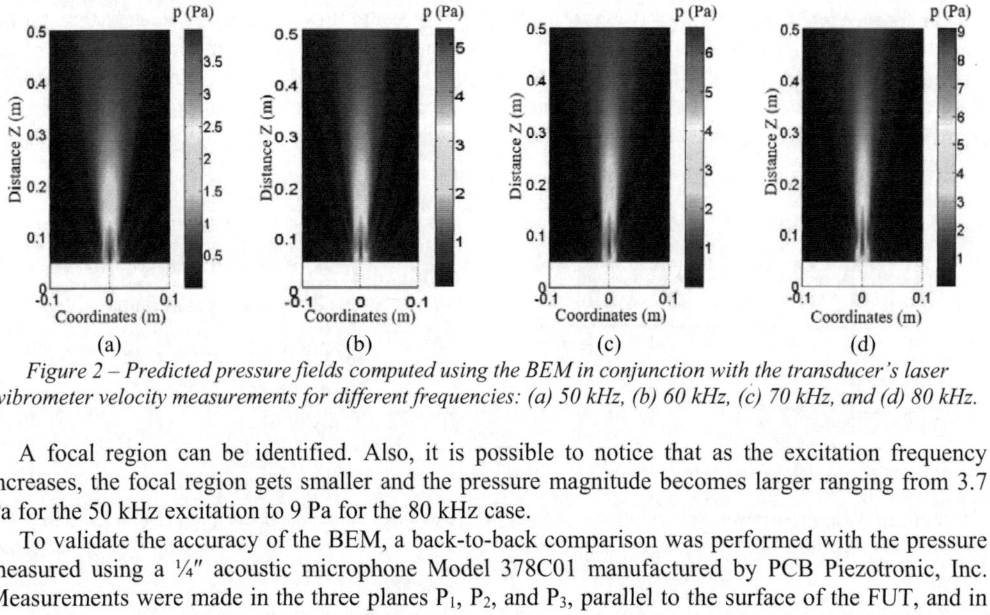

Figure 2 – Predicted pressure fields computed using the BEM in conjunction with the transducer's laser vibrometer velocity measurements for different frequencies: (a) 50 kHz, (b) 60 kHz, (c) 70 kHz, and (d) 80 kHz.

A focal region can be identified. Also, it is possible to notice that as the excitation frequency increases, the focal region gets smaller and the pressure magnitude becomes larger ranging from 3.7 Pa for the 50 kHz excitation to 9 Pa for the 80 kHz case.

To validate the accuracy of the BEM, a back-to-back comparison was performed with the pressure measured using a ¼″ acoustic microphone Model 378C01 manufactured by PCB Piezotronic, Inc. Measurements were made in the three planes P_1, P_2, and P_3, parallel to the surface of the FUT, and in one horizontal plane P_4, perpendicular to the surface of the FUT. P_1, P_2, and P_3 were 140, 150, and 160 mm away from the FUT surface, while plane P_4 extended in a region spaced 140 and 160 mm from the FUT. Figure 3 shows an example of the pressure fields measured using the microphone and those simulated using the BEM. For the sake of brevity, only results in plane P_1 and P_4 for a driven frequency of 50 kHz are shown here. Experimental data and numerical simulations are in excellent agreement. The comparison indicates that the pressure values are close and the pressure profiles are similar.

IOP Conf. Series: Journal of Physics: Conf. Series **1149** (2018) 012001 doi:10.1088/1742-6596/1149/1/012001

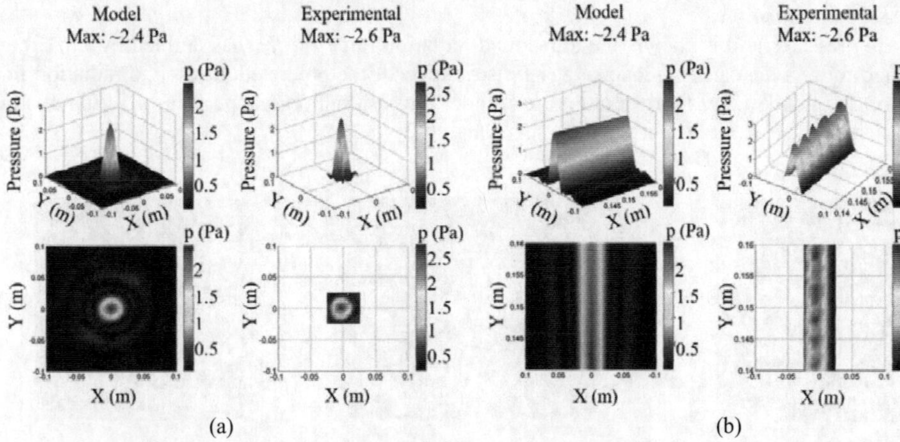

Figure 3 – Comparison of model and experimental ultrasound pressure when the FUT is excited at ~25 V_{rms} and 50 kHz for two measurement plane slices: (a) P_1 = 140 mm, (b) P_4 = 140 - 160 mm.

The simulation results seem to underestimate the actual value of the maximum acoustic pressure value, but the average error for the four planes considered is equal to 0.15 Pa and can be considered relatively small.

3.1. *FUT pressure field simulation in the higher frequency range*

With the availability of the validated acoustic BEM model, the ultrasound pressure field in the higher frequency range (i.e., the carrier frequency range of the FUT) can be predicted. Vibrational FRF testing using an SLDV was conducted to determine the velocity profile U of the FUT in the 300 - 400 kHz frequency range. The full ultrasound pressure field at three different frequencies (i.e., 325, 362.5, and 400 kHz) for the plane P_1 is shown in Figure 4.

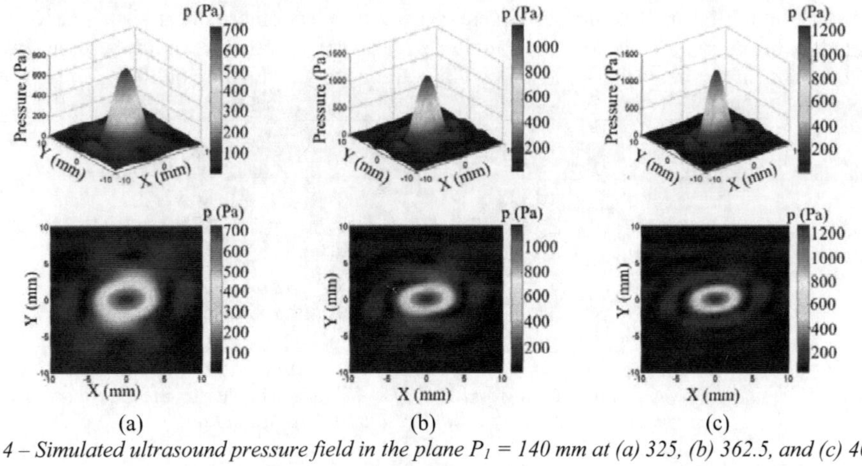

Figure 4 – Simulated ultrasound pressure field in the plane P_1 = 140 mm at (a) 325, (b) 362.5, and (c) 400 kHz.

Compared with the results at lower frequencies, the pressure value at carrier frequency level is approximately three orders of magnitude higher. The maximum value of the ultrasound pressure increases as frequency increases reaching peaks higher than 1200 Pa as the 400 kHz excitation is considered. Moreover, it is possible to observe that a highly focused acoustic spot with a diameter of ~3 mm is produced.

IOP Conf. Series: Journal of Physics: Conf. Series **1149** (2018) 012001 doi:10.1088/1742-6596/1149/1/012001

3.2. Force quantification

Once the pressure field is known, the ultrasound radiation force $F_{\Delta f}(\mathbf{r}, t)$ as defined by Eq. (3), can be estimated for a given carrier frequency. The distribution of the predicted ultrasound radiation force for 325, 362.5, and 400 kHz is shown in Figure 5 under the assumption of a perfect reflection of sound waves by test article (i.e., drag coefficient $\mathbf{d_r}$ sets equal to 2).

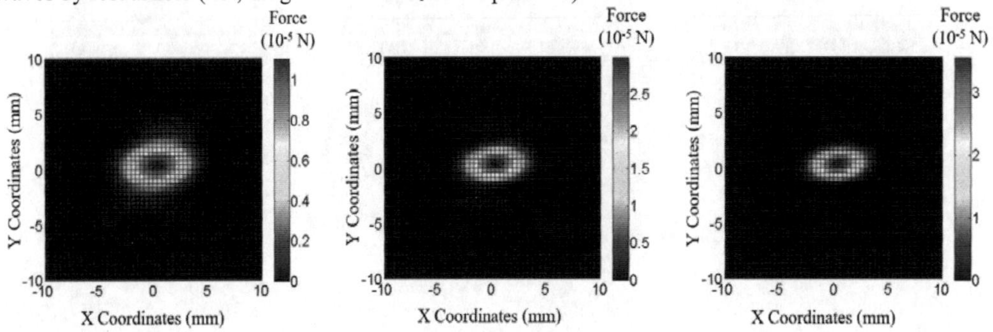

(a) (b) (c)

Figure 5 – Quantification of ultrasound radiation force $\mathbf{F}_{\Delta f}(\mathbf{r}, t)$ acting on a test article140 mm away from the FUT for three frequencies: (a) 325 kHz, (b) 362.5 kHz, and (c) 400 kHz.

It can be found that the majority of ultrasound radiation force is confined within a range of ~3 mm in diameter around the geometrical center of the FUT and that it has maximum values of ~$1.3\text{x}10^{-5}$ N, $2.7\text{x}10^{-5}$ N, and $3.5\text{x}10^{-5}$ N for the three frequencies considered.

4. Experimental modal analysis

To validate the capability of the FUT as an excitation technique for modal analysis, a back-to-back comparison with a modal test performed using a traditional excitation system (i.e., shaker) was completed. The structural response of the upright structure shown in Figure 6 was measured using a Polytec SLDV as the beam structure was excited using an electromechanical shaker driven with a pseudo random signal from 0 Hz to 4 kHz applied at the center of the mounting base, and with the FUT set to have a carrier frequency of 359 kHz with a frequency modulation of 20 Hz.

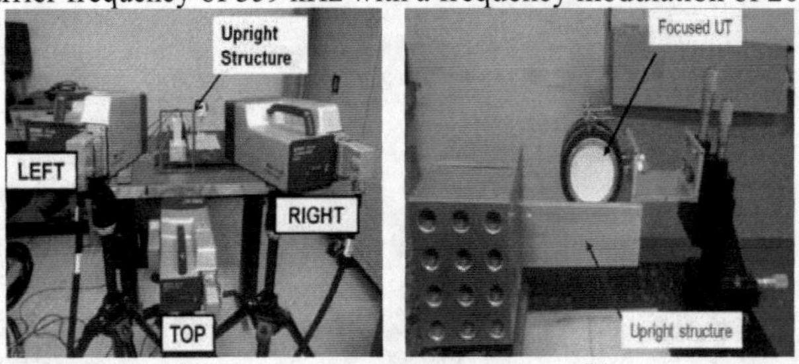

(a) (b)

Figure 6 – Experimental setup of modal testing and SLDV for displacement measurements: (a) using shaker excitation; (b) using ultrasound radiation force excitation.

The results of the modal test performed as the shaker excitation is used are shown in Figure 7 and Table 1, where the values of the first four natural frequencies for the test object are summarized. In this test, 45 scan points were used as measurement pattern by the SLDV set-up in a three-dimensional configuration. Using the modal information, the dynamic characteristics of the target structure can be determined. It can be seen that the first flexible mode of the upright plate structure is its first bending mode at ~371 Hz, the second flexible mode is its first torsional mode at ~1585 Hz, the third flexible mode is its second bending mode at ~1966 Hz, and the fourth flexible mode is its first in-plane bending mode at ~2499 Hz. In this study, the rigid body modes in low-frequency range are neglected and only the flexible modes are assessed.

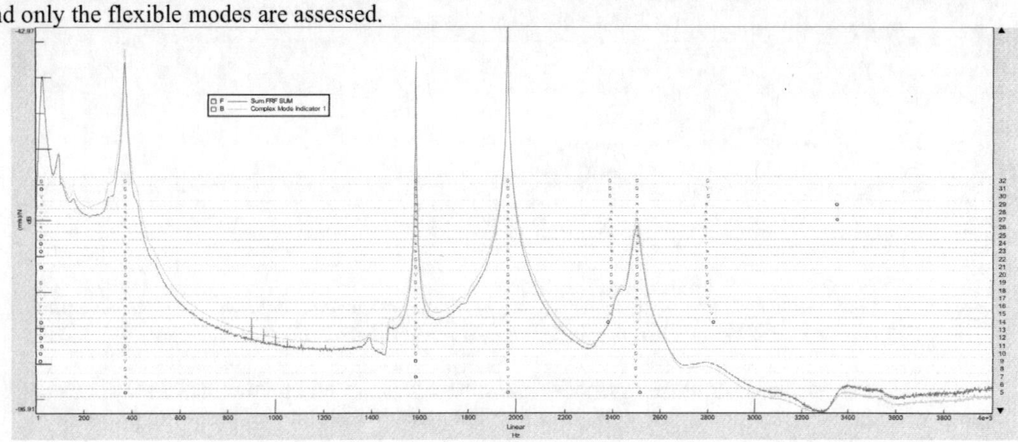

Figure 7 – Stabilization diagram obtained as the upright structure is excited using the shaker.

Table 1: Natural frequencies determined from modal analysis using the shaker.

Mode number	Mode name	Frequency
(-)	(-)	(Hz)
1	1st bending	371.4
2	1st torsional	1584.5
3	2nd bending	1965.8
4	1st in-plane bending	2498.9

After that, the FUT was used to excite the first three flexible modes of the structure, the FRFs were measured, and the modal parameters were extracted. In this testing, a DSB-SC AM signal with a carrier frequency of 359 kHz with a random variation of 10 cycles at 10 kHz was used as an excitation signal. To avoid damage to the FUT, the bandwidth was limited to 20 Hz to prevent it from getting hot. Due to the defocusing effect, only the first three flexible modes were tested. Therefore, based on a previously developed finite element model of the structure, the excitation frequencies were defined as 360-380 Hz, 1570-1590 Hz, and 1970-1990 Hz. Two-dimensional vibration measurements were made using the SLDV with a frequency resolution of 1 Hz, and only 21 points were scanned to limit the usage cycle causing heating of the transducer. The identified modes and natural frequencies are shown in Table 2.

Table 2: Natural frequencies determined from modal analysis using the FUT.

Mode number	Mode name	Frequency
(-)	(-)	(Hz)
1	1st bending	371.1
2	1st twisting	1579.3
3	2nd bending	1980.6

It can be seen that first mode is the first bending mode at 371.1 Hz, the second mode is the first twisting mode at 1579.3 Hz, and the third mode is the second bending mode at 1980.6 Hz.

Excellent correlation is also observed when the mode shapes are compared using the Modal Assurance Criterion (MAC). From data reported in Figure 8, it is possible to observe that MAC values for the first three flexible modes, are equal to 98.9%, 94.9%, and 84.2% respectively. The good MAC values imply a good correlation of mode shapes between the shaker testing and ultrasound radiation force excitation testing. The correlated mode shapes for the first bending mode, first torsional mode, and second bending mode are also presented in Figure 8b.

MAC (%)		Test Mode - Shaker		
		1	2	3
Test Mode - Ultrasound	1	98.9	0.2	7.1
	2	0	94.9	.0.0
	3	0.1	0	84.2

(a)

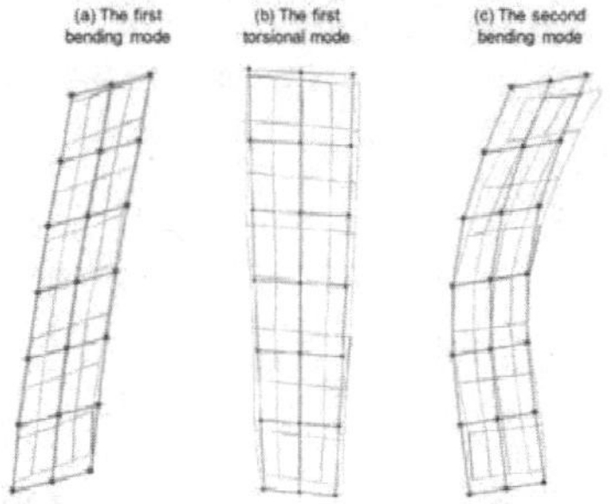

(b)

Figure 8 – Analysis of the results: (a) MAC values; (b) mode shapes correlation between the shaker testing and ultrasound radiation force excitation.

Overall, the excellent correlation of both resonant frequencies and mode shapes indicates that the focused ultrasound radiation force excitation technique is helpful in performing modal tests. Differences in the MAC at the higher frequency modes may be attributed to the coupling and influence that the attached shaker has on the structure.

5. Conclusions

In this study, the acoustic pressure field and the acoustic force generated by the focused ultrasonic transducer are determined and simulated using a boundary element model for carrier frequencies higher than 300 kHz. The results indicate that the radiation pressure field has a spot size of ~3 mm in diameter and the radiation force has a maximum intensity equal to 2.5 10^{-5} N.

The effectiveness of the dynamic ultrasound radiation force as an excitation technique was analysed. Two different modal analysis methods, shaker excitation, and dynamic ultrasound radiation force excitation are presented to perform modal analysis on a target beam structure. Structural dynamic characteristics including resonant frequencies and mode shapes are compared and correlated. The percent difference in resonant frequencies is less than 2%, while MAC values are reasonably high. The results demonstrate that both the resonant frequencies and mode shapes are in good agreement. Overall, the excellent correlation indicates that the dynamic focused ultrasound radiation force excitation is a valid and useful non-contact excitation option in performing modal tests for very small or lightweight structures.

Acknowledgement

This work was funded by the NSF Civil, Mechanical and Manufacturing Innovation (CMMI) Grant No. 1266019 entitled "*Collaborative Research: Enabling Non-contact Structural Dynamic Identification with Focused Ultrasound Radiation Force*". Any opinions, findings, and conclusions or recommendations expressed in this material are those of the authors and do not necessarily reflect the views of the particular funding agency.

References

[1] P. Avitabile, Experimental modal analysis, Sound and vibration, 35(1), (2001) 20-31.
[2] D. Cloutier, P. Avitabile, R. Bono, R., M. Peres, Shaker/stinger effects on measured frequency response functions, (2009), In Proceedings of the XXVII International Modal Analysis Conference, 9-12 February, Orlando, FL.
[3] P.J. Westervelt, The theory of steady forces caused by sound waves, The Journal of the Acoustical Society of America, 23(3) (1951) 312-315.
[4] Y.F. Xu, W.D. Zhu, Operational modal analysis of a rectangular plate using noncontact acoustic excitation, in: E. B. Zimmerman (Ed.), Rotating Machinery, Structural Health Monitoring, Shock and Vibration, Volume 5, Springer, New York, 2011, pp. 359-374.

[5] Y.F. Xu, W.D. Zhu, Operational modal analysis of a rectangular plate using non-contact excitation and measurement, Journal of Sound and Vibration, 332(20) (2013) 4927-4939.
[6] Y. Zhou, Acoustic power measurement of high-intensity focused ultrasound transducer using a pressure sensor, Medical engineering & physics, 37(3), (2015) 335-340.
[7] P.L. Marston, Shape oscillation and static deformation of drops and bubbles driven by modulated radiation stresses—Theory, The Journal of the Acoustical Society of America, 67(1), (1980) 15-26.
[8] C. Chen, J. Wang, B. Jia, F. Li, Design of a noncontact spherical bearing based on near-field acoustic levitation, Journal of intelligent material systems and structures, 25(6) (2014) 755-767.
[9] K.Y. Jhang, Nonlinear ultrasonic techniques for nondestructive assessment of micro damage in material: a review, International journal of precision engineering and manufacturing, 10(1) (2009) 123-135.
[10] K. Nightingale, K, Acoustic radiation force impulse (ARFI) imaging: a review. Current medical imaging reviews, 7(4) (2011) 328-339.
[11] T.M. Huber, M. Fatemi, R. Kinnick, J.F. Greenleaf, Noncontact modal analysis of a pipe organ reed using airborne ultrasound stimulated vibrometry, The Journal of the Acoustical Society of America, 119(4) (2006) 2476-2482.
[12] T.M. Huber, D. Calhoun, M. Fatemi, R.R. Kinnick, J.F. Greenleaf, J. F. Noncontact modal testing of hard-drive suspensions using ultrasound radiation force. Journal of the Acoustical Society of America, 118(3) (2005) 1928-1928.
[13] T.M. Huber, B.C. Abell, D.C. Mellema, M. Spletzer, A. Raman, A. Mode-selective noncontact excitation of microcantilevers and microcantilever arrays in air using the ultrasound radiation force, Applied physics letters, 97(21) (2010) 214101.
[14] M. Fatemi, J.F. Greenleaf, Ultrasound-stimulated vibro-acoustic spectrography, Science, 280(5360) (1998) 82-85.
[15] G.T. Silva, S. Chen, and L. P. Viana, Parametric amplification of the dynamic radiation force of acoustic waves in fluids. Physical review letters, 96(23) (2006) 234301.
[16] Non-contact ultrasound (NCU), Phenomenally high-efficiency transducers – 50kHz to >50MHz, Available online: http://www.ultrangroup.com/Transducers/Catalogs/NCU.pdf. (accessed 31.07.17).

IOP Conf. Series: Journal of Physics: Conf. Series **1149** (2018) 012002 doi:10.1088/1742-6596/1149/1/012002

Characterization of Damping Properties in 3D Printed Structures

Jenna Gietl[1], Joseph Vignola[1], Ph.D, John Sterling[1], Teresa Ryan[2], Ph.D

[1]Center for Acoustics, Vibrations and Structures, Dept. of Mechanical Engineering, The Catholic University of America, Washington, DC
[2]East Carolina University, Department of Engineering
Greenville, NC 27858

E-mail: 02gietl@cua.edu

Abstract.
Current research on the effectiveness of a subordinate oscillator array (SOA) as a broadband mechanical filter relies on adequate knowledge of the SOAs material properties. Recent studies have shown a high sensitivity of these structures to disorder. A desire to produce large numbers of arrays to test this sensitivity to disorder motivated a transition from metal to 3-dimensional printed plastic SOAs. Irregularities associated with the curing process of the 3D printed polymers, as well as a general inconsistency of material properties of plastics, in turn highlighted the need for characterization of properties of 3D printed materials, especially those properties related to damping. As part of this study, several 3D printed plastic cantilevers, varying in material, printing technique, and printing orientation, are measured. Quantities of interest include the Young's modulus and density, as well as phenomenological properties, like the quality factor of specific designs. An ASTM standard test method for property determination is implemented with a laser Doppler vibrometer (LDV) to test each polymer. In addition to the ASTM protocol, tests are conducted in vacuum to distinguish internal damping mechanisms of the cantilever from external fluid mechanisms. Results are compared to both analytic and numeric predictions and published theory.

1. Introduction

Three-dimensional printing is increasing in popularity as new economical and efficient technologies are invented. Often engineers and architects utilize 3D printing because of its combination of high resolution and rapid prototyping. Beyond the scientific world, 3D printing has found its use in jewelry molds, custom manufacturing fixtures, and prop making [1]. As a relatively new technology, however, a thorough understanding of the properties of the materials produced by 3D printing is still emerging.

The authors have conducted a number of studies to understand and quantify different aspects of acoustic and vibrational damping [2, 3]. More recently, the apparent damping effect publicized by Soize [4] and later observed by Strasberg and Feit [5] is being investigated at CUA through the design of subordinate structures that, once attached to a primary structure, transfer energy[6, 7]. These subordinate structures are designed such that once the energy is transferred downscale away from the primary structure, it is dissipated in the subordinate array, essentially trapping the energy from the primary structure. The design of the subordinate array structure relies on thorough knowledge of its material properties, specifically the Young's modulus and density, and a way to predict mode frequencies and damping for the system.

The literature on material properties, particularly those related to damping, of most 3D-printed polymers is largely inconsistent or lacking altogether. Therefore, further investigation of the effectiveness of 3D-printed mechanical filters requires quantification of the internal friction of the material. The purpose of this work is to use a standard testing protocol to determine material properties, especially those related to damping, of varying 3D-printed polymers. Specifically, modulus of elasticity, density, and properties related to damping will be addressed in this work. By measuring the resonance frequency and the density, the Young's modulus can be determined. The quality factor,Q, which is a measure of damping, can be calculated using the half-power bandwidth method.

In 3D printing, particularly stereolithography (SL), there are a number of fabrication variables both during and after material production that may contribute to inconsistencies in the final product. For example, printing technique, sample orientation during printing, printing temperature, post-printing curing time and wavelength, and post-printing wash time can all affect the material properties of a sample. Formlabs, a company that produces the stereolithograph apparatus (SLA) and the SLA used for this work, has investigated the effect of some of these variables, including post-curing UV wavelength [1] and geometric isotropy [8]. Each engineering resin produced by Formlabs was found to harden appropriately at a defined UV wavelength. Formlabs also claims that printing orientation does not affect material properties, particularly the Young's Modulus and tensile strength. In other words, unlike the anisotropy of the fused deposition modeling (FDM) technique, material properties are not direction-dependent.

2. Background

This study is motivated by previous and current research on damping by the authors. More recently, examination of the apparent damping effect[7] combined with piezoelectric control [9] has highlighted the need for characterization of damping of 3D printed materials.

Apparent damping is the dissipation of energy observed in a primary or master structure caused by subordinate attachments that draw energy away from the primary. This effect is seen in a tuned-mass-damper, in which a small mass is elastically attached to a larger primary mass and, if designed to have nominally the same resonant frequency as the primary, causes an effect similar to an electrical notch filter. Then, if several subordinate elements (rather than one) are attached to a primary, each designed to have an isolated natural frequency within a range of frequencies surrounding the resonant frequency of the primary, the apparent dissipation of energy from the primary is broadened across the range. The effect of attaching these subordinate elements for a specified mass ratio is analogous to applying an electrical band-rejection filter [6]. The subordinate oscillators are referred to here as a subordinate oscillator array, or SOA.

Counteracting the disorder sensitivity of these structures can be done by refining the design and production of each part. Design parameters that cause disorder are not discussed in this work, but production sensitivity is evaluated for various printing techniques. Using stereolithography, one of the many additive manufacturing/3D printing technologies available, small parts can be fabricated within a tolerance of one-hundredth of a millimeter, which is sufficient to counter disorder sensitivity. Stereolithography (SL) employs photopolymerization of liquid resins to build a solid polymer layer-by-layer. Once complete, ultraviolet curing fully hardens the structure [10]. This kind of additive manufacturing with curing produces a plastic with unique properties that vary with machine type, curing time and wavelength, and resin choice. Although material properties are provided for each resin, these fabrication variables result in large inconsistencies in the properties. Results from preliminary 3D printed SOA samples manufactured as print protocols were being developed showed up to 40% variation in

Young's modulus and quality factor from design to fabrication indicating the need for strict processing control and the possibility for sweeping variation when such processing (such as cure parameters) are not.

2.1. Samples

For this study, stereolithographic (SL) and fused deposition modeling (FDM) printing were the print methods used to produce samples. The primary difference between SL and FDM printing lies in the material state during printing. Stereolithography is a vat photopolymerization technique that uses UV light to partially cure material as each layer is laid down. Once the sample is fully printed, the material is in a green, or uncured, state and requires further UV curing to fully harden [10]. By contrast, FDM printing produces a part in its fully hardened state. Each of these methods differs with elements like strength of material, building volume, printing time, and material availability. One specific difference this work explores is isotropy of the printed material.

A secondary question addressed in this work is the effect of SL printing orientation on these properties. The material isotropy of stereolithography as discussed earlier is different than most 3D printing techniques. The FDM prints are anisotropic, or directionally dependent, because they are printed layer-by-layer without a post-print hardening process. With this technique, the bonds between layers are not as strong as those along a plane. However, because the layers of SL resins are not fully bonded before being cured, the bonds between layers harden at the same time as those along the plane. This technique produces an isotropic material. Therefore, the material properties are not expected to vary with respect to printing orientation. In the study of isotropy by Formlabs for their Form 2 printer, several specimens with an ASTM standard geometry were printed at orientation angles of 15 degree increments with respect to the xy-plane. Standard tensile tests indicated low error in Young's Modulus and yield strength between each sample [8]. The study by Formlabs concluded that strength of material is independent of printing orientation. The present work will compare results for vibrations samples and expand the study to determine whether damping is affected by build angle. The Young's modulus as well as resonance and damping properties will be compared for various printing orientations of SL samples.

2.2. Property Determination

One purpose of this work is to compare measured properties of seven 3D printed materials to the values provided by the manufacturer. Specifically, material properties, like the Young's Modulus and density, and phenomenological properties, such as the natural frequency and damping, of each sample will be determined experimentally. Each sample is designed through a finite-element, multiphysics simulation software, COMSOL, in which the dimensions and material properties are inputs. Combining our design with manufacturer-given material properties, this software predicts the eigenvalues, or bending modes, for each simple cantilever sample. These values are then used to assess the natural frequency of the part. As phenomena like the natural frequency and quality factor are dependent on these material properties, a comparison will be made between the given properties and predicted physical phenomena to measured values.

The density of each sample can be determined by the mass-volume ratio. Mass is measured on a scale, and the dimensions of the cantilever are measured with a digital caliper to calculate volume. The properties as designed are compared to those as printed in Table 1. These values indicate the accuracy of the printers by the percent difference, and the precision by the standard deviation.

The Young's modulus is often determined using well-known mechanical tests, such as the uniaxial tensile test or a three-point bending test. In such tests, Hooke's law is used to extract the modulus of elasticity in the elastic region on a stress-strain curve. For plastics, the elastic region is very small; therefore, in this work, the Young's modulus will be determined by measuring the natural frequency. The relationship between natural frequency, density, and Young's modulus is shown in Eq. 1, where ω_n is the measured natural frequency of the cantilever, I_z is the mass moment of inertia, ρ is the density of the material, L and t are the length and thickness of the cantilever, $k_n L$ is the spatial frequency or wave number, and A is the cross-sectional area.

$$\omega_n = \frac{1}{L^2}\sqrt{\frac{EI_z}{\rho A}}(k_n L)^2 \tag{1}$$

Solving Equation 1 for the Young's modulus for a cantilever beam yields

$$E = \frac{1}{(k_n L)^2}\frac{12\rho L^4}{t^2}(\omega_n)^2 \tag{2}$$

Phenomenological properties, particularly damping, are the most difficult to predict because of the contributions of external factors. The quality factor is a measure of all damping of the material, both internal damping and external mechanisms, so damping itself is not a material property. However, by reducing damping to primarily internal mechanisms, a property-like attribute for a defined geometry and material can be determined. Damping can be quantified in various ways, so the quality factor will be used as the measure of damping for this thesis. A value representing damping can be extracted by measuring the frequency response of a beam and using the half-power bandwidth method for the first bending mode. Judge *et al.* define the quality factor of a damped material to be proportional to the ratio of the mechanical energy stored in the device to the energy shed per cycle of oscillation[2]. The shed energy in this case comes from a variety of internal and external mechanisms that draw energy from the resonator. External damping mechanisms include energy loss by attachment and a fluid environment. Attachment loss depends on the material, mass, and geometry of the affixed object and can be minimized by attaching a resonator with no common modes of energy. Fluid loss is present in different forms, depending on the fluid and geometry of the resonator. Fluid damping can be eliminated by placing the object in a vacuum.

Three dominant fluid mechanisms presented by Judge *et al.* include viscous damping, acoustic radiation loss, and squeeze film damping. Viscous loss, proportional to the square root of the fluid's pressure as seen in Eq. 3, dominates in denser fluids or in small enclosed environments. Acoustic loss, shown in Eq. 4 to be proportional to the fluid's pressure, also depends on fluid sound speed, resonator geometry, and the wavelengths of each. As its name suggests, squeeze film damping is significant for very narrow gaps between objects[11].

$$Q_{\text{viscous}} = \frac{t\rho_s}{3}\sqrt{\frac{\omega_n}{2\eta\rho}} \tag{3}$$

$$Q_{\text{acoustic}} = \frac{t\rho_s\omega_n}{2\sigma_r\rho c} \tag{4}$$

These external mechanisms in conjunction with internal damping form the total quality factor of the resonator. To isolate internal friction from external damping, each of these mechanisms is targeted in turn by different experimental approaches. Attachment loss is minimized by clamping each cantilever tightly to create a high impedance boundary. By placing the cantilever in a vacuum chamber with pressure on the order of milliTorr, fluid loss is reduced until it is insignificant.

IOP Conf. Series: Journal of Physics: Conf. Series **1149** (2018) 012002 doi:10.1088/1742-6596/1149/1/012002

3. Experimental Methods

3.1. Density

To determine the density of each material, the total mass of each sample is measured on an analytical balance. Using a digital caliper, the length, width, and thickness of both the cantilever beam and the head that holds the bolt hole are measured. The basic shape of the head and the cantilever is seen in Fig. 1. The volume of each sample is then extracted from these measured dimensions of the geometry; the sum of the volume of the cantilever and that of the head minus the bolt hole yield the total volume. Then, the ratio of mass to volume is compared to given values of density by the manufacturer.

3.2. Young's Modulus and Quality Factor

The Young's modulus and quality factor are determined from the same experiment, as both are extracted from measured phenomenological properties. The measured resonant frequency and its half-power bandwidth are used to calculate these two properties. By exciting each cantilever with energy in targeted frequency bands and measuring its response velocity with a laser Doppler vibrometer (LDV), damping of these 3D printed materials is characterized. Measuring 100 averages of the cantilever response reduces the signal to noise ratio and distinguishes the resonant frequency peak in the frequency domain. Then in MATLAB, a non-linear least-squares algorithm is used to fit the data to a Lorentzian function. This yields the center frequency, half-power bandwidth, and the quality factor, Q. An example of this curve fit is shown for an SL sample in Fig. 2.

3.3. ASTM Standard Requirements

ASTM Standard E756-05, the Standard Test Method for Measuring Vibration-Damping Properties of Materials, is implemented in this study [12]. Parameters regarding the shape, attachment, and excitation force of the samples are defined in this standard. The standard designates the use of a cantilever beam with a thickness to length ratio of, at minimum, 1:10. Two transducers are used in the experiment: one to provide a non-contact excitation force and one to measure (also without contact) the response. The non-contact force transducer in this experiment is an electrostatic force that is produced by sending a large DC offset voltage across

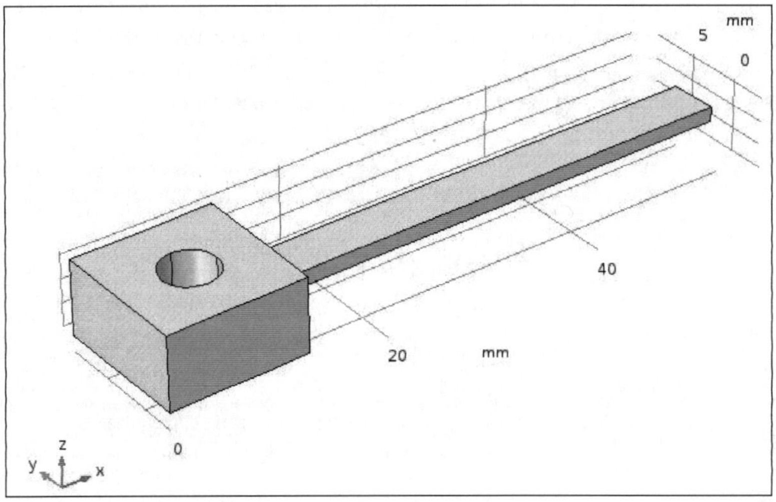

Figure 1: COMSOL is used to design the cantilevers with an attachment hole.

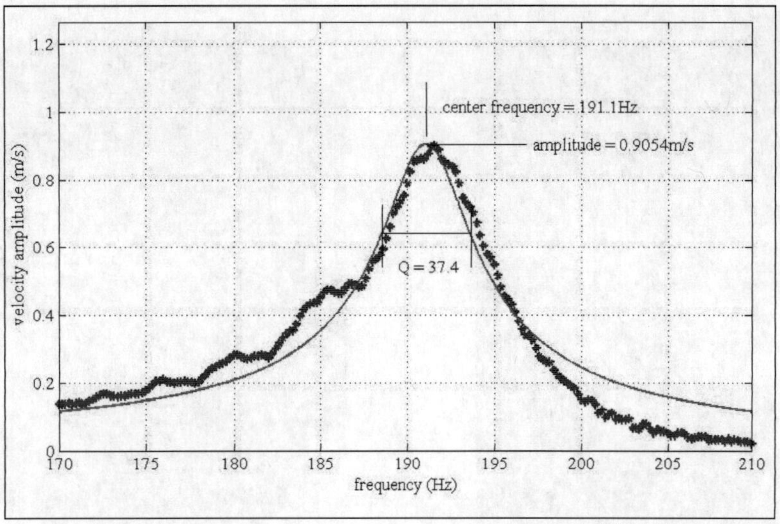

Figure 2: This figure shows the response of a gold-coated stereolithography cantilever to a 120-200 Hz chirp excitation.

a small gap between two conducting plates. One of the plates in this case is the cantilever, and the other is a thin plate of aluminum. For materials that are non-conductive or nonferrous, the standard suggests attaching "small bits of magnetic material...to the base beam side of the specimen to achieve specimen excitation and measurable response" [?]. To that end, sputter deposition was used to ionically bond a 35 nm layer of gold to the underside of the cantilever. As the aluminum sample is conductive, no gold was added.

3.4. Electrostatic force - F_s

An electrostatic force, F_s, is applied to the beam by grounding a plate that is positioned a very small distance below the beam. The cantilevers, which have been mutualized with a very thin layer of gold on the surface facing the grounded plate, are bolted to a high impedance mass. A voltage signal that has both an AC and DC component is then applied to the plate to produce the dynamic force in accordance with

$$F_S = \frac{\mathrm{A} \cdot \epsilon_o}{2} \frac{(V_{ac} + V_{dc})^2}{r^2} \tag{5}$$

If only the static portion is considered, the remaining expression is

$$F_S = \mathrm{area} \cdot \epsilon_o \frac{V_{ac} V_{dc}}{r^2} \tag{6}$$

where the permittivity of free space is $\epsilon_o = 8.854 \times 10^{-12} \frac{C^2}{Nm^2}$, r is the gap between the beam and electrode plate, and V is the total voltage [13].

3.5. Experimental Apparatus

As the focus in this experiment is to determine damping properties of each specimen, external fluid damping mechanisms will be reduced or eliminated by testing each sample within a vacuum chamber with pressure under 1 Torr. The sample, attached to a high impedance mass to reduce attachment loss, is bolted so that the majority of the cantilever hovers less than a millimeter

IOP Conf. Series: Journal of Physics: Conf. Series **1149** (2018) 012002 doi:10.1088/1742-6596/1149/1/012002

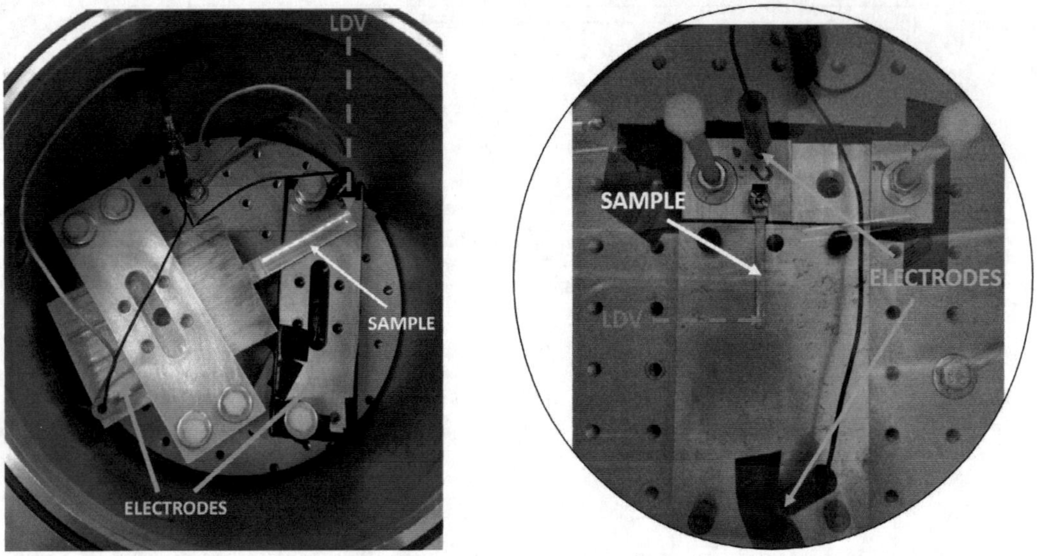

Figure 3: Bird's eye view of the vacuum chamber in which an aluminum (on the left) and then 3D printed (on the right) cantilever is attached to a high impedance mass and excited electrostatically.

above the conducting plate. Nylon non-conducting bolts secure the sample to the base plate of the vacuum chamber, and electrical tape separates the mass holding the cantilever from the base. The sample and high impedance mass act as a positive electrode, while the conducting plate and chamber are the ground component. Wires attached to each electrode connect the samples within the chamber to an external AC/DC amplifier. The electrostatic force that excites the sample is produced by a generated 1 V signal amplified with a gain of 14 dB and an offset of 200 V DC. The high impedance voltage between the sample and the conducting plate perturbs the sample enough that the laser Doppler vibrometer (LDV) can detect its vibration. The excitation signal used is a chirp for which the range of frequencies depends on the sample being tested. Using Eq. 2, the measured dimensions of each cantilever, and approximate values for density, an estimate of the natural frequency for each sample is calculated. The chirp is then designed with a 100-200 Hz range around the fundamental natural frequency to concentrate the energy in the first bending mode. Fig. 3 displays the two sample configurations; the first shows the aluminum cantilever machined from the large mass clamped to the vacuum base by a second plate. The second displays the 3D printed cantilever bolted to an aluminum block that is secured to the vacuum base. The cantilever hovers just above a plate whose four corner screws are attached to springs so that the plate can be adjusted to best fit the shape of the cantilever.

The beam from the laser Doppler vibrometer (LDV) shines directly onto the end of the cantilever through the glass lid of the vacuum chamber. Velocity data from the LDV measurement is collected by a National Instruments PCI-6221 with a sampling frequency of 10,000 samples per second and 100 averages of the signal. Time history and frequency spectra of the cantilever response are recorded. The schematic in Fig. 4 portrays the input and output of the signal and their connection to the physical components.

4. Results

The aluminum sample was tested first, as its material properties are well known. Using the same setup, the aluminum sample was tested with and without vacuum. The center frequency was estimated with material properties given by the manufacturer and measured dimensions

IOP Conf. Series: Journal of Physics: Conf. Series **1149** (2018) 012002 doi:10.1088/1742-6596/1149/1/012002

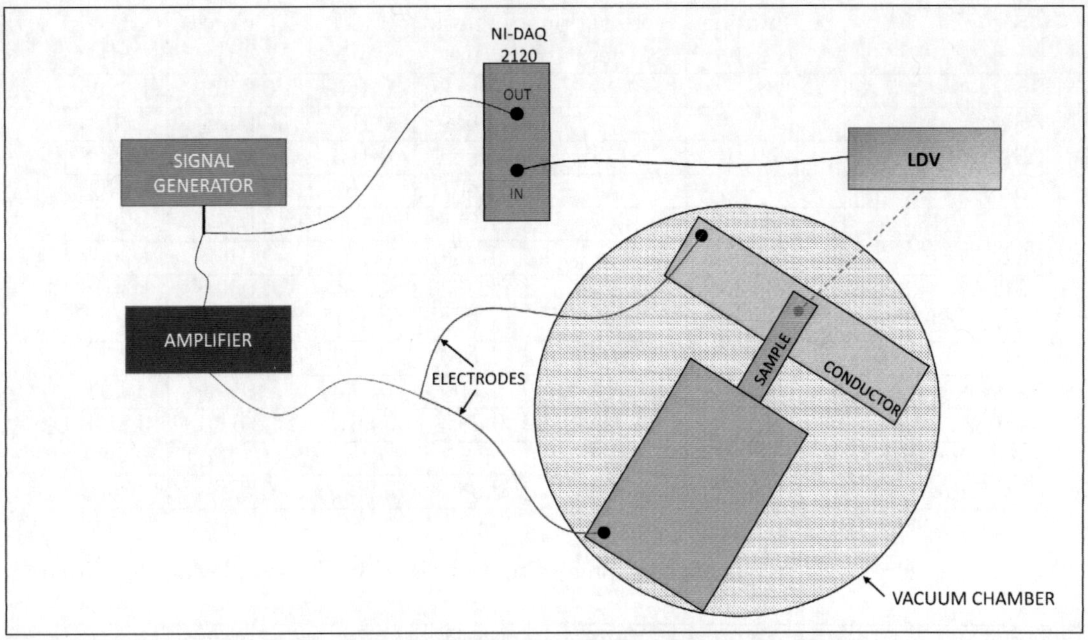

Figure 4: Top view of the experiment schematic indicating a generated input to excite the cantilever and a measured output from the LDV.

to have a value of 176 Hz. Therefore, a 120-200 Hz chirp was generated to concentrate the energy around the resonant peak. From the frequency response of the aluminum cantilever, the resulting center frequency of its resonant peak was found to be approximately 170 Hz. After curve fitting, the quality factor was found to be about 244 at ambient pressure (760 Torr), seen in Fig. 5, and 668 at a pressure of 900 milliTorr. The drastic increase in quality factor indicates that fluid damping plays a significant role for this oscillator.

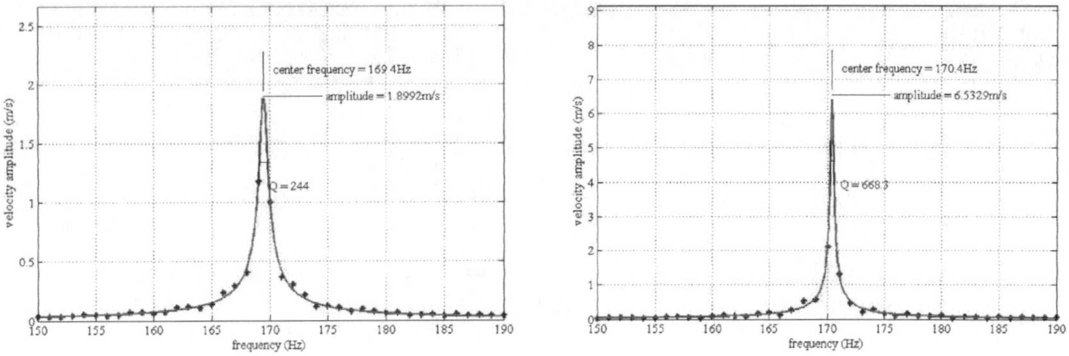

Figure 5: Each sample was tested and processed with identical excitation and curve-fitting limits in between trials with and without vacuum.

Sample	X Rotation (deg)	Y Rotation (deg)	Length (in)	Width (in)	Thickness (in)	Mass (g)
SL 1	0	45	1.672	0.163	0.0402	1.2805
SL 2	15	45	1.669	0.163	0.0488	1.3025
SL 3	30	45	1.660	0.163	0.0461	1.2908
SL 4	45	45	1.661	0.162	0.0476	1.3001
SL 5	60	45	1.665	0.163	0.0469	1.2877
SL 6	75	45	1.664	0.166	0.0469	1.3069
SL 7	45	15	1.663	0.162	0.0504	1.3019
SL 8	45	30	1.672	0.161	0.0472	1.2872
SL 9	45	60	1.670	0.163	0.0453	1.2751
Average	NA	NA	1.666	0.163	0.0466	1.293
DFA	NA	NA	0.28%	0.86%	%6.1	0.85%
As Designed	NA	NA	1.67	0.167	0.050	NA
% Difference	NA	NA	0.03%	2.28%	6.82%	NA

Table 1: Measured dimensions of the geometry and mass of the SL samples are displayed.

Sample	Density (kg/m^3)	Young's Modulus (GPa)	Center Frequency (Hz)	Q (no vacuum)	Q (vacuum)
SL 1	1086.9	2.22	180.4	20.7	23.8
SL 2	1043.3	1.39	179.1	20.6	23.0
SL 3	1051.4	1.56	176.0	20.9	22.0
SL 4	1036.2	1.44	178.3	19.5	23.8
SL 5	1043.2	1.55	179.6	19	25.1
SL 6	1063.2	1.58	180.6	21.2	24.7
SL 7	1034.3	1.16	171.1	18.1	21.1
SL 8	1074.8	1.44	171.1	20.3	22.3
SL 9	1057.2	1.67	175.8	19.7	22.5
Average	1054.5	1.56	176.9	20.0	23.1
DFA (%)	1.69	18.4	2.09	5.04	5.67
Calculated	1105	2.80	157	NA	NA
DFM (%)	4.57	270%	12.7%	NA	NA

Table 2: Material and phenomenological properties are determined for each printing orientation.

4.1. SL Samples

The measured dimensions and mass followed by calculated values of density, Young's modulus, center frequency, and quality factor are found in Tables 1 and 2. The results in Table 1 were used to determine the density and Young's Modulus of each sample given in Table 2. Fabrication error became increasingly evident as dimensions decreased–the thickness of the SL samples varied as much as 6.8% from the designed dimension. Compared to the manufacturer's values in Table 5, the average density of the tested samples had a percent error of 4.6% and the Young's modulus had an error of approximately 100%.

Table 2 displays the results of the properties found for each sample. Two kinds of error are observed here, which are defined as Deviation From Average (DFA) and Deviation From Manufacturer (DFM). The DFA is only considered for the SL samples as there are multiple

Sample	Length (in)	Width (in)	Thickness (in)	Mass (g)
Aluminum	3.0545	0.505	0.052	NA
ABS	4.145	0.410	0.132	17.179
CFN	4.163	0.407	0.143	18.153
FGN	4.157	0.405	0.135	16.403
PCF	4.153	0.405	0.131	19.255
PETG	4.194	0.413	0.139	21.907
PLA	4.180	0.409	0.139	19.294
Average	4.167	0.408	0.137	NA
DFA (%)	2.22%	0.81%	2.58%	NA
As Designed	4.167	0.417	0.125	NA
DFM (%)	0.003%	2.05%	9.24%	NA

Table 3: Measured dimensions and mass of the FDM samples are compared to those as designed.

samples printed from one material. For DFM, the mean of the property is compared to the property given by the manufacturer. This is the error observed for the FDM samples.

Overall, density for the SL cantilevers varied from the average (DFA) by 1.69%, Young's Modulus varied by 18.4%, resonant frequency varied by 2.09%, and quality factor varied by about 5%. Dividing the SL samples into two groups based on their printing orientation, those rotated in the X-direction and those rotated in the Y-direction, the same quantities are compared. For the six samples rotated about the x-axis, the density varied by 1.76%, Young's modulus varied by 19%, resonant frequency varied by 0.63%, and quality factor varied by 4.75%. Similarly, for the 4 samples rotated about the y-axis, density varied by 1.82%, Young's Modulus varied by 12.4%, resonant frequency varied by 0.63%, and quality factor varied by 4.91%. The modulus of elasticity is the most inconsistent of the properties but does not indicate orientation-based trends. It may be noted that the vacuum levels vary between 300 mTorr and 5 Torr. Constant peak vacuum was unattainable; however, as the quality factor of most of the samples only decrease by about 5-10% in most cases, it is reasonable to assume that fluid loss is not a dominant damping mechanism for these 3D printed plastics.

These results indicate that density, resonant frequency, and quality factor are sufficiently independent of printing orientation. The Young's modulus clearly varies per sample, but whether or not this can be attributed to printing orientation or sensitivity to fabrication error is unclear.

4.2. FDM Samples

The same tests imposed on the larger FDM samples produced more varied results. With only one sample per material, percent variation between samples cannot be estimated (DFA), but the percent error between the theoretical or estimated values and the measured values were compared (DFM). Table 3 displays the DFA and DFM for the FDM samples, both of which present error under 10% for the dimensions and mass. The error in these larger samples compared to that for the SL samples is significantly smaller. It is probable that the larger size of the samples allows for more accurate printing and therefore lower fabrication error.

The percent error between manufacturer and experimental values given in Table 4 indicate nearly 100% variation in Young's modulus and 10-50% error in density. Without a predicted value for any kind of damping per material, the quality factor is simply compared before and after vacuum. The percent difference in quality factor for four of the six samples is very low, at approximately 3-5%. However, for the ABS and PLA thermoplastics, Q varies more than 10% in vacuum, indicating a slightly higher sensitivity to fluid damping mechanisms.

IOP Conf. Series: Journal of Physics: Conf. Series **1149** (2018) 012002 doi:10.1088/1742-6596/1149/1/012002

Sample	Density (%)	Young's Modulus (%)	Q (%)
ABS	16.3	94	32.4
CFN	34.9	99	4.9
FGN	48.1	98	3.4
PCF	NA	100	4.0
PETG	11.8	95	4.7
PLA	22.5	95	11.1

Table 4: Density and Young's modulus DFM as compared to values provided in Table 5 are shown with the percent difference in quality factor with and without vacuum.

Sample	Density (kg/m^3)	Young's Modulus (GPa)
Aluminum	2700	69
ABS	1040	2.2
CFN	1400	54
FGN	1600	21
PCF	NA	54
PETG	1250	2.11
PLA	1270	3.5
SL Resin	1150	2.8

Table 5: Manufacturer-given values as available for density and Young's modulus. Found in material index on 3D Hubs webpage and in datasheets for each printer.

5. Conclusion

These results imply that the effect of printing orientation on stereolithographic prints is minimal. The Young's modulus for SL samples varied greatly from the manufacturer value. The large error in Young's modulus does not follow a trend with printing orientation, so more experiments must be done to determine the cause. Further investigations of these material properties, perhaps in which all fabrication variations are held constant, may help to explain the inconsistencies. Another study of interest would be to measure the frequency response as a function of vacuum level, particularly for the oscillators that had relatively large changes in damping when in vacuum.

The results of the fused deposition modeling samples indicated lower variation between designed and printed dimensions compared to the stereolithographic samples, which is likely due to the larger size of the FDM samples. The properties measured in this work of density and Young's modulus differed from manufacturer values significantly. The density was extracted directly from the measured dimensions and weight of each sample and diverged from literature up to 50% for some samples. The Young's modulus, which was calculated from the measured dimensions and the center frequency of each sample, had an even more extreme error of 90-100%. However, inconsistent data on material properties of 3D printed materials by different manufacturers makes this error less incriminating. The percent difference in quality factor with and without vacuum for the FDM samples is less than 5% for most of the materials, indicating low sensitivity to fluid damping.

For future work, continuing the study of 3D printed materials and their intrinsic properties by isolating specific aspects of the printing process will formulate a fuller characterization of these materials. Printing a large number of samples with constant design specifications would target fabrication disorder studies more directly. Once a baseline for fabrication disorder is established (per printer, per material), investigations in unique 3D printing fabrication variables can be considered.

Acknowledgement
This work was supported in part by the NASA DC Space Grant Consortium Award 31377.

References

[1] Formlabs. Form 2 tech specs.
[2] John A. Judge, Douglas M. Photiadis, Joseph F. Vignola, Brian H. Houston, and Jacek Jarzynski. Attachment loss of micromechanical and nanomechanical resonators in the limits of thick and thin support structures. *Journal of Applied Physics*, 101(1):013521, January 2007.
[3] John A. Judge, Joseph F. Vignola, and Jacek Jarzynski. Dissipation from microscale and nanoscale beam resonators into a surrounding fluid. *Applied Physics Letters*, 92(12):124102, March 2008.
[4] C. Soize. Probabilistic structural modeling in linear dynamic analysis of complex mechanical systems, I - Theoretical elements. *La Recherche Aerospatiale (English edition)*, 5(-):Pages: 23–48, 1986.
[5] Murray Strasberg and David Feit. Vibration damping of large structures induced by attached small resonant structures. *J. Acoust. Soc. Am.*, 99(1):335–344, 1996.
[6] J.F. Vignola, J.A. Judge, and A.J. Kurdila. Shaping of a system's frequency response using an array of subordinate oscillators. *Journal of the Acoustical Society of America*, 126:129–139, 2009.
[7] Joseph Vignola, Aldo Glean, John Judge, and Teresa Ryan. Optimal apparent damping as a function of the bandwidth of an array of vibration absorbers. *The Journal of the Acoustical Society of America*, 134(2):1067–1070, August 2013.
[8] Formlabs. Validating Isotropy in SLA 3d Printing, October 2016.
[9] Sai Tej Paruchuri, John Sterling, Andrew Kurdila, and Joseph Vignola. Piezoelectric composite subordinate oscillator arrays and frequency response shaping for passive vibration attenuation. In *Control Technology and Applications (CCTA), 2017 IEEE Conference on*, pages 702–707. IEEE, 2017.
[10] Formlabs. Ultimate Guide to Stereolithography.
[11] Joseph F. Vignola and John A. Judge. Architectural considerations of micro- and nanoresonators for mass detection in the presence of a fluid. *Journal of Applied Physics*, 104(12):124305, December 2008.
[12] ASTM E756-05, Standard Test Method for Measuring Vibration-Damping Properties of Materials, 2017.
[13] Subrahmanyam Gorthi. *Characterization of MEMS Electrostatic Actuators Beyond Pull-In.* PhD thesis, Indian Institute of Science, Bangalore, January 2006.

IOP Conf. Series: Journal of Physics: Conf. Series **1149** (2018) 012003 doi:10.1088/1742-6596/1149/1/012003

Toward fully anisotropic viscoelastic material models using an automated high-speed optical rig

Luca Manzari[1], Peter Göransson[1], Jacques Cuenca[1,2] and Ines L Arteaga[1]

[1] Aeronautical and Vehicle Engineering, KTH Royal Institute of Technology, Teknikringen 8, SE-10044 Stockholm, Sweden
[2] Siemens Industry Software, Interleuvenlaan 68, B-3001 Leuven, Belgium

E-mail: manzari@kth.se

Abstract. This paper describes some of the algorithms that are employed in the post-processing of the data obtained from an automated high-speed optical rig. The setup, entirely designed by some of the authors, is used for measuring the full field displacement in time of anisotropic viscoelastic solids undergoing controlled dynamic excitation. The setup main goal is to provide complex transfer functions relating the three-dimensional sinusoidal displacement on the surface of the sample to the sinusoidal motion of the foundation, suitable for the inverse estimation of the dynamic properties of the material under analysis. The complete automation of the measurement process aims at maximizing repeatability and signal-to-noise ratio. In particular, a stereo high-speed camera system is responsible for the actual measurement, while a laser Doppler vibrometer is used for a preliminary autoranging procedure and for data validation.

The camera system does not directly output the high-quality data that is needed for a successful, physically meaningful inverse estimation. Hence, here the algorithms described in this paper come into play. Since digital image correlation (DIC) may fail in some frames or in some regions of the pictures, a mask is applied and calculated for every measurement run to preserve information only where the success rate of the DIC algorithm is higher than a specified threshold. The sample is mounted on a foundation whose sides are flat: random sample consensus (RANSAC) and principal component analysis (PCA) are used to respectively eliminate outliers and find the best fitting plane to the foundation side – the measured displacement field can then be expressed in a local coordinate system, truly relative to the foundation. The transfer functions are now obtained in two steps: first an optimizer is used to fit a cosine to all the measured displacement time-series, then the complex amplitudes representing the motion of the sample faces are divided by the average of the complex amplitudes representing the motion of the foundation.

1. Introduction

3D digital image correlation (3D-DIC) proved itself as a viable alternative to more conventional accelerometer- and laser-based measurements to validate finite element models and to provide qualitative information about the resonant behavior of simple structures [1, 2, 3, 4, 5]. Typical limitations of the method (e.g. the evaluation of displacements that the human eye cannot perceive when observing the image sequence) are being overcome with the integration of phase-based video magnification techniques [6]. Still, if the data obtained from 3D-DIC is meant to be the target of an inverse estimation process aimed at determining material parameters, it is necessary to develop a robust post-processing chain able to automatically deal with the mishaps

IOP Conf. Series: Journal of Physics: Conf. Series **1149** (2018) 012003 doi:10.1088/1742-6596/1149/1/012003

that inevitably occur when repeating a measurement a large number of times: failures of the DIC algorithm for some regions of the image or a misalignment of the test object with respect to the coordinate system used by the measurement equipment.

In this particular case, a cubic sample of a viscoelastic anisotropic material is to be dynamically characterized. The measurement of the displacement undergone by the outer faces of the sample, as well as the displacement of the driving foundation on which the sample is mounted, is computed using 3D-DIC and a pair of monochromatic high-speed cameras. The setup is briefly described in section 2, and yields three-dimensional full field displacement information in time for both the four sample side faces and the foundation. The data output by the 3D-DIC algorithm and in particular the flaws that render it not directly usable for a computation of the transfer functions are discussed in section 3. The solutions to such issues are discussed in sections 4 to 6 and represent the main contribution of this work. Results are shown for each algorithm, using both synthetic data for ease of understanding and real data to show the efficacy of the methods.

2. Brief setup description

The setup has been extensively described in [7], a brief overview is here given for the sake of self-containedness. Many details considered irrelevant for the current work are here omitted, the reader is again pointed at [7] for a much more detailed exposition. A cubic sample of the material to be tested is glued to a foundation with speckled sides, driven in the vertical direction by a uniaxial shaker imposing a sinusoidal motion to the foundation. A tuned seismic mass is glued to the top of the sample, and a couple of high-speed cameras record the motion of the sample and of the foundation for the whole duration of the experiment. The sample, along with the foundation and the seismic mass, is represented in figures 1 and 2.

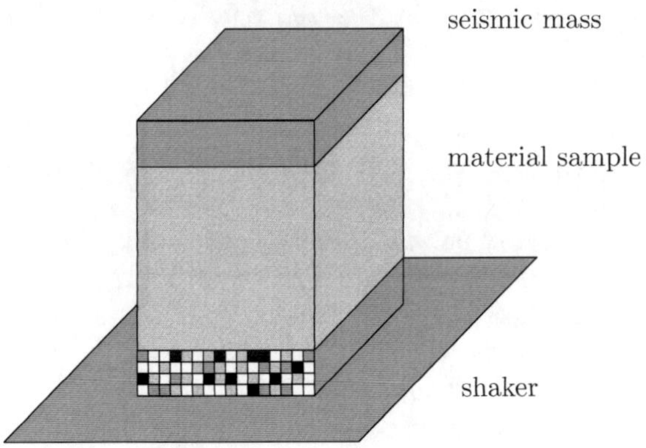

Figure 1: Schematic representation of a sample ready to be measured.

Figure 2: Picture of a real material sample.

The whole assembly (shaker, foundation, sample and seismic mass) is placed in a vacuum chamber on top of a rotation stage that allows for the rotation of the sample without opening the vacuum chamber. This avoids the need to re-calibrate the camera system each time the sample is rotated, since the pressure difference deforms the window through which the cameras look at the sample. Once the sample has been set up and the vacuum chamber depressurized, the measurement proceeds as follows:

(i) the sample is rotated in such a way that both cameras see one of its faces;

(ii) the shaker is driven with a pure sine of a randomly chosen frequency among the frequencies of interest;

(iii) the amplitude is chosen by an iterative auto-ranging procedure (explained in [7]);

(iv) once the system reaches steady state, the cameras acquire a stereo video recording of the sample and the foundation;

(v) points (ii) to (iv) are repeated until all frequencies of interest are spanned;

(vi) points (i) to (v) are now repeated for the three remaining faces of the sample.

3. Data obtained from the 3D-DIC camera system

The 3D-DIC is performed by a proprietary software bundled with the cameras. For each region of the image on which 3D-DIC is performed (two in our case, the foundation and the face of the sample), two outputs are obtained:

- N files, one per stereo frame, containing the surface height map of the measured area;
- N files, one per stereo frame, containing the coordinates of each measurement facet as computed in the reference image along with its three-dimensional displacement with respect to the aforementioned reference.

In order to compute the sought after transfer functions, it is necessary to know the amplitude and phase of each facet per each frequency and per each coordinate – assumed a linear elastic behavior of the sample and a three-dimensional cartesian coordinate system such that:

- the y axis is vertical, parallel to the direction in which the shaker excites the foundation;
- the face of the foundation currently observed entirely lies in the xy plane;
- the origin of such coordinate system is not important, since zero-mean signals are of interest for the current application.

Sadly, the DIC algorithm does not necessarily return valid results for all frames nor for all measurement facets. This can happen between the frames in a stereo pair (e.g. specular reflections, appearing as extremely bright points in only one of the two frames) as well as between stereo frames captured at different times (a facet laying on the boundary of the sample would be particularly tricky to match). To automatically obtain the transfer functions it is then necessary to exclude the facets that are not consistently tracked. A procedure for doing so is described in section 4.

Due to the need of measuring in vacuum, a thick sheet of acrylic glass separates the cameras from the sample. The calibration routine adopted by the software forces the coordinate system to be aligned with the calibration plate [8, 9]. This implies that if the face of the sample under observation is not perfectly parallel to the calibration plate (which is the real scenario, given the unlikeliness of achieving a perfect alignment) then the displacement fields obtained need to be expressed in a new coordinate system before they can be used to compute transfer functions. This scenario is dealt with in section 5.

Once the valid displacement time series have been transformed in the new coordinate system, the amplitude and phase of each facet per each frequency has to be evaluated. This is done by fitting a cosine to each displacement time series, as described in section 6.

4. Discarding invalid measurement facets

The algorithm here described is operated on the data output of each individual video recording, that is for each individual frequency and for each face of the sample and of the foundation. A description of the algorithm in natural language follows, an example from real measurement data being presented in figure 3.

(i) the number of valid DIC occurrences per facet is counted;

(ii) the facets boasting less valid DIC occurrences than the average number of valid DIC occurrences are discarded;

IOP Conf. Series: Journal of Physics: Conf. Series **1149** (2018) 012003 doi:10.1088/1742-6596/1149/1/012003

(iii) the biggest rectangle enclosing only non-discarded element is found;

(iv) only the points belonging to such rectangular domain are considered valid and used for the subsequent analysis.

The choice of a rectangular domain is done for mere convenience: on such a regular domain it is easier to perform a variety of operations (e.g. a one- and two-dimensional FFT), should it be needed in the future. Moreover, as the size of the facets is reduced, the shape of the rectangular domain better approximates the shape of the sample.

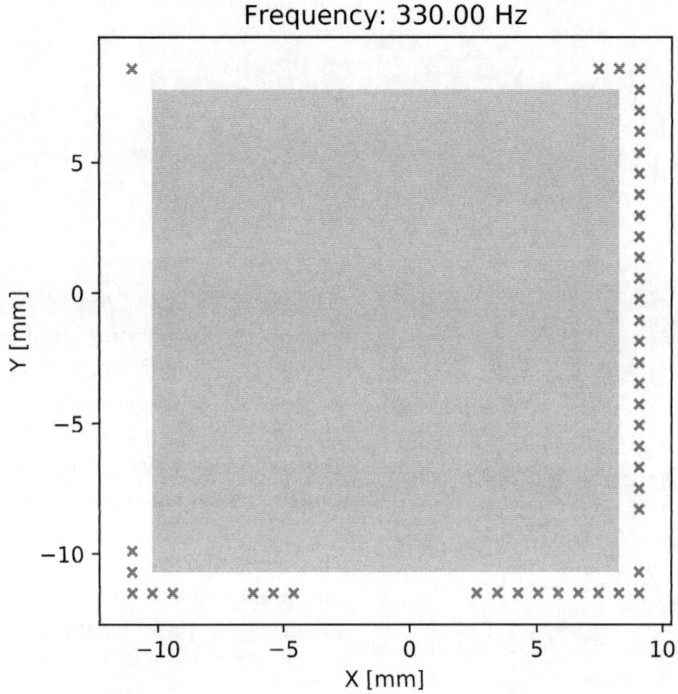

Figure 3: A graphical walk-through of the algorithm that select a valid domain based on the number of valid DIC occurrences per facet. Each light blue dot represents a facet: the darker the dot, the higher the number of valid DIC occurrences for that facet. Facets with too few valid DIC occurrences are rejected (marked with a red "x"), and the biggest rectangle containing only non-discarded facets (in light blue) is chosen as the valid domain for all subsequent computations.

5. Changing the coordinate system

As it can be observed in figure 3, the origin of the coordinate system does not overlap with the physical center of the sample face. While this would not be a problem in itself, it is a symptom of the misalignment of the sample and the calibration plate. The foundation on the other hand (painted with a speckle pattern and clearly visible in figure 2) is aligned with the sample and provides for a useful reference to express the displacement data in a more convenient measurement system. In particular the foundation height map is a point cloud that should ideally fit into a plane: such a plane would provide the xy-plane of the new coordinate system, relative to the foundation and with the y-axis aligned with the direction of excitation. In reality, the foundation height map is not a perfect plane because of two different phenomena:

(i) the resolving power of the cameras is limited (as it is for all digital equipment), hence the true location of the points oscillates around the true value;

(ii) outliers, due to imperfect manufacturing of the foundation and errors in the matching algorithm, pollute the point cloud.

Keeping these phenomena in mind, it is necessary to remove the outliers before using the point cloud to determine the best fitting plane hence deciding on a coordinate system.

The outlier detection is operated using the random sample consesus (RANSAC) paradigm, an iterative method well suited to estimate model parameters from data that is contaminated with outliers [10]. RANSAC requires the setting of some confidence parameters, of which the most critical is the threshold value for determining whether a data point fits the model or not. The determination of this threshold value is the only operation that requires some human intervention, since it is sensitive to the surface roughness of the material observed and to the quality of the camera calibration. After RANSAC has determined the set of inliers, principal component analysis (PCA) can be used to determine the eigenvectors that will transform the displacement fields to the new coordinate system [11]. The procedure is illustrated using synthetic data for the sake of simplicity in figure 4.

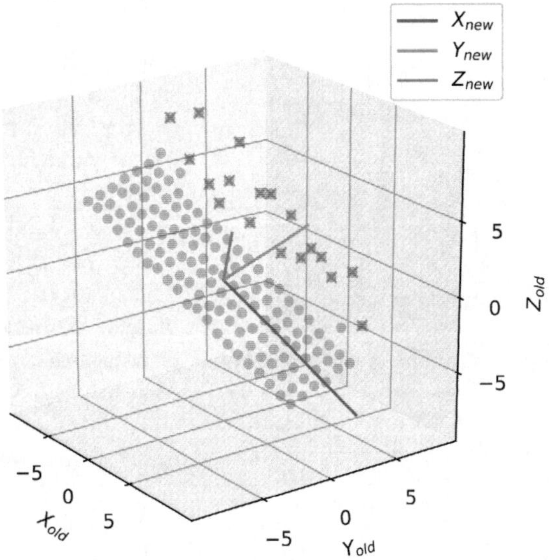

Figure 4: The blue dots are points in a point cloud which could be modeled as a plane. RANSAC excludes the outliers (marked with a red "x"), and PCA is run on the non-excluded points to finds the eigenvectors representing the major axis of the plane. Such eigenvectors effectively represent the new coordinate system.

6. From time series to amplitudes and phases

There are many different ways to extract an amplitude and a phase from an ideally sinusiodal signal, with the Discrete Fourier Transform being among the most common tools in the arsenal. Still, the main drawback of such a method is that (all the other parameters being fixed) high frequency resolution requires a high number of samples [12], a computationally expensive luxury when dealing with 3D-DIC. The approach used relies instead on solving the nonlinear least-squares problem of fitting a cosine to the time series, a well-posed problem given reasonable bounds and initial guesses [13]. A one-dimensional linear interpolation is used to recover information for missing data points, where the DIC might have failed. Then amplitude, phase, offset and frequency are the four parameters that are obtained for each uni-directional displacement of each facet. An example with real measurement data is provided in figure 5.

While such an approach might mask any non-linear behavior of the system, the material model used in subsequent stages for the estimation of the dynamic properties is a linear one: as such, it would be unable to incorporate nonlinear phenomena in its current state. The estimated covariance matrices of the optimal values resulting from the solution of the nonlinear least-squares are nevertheless an indicator of the quality of the fit.

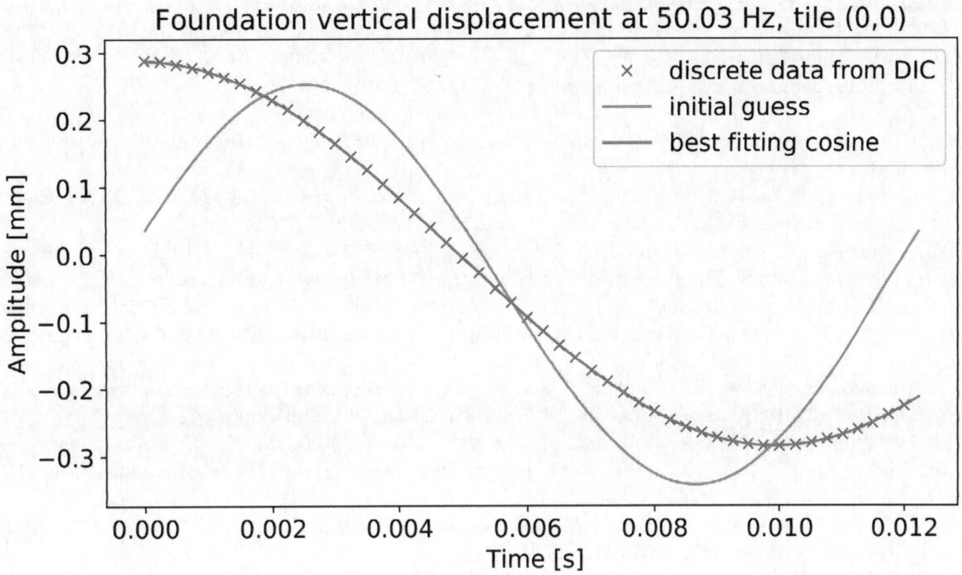

Figure 5: Vertical displacement of one of the foundation facets: the DIC data from 50 stereo frames, the initial guess and the best fitting cosine are shown. The method works even with non-periodic data, and in the case of frequencies of longer time period than the duration of the observation.

7. Conclusions

Part of a much broader project, the work presented in this paper allows for the fully automatic post-processing of the data produced by a 3D-DIC system meant for the characterization of viscoelastic anisotropic materials: the outcome of the process is a series of three-dimensional displacement transfer functions. Future work may include (but not be limited to):

- uncertainty analysis of the measured quantities;
- an automatic, smart choice of the RANSAC threshold value (e.g. MLESAC [14]);
- a high-performance, highly parallel implementation of the above routines;
- development of an own 3D-DIC solver.

Acknowledgments

Part of the funding for this work is received under the Swedish Research Council, Grant No. 2015-04258. The authors also acknowledge the funding provided by the Knut and Alice Wallenberg Foundation, in the form of the the MUSCLE project awarded to the Odqvist Laboratory in which the equipment used in the current research has been funded.

References

[1] Beberniss T J and Ehrhardt D A 2017 *Mechanical Systems and Signal Processing* **86** 35 – 48 ISSN 0888-3270 full-field, non-contact vibration measurement methods: comparisons and applications URL `http://www.sciencedirect.com/science/article/pii/S0888327016300528`

[2] Reu P L, Rohe D P and Jacobs L D 2017 *Mechanical Systems and Signal Processing* **86** 2 – 16 ISSN 0888-3270 full-field, non-contact vibration measurement methods: comparisons and applications URL `http://www.sciencedirect.com/science/article/pii/S0888327016000339`

[3] Siddiqui M Z 2014 *Aerospace Science and Technology* **39** 128 – 136 ISSN 1270-9638 URL `http://www.sciencedirect.com/science/article/pii/S1270963814001795`

[4] Wang W, Mottershead J E, Ihle A, Siebert T and Schubach H R 2011 *Journal of Sound and Vibration* **330** 1599 – 1620 ISSN 0022-460X URL `http://www.sciencedirect.com/science/article/pii/S0022460X10007224`

[5] Sutton M A, Orteu J J and Schreier H 2009 *Image Correlation for Shape, Motion and Deformation Measurements: Basic Concepts,Theory and Applications* (Springer Science & Business Media) ISBN 978-0-387-78747-3

[6] Javh J, Slavič J and Boltežar M 2018 *Mechanical Systems and Signal Processing* **98** 344 – 351 ISSN 0888-3270 URL `http://www.sciencedirect.com/science/article/pii/S0888327017302637`

[7] Manzari L, Göransson P, Cuenca J and Lopez Arteaga I 2018 *Proceedings of NOVEM 2018 Noise and vibration emerging methods, The 6th conference, Ibiza (Balearic Islands), Spain* 175109

[8] LaVision GmbH 2015 *Product-Manual for DaVis 8.3: Imaging Tools* (LaVision GmbH) document name: 1003012_ImagingTools_D83.pdf URL `https://www.lavision.de/en/downloads/manuals/systems.php`

[9] LaVision GmbH 2015 *Product-Manual for DaVis 8.3: StrainMaster* (LaVision GmbH) document name: 1003017_StrainMaster_D83.pdf URL `https://www.lavision.de/en/downloads/manuals/systems.php`

[10] Fischler M and Bolles R 1981 *Communications of the ACM* **24** 381–395 ISSN 0001-0782

[11] FRS K P 1901 *The London, Edinburgh, and Dublin Philosophical Magazine and Journal of Science* **2** 559–572 (*Preprint* `https://doi.org/10.1080/14786440109462720`) URL `https://doi.org/10.1080/14786440109462720`

[12] Manolakis D G and Ingle V K 2011 *Applied Digital Signal Processing : Theory and Practice* (Cambridge: Cambridge University Press) ISBN 9780511835261

[13] Branch M, Coleman T and Li Y 1999 *SIAM J. SCI. COMPUT* **21** 1–23 ISSN 1064-8275 URL `http://search.proquest.com/docview/27112900/`

[14] Torr P and Zisserman A 2000 *Computer Vision and Image Understanding* **78** 138–156 ISSN 1077-3142

IOP Conf. Series: Journal of Physics: Conf. Series **1149** (2018) 012004 doi:10.1088/1742-6596/1149/1/012004

Determination of acoustic behaviour of locally modified structures by Scanning LDV

A. Lohrengel, W. Rolshofen, G. Schäfer, D. Thoden

TU Clausthal, Institute of Mechanical Engineering, Clausthal-Zellerfeld, Germany

E-mail: schaefer@imw.tu-clausthal.de

Abstract. Acoustic properties become more and more important in product design. To change the vibration characteristics additional masses or damping systems are common, but not lightweight. Integrated fins and the damping performance of joints have no negative impact on the weight. Another new advanced method is the local change of material properties by bake hardening. A very effective way to quantify the acoustic influence is the measurement with scanning Laser Doppler Vibrometry, also for the quality assurance.

1. Design of low noise products

Besides the main functions and the costs of a product, acoustic behavior becomes more and more important. In general a customer expect a noiseless use, e.g. lift, air condition, premium vehicles, household appliance and so on. Only a few products like motorcycles or sports cars are additionally sound engineered to have a typical acoustic appearance. But the doors, steering system and drive shaft also have to be quiet. Especially products and tools for the use in working environment have to have a minimum vibration and noise emission.

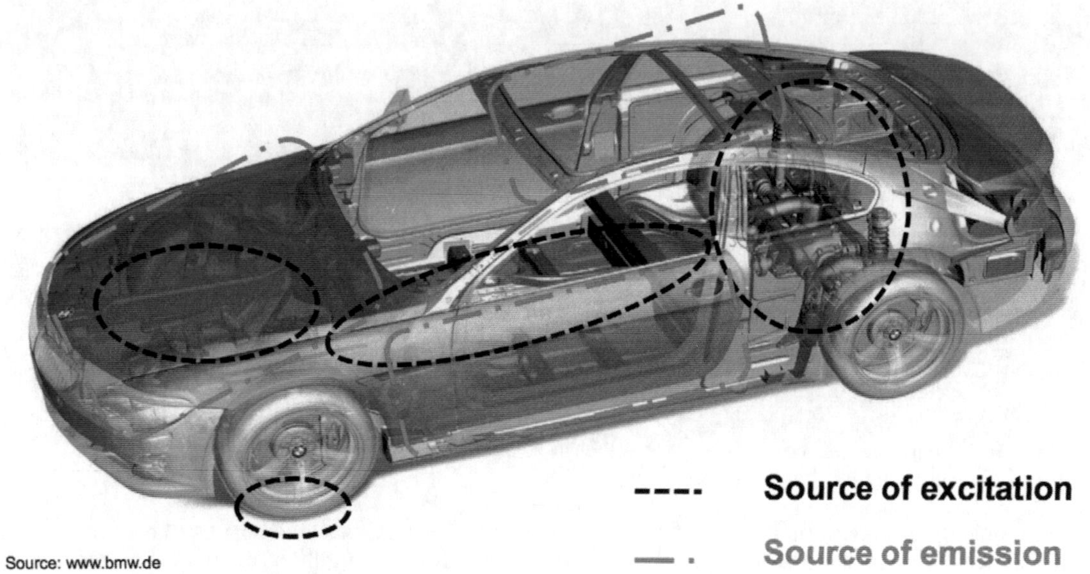

Figure 1: Sources of excitation and emission on a car

To design a low noise product, Dietz and Gummersbach [1] collected a huge number of design rules and defined three basic acoustic elements, see figure 1:

- the origin of vibrations (source of excitation)
- the path between the excitation and emission
- the surface (source of emission), where vibrations are changed into airborne noise

The following chapter 2 deals with some solutions regarding the second and third acoustic element to design a quiet product. Chapter 3 contains some suggestions of the numerical simulation and simple measurements of the acoustic behaviour. Chapter 4 shows some results of the advanced non-contact measurement using a scanning Laser-Doppler-Vibrometer (LDV).

2. Noise reduction by design

Often additional masses to change the natural frequency or damping systems are used to make a product less noisy. The problem of these straight ways to the quiet product is the increasing weight of the product. Regarding the idea of lightweight design, other solutions are necessary, which are focussed on changing the stiffness or using the damping capacity of existing junctions.

2.1. Noise reduction by integrated fins

Figure 2 shows some common versions to influence the stiffness of surface areas, which are causal for airborne noise emission. The integrated fins, which are realised by cold forming, converted the 2D sheet metal into a 3D structure with certain beam stiffness. The geometrical modification works well to the vibration characteristics, without increasing the weight. The disadvantage is the geometry its self, because most of the customers don´t want to have these patterns on the visible surface.

Figure 2: Noise reduction by integrated fins

Reasonable examples for the use of integrated fins are the roof of small lorries, the floor section and bulkheads of cars and the left, right, back and floor section of household appliance, in general all invisible plates.

IOP Conf. Series: Journal of Physics: Conf. Series **1149** (2018) 012004 doi:10.1088/1742-6596/1149/1/012004

2.2. Noise reduction by junctions
Another kind of noise reduction without additional weight is the use of junctions with high damping characteristics. Junctions mostly necessary in products and so they can be used to influence the acoustic product quality. Figure 3 shows four different local junctions for sheet metal. A complete welding seam is more or less useless for damping. Between two spot welds exists an area with a frictional contact, this leads to the frictional damping of a spot weld junction.

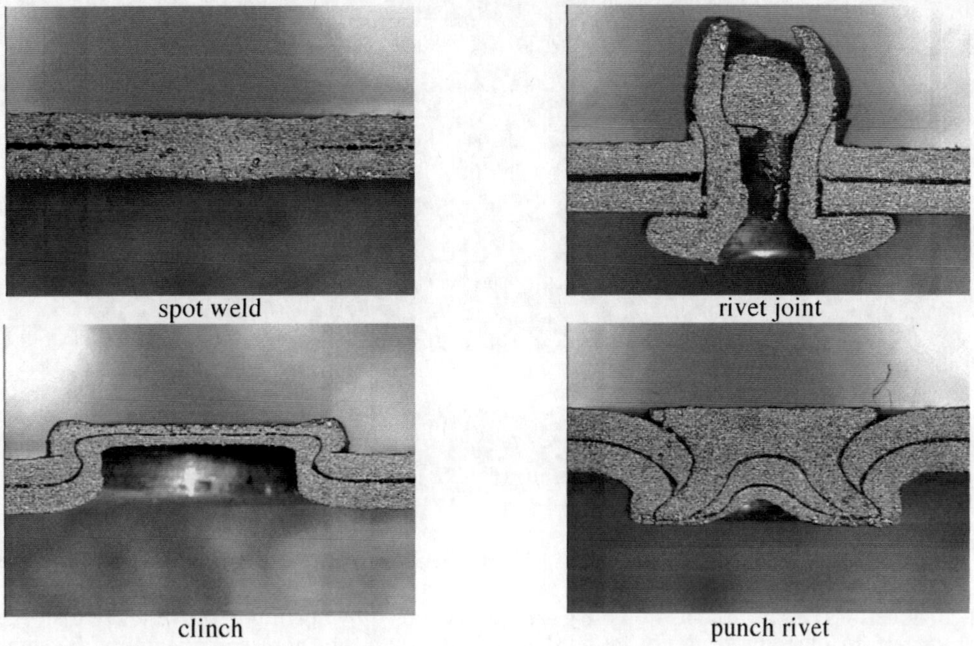

Figure 3: Noise reduction by junctions

Käferstein [2] investigated the damping properties of different sheet metal junctions. The punch rivet showed the highest damping factor, followed by the clinch and rivet joint. The spot weld showed the lowest damping factor of the four junctions in figure 3. In simple words, the damping factor is proportional to the frictional contact area of the junction. In [2] values for damping in junctions up to 800 Hz are determined, also references for FE-modelling of such junctions are given.

2.3. Noise reduction by additional reinforcing fins or local material modification
The performance comparison of the integrated fins and junctions on the acoustic behaviour showed a benefit for the integrated fins. To increase this property and overcome the pattern problem, the use of additional reinforcing fins is very common in car industry, see figure 4. By these additional fins the weight of engine hood increases slightly, but the stiffness and acoustic performance is obvious better.

Analysing the way such fins worked, it becomes clear, that the structured local change of material properties, e.g. stiffness and stability, is able to influence the acoustic behaviour.

The awareness to use a local change of material properties leads to the idea of a structured local heat treatment of sheet metal e.g. by laser. Using the correct laser parameters, there will be no visible pattern and of course no additional weight. This idea in principle is highlighted with yellow laser tracks in figure 4.

After pre-stressing the material by forming the hood, the laser heated up the hood very slightly and structured. This process is well none in general as Bake-Hardening. The local Bake-Hardening should be used instead of the additional fins.

Figure 4: Noise reduction by additional reinforcing fins or local material modification

2.3.1. Material modification "Bake-Hardening". The Bake-Hardening is a combined mechanical and thermal process to increase the stability of steel. The first step is the mechanical pre-stressing to generate dislocations. In the second step the C-atoms are forced by heating to move to the dislocations. The aggregation of C-atoms leads to the hardening effect with an increase of stability, as shown in figure 5.

Figure 5: Modification of material properties by Bake-Hardening

IOP Conf. Series: Journal of Physics: Conf. Series **1149** (2018) 012004 doi:10.1088/1742-6596/1149/1/012004

3. Numerical simulation and first measurements

The first step for the verification and dimensioning of the structured local bake hardening was the set-up of a simple geometry model according to Hambric [3] with a single line of bake hardening and two measurement points on each side left and right of the line, see figure 6. With this simple model the transfer function with and without bake hardening was measured. Therefore an impact excitation and piezo sensors were used.

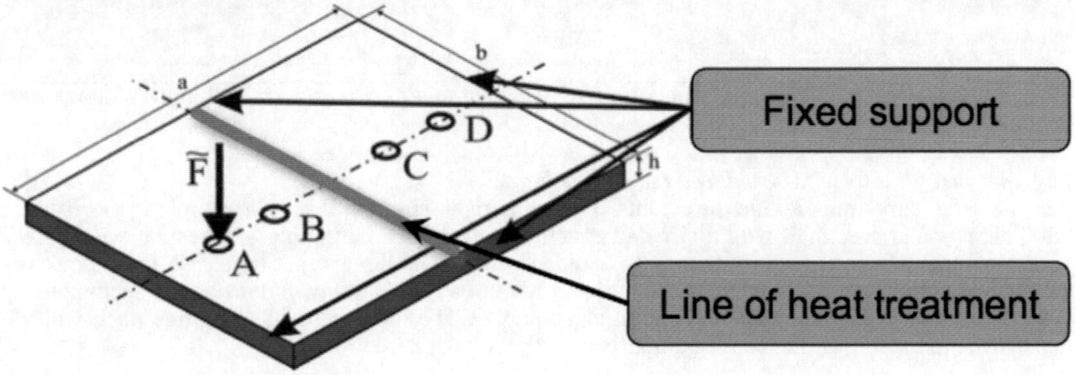

Figure 6: Simple geometry model

For the future numerical dimensioning of structured local bake hardening a valid Finite Element Model was necessary. Equivalent to the simple geometry model in figure 6 an FE-model was generated, see figure 7. On the left side of figure 7 a cross section of the heat affected zone is shown. The first simple seam geometry A was taken from a classic welding seam. The height of the heat treated zone was equal to the sheet metal thickness. The accordance between the measurement and numerical simulation was poor. Having a look to the cross section in figure 7 left, the seam geometry B was generated, which leads to a bit better accordance. Within a third stop of refining the numerical model, the seam geometry C was established. Seam geometry C represents all three areas in figure 7 left, with different modulus of elasticity.

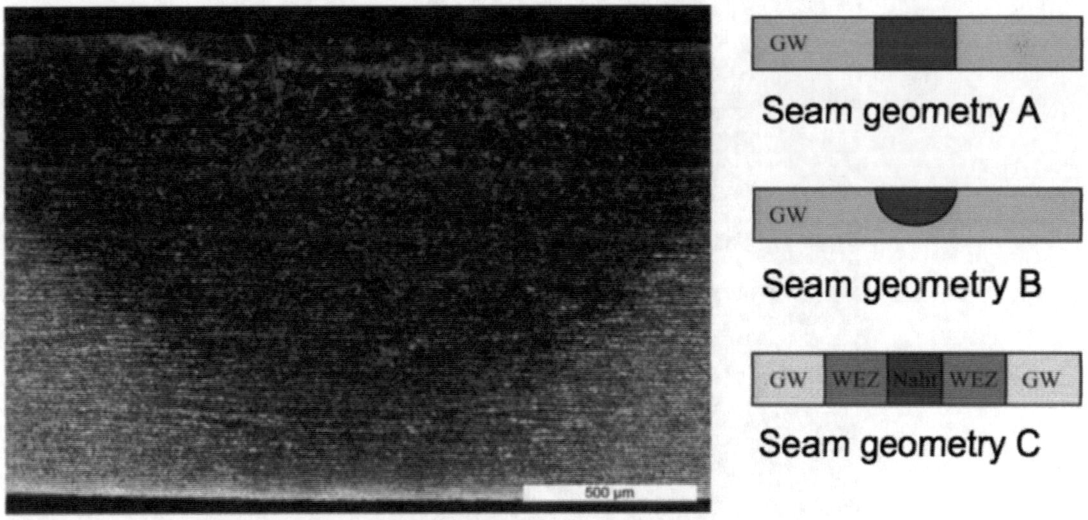

Figure 7: Image of heat treated zone and variations for FE-model, Naht = heat treated zone, WEZ = heat affected zone, GW = basic material

After the first successful matching of numerical and experimental results a second test setup, see figure 9, was produced in sheet metal and modelled for the numerical simulation. First results for the mode simulation without bake hardening are shown in figure 8.

IOP Conf. Series: Journal of Physics: Conf. Series **1149** (2018) 012004 doi:10.1088/1742-6596/1149/1/012004

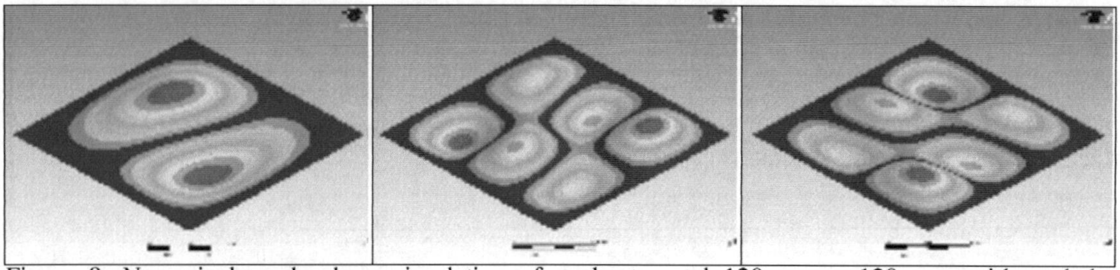

Figure 8: Numerical mode shape simulation of a sheet metal 120 mm x 120 mm without bake hardening

3.1. Experimental tests with structured local bake hardening

After the first experimental measurements and numerical simulations with simple geometries, a DP800 sheet metal specimen with a circular structured local bake hardening was tested, see figure 9. The measurement was carried out with two piezo sensors in three repetitions, so that six sensor positions were used to get a good local resolution without having to many masses on the specimen.
On the lower left corner a shaker induced a chirp signal. The measurement was done once with the circular bake hardening and once for comparison without bake hardening.

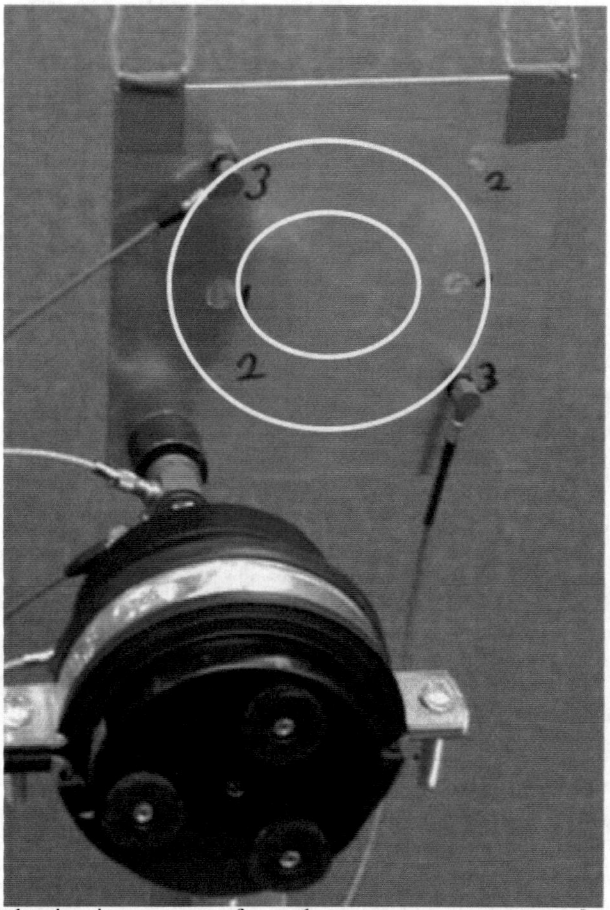

Figure 9: Set-up with classic piezo sensors for a three step measurement of two DP800 sheet metal plates 120 mm x 120 mm

The results for the sensor position 3, like it is shown in figure 9, are compared in figure 10. Around 4 kHz a significant positive influence of the bake hardening is observable. The acceleration peak is decreased to less than 50%.

IOP Conf. Series: Journal of Physics: Conf. Series **1149** (2018) 012004 doi:10.1088/1742-6596/1149/1/012004

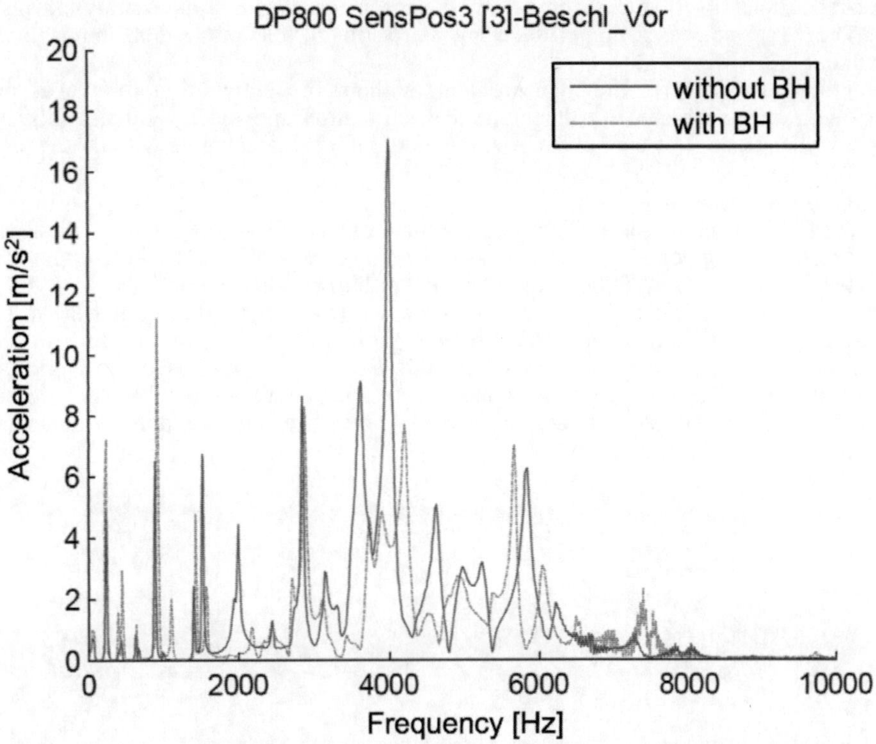

Figure 10: Comparison of two DP800 sheet metal plates, with and without bake-hardening (BH) traces by electron beam

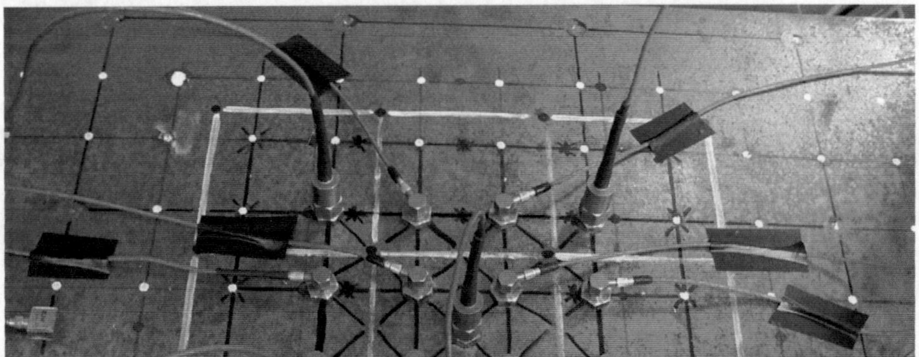

Figure 11: Set-up with classic piezo sensors for a one step measurement of a sheet metal plate 500 mm x 500 mm

The three repetitions for the measurement in six sensor positions and their analyses was very time consuming. The next step was a bigger sheet metal specimen, 500 mm x 500 mm with full sensor equipment, like it is shown in figure 11.

The expenditure of time for the measurement, without the preparation time, was ok, but the comparison of the measurement and the numerical simulation showed a dramatic influence of the sensor masses. The influence of the bake hardening would not be measurable with theses sensors.

4. Non-contact measurement

As result of the full contact sensor measurement, a non-contact measuring device with high local and frequency resolution was necessary. After a market analysis a scanning Laser Doppler Vibrometer (LDV) (Polytec PSV-400 + OFV-5000) was used for the following measurements.

Using the laser beam for the measurement, it is a reactionless method to gather the information's, which are necessary for structural acoustics analyses regarding the local material properties [4].

Figure 12 shows the measuring grid of the scanning LDV, which was used for the 500 mm x 500 mm sheet metal plate. The measuring grid has an increased resolution with 287 measuring dots compared to the 17 piezo sensors in figure 11. The LDV measurement was done with a sampling rate of 2,5 MHz to increase the time resolution.

Figure 12: Measuring grid for the LDV-Measurement with 287 dots on a sheet metal 500 mm x 500 mm

One measuring result, the wave shape in time domain of a sheet metal with circular structured bake hardening is shown in Figure 13. The excitation was realised from the backside with an acoustic radiator aligned to the centre of the plate. For other measurements an impact hammer or shaker, connected to the centre of the plate or one of the lower corners, was used.

IOP Conf. Series: Journal of Physics: Conf. Series **1149** (2018) 012004 doi:10.1088/1742-6596/1149/1/012004

Figure 13: Wave shape in time domain of a circular reinforced bake hardening sheet metal 500 mm x 500 mm

4.1. Modal analysis

The scanning LDV data were used for an experimental modal analysis with the software ME´Scope VES. Besides the experimental analysis a numerical modal analysis was done with ANSYS.
Within the frame of investigations not only the sheet metal with one layer with and without bake hardening were measured, there are also sandwich plates with two sheet metal and between one polypropylene-polyethylene copolymer (PP-PE) layer measured. The two sheet metal layers allowed a different pattern and orientation of the structured bake hardening which provides a perfect adaptability to the acoustic requirements. The PP-PE layer allows also additional structured reinforcement, so that there is a very wide area of future research work.

For the first test a bake hardening pattern of radial lines with 45° pitch on both sheet metal layers was used compared to a pure sandwich plate without stiffeners. Figure 14 shows the results of the experimental and numerical modal analysis [5]. From left to right the influence of the material modification is perfect observable in both analyses. The experimental analysis of the natural frequency showed also a significant shifting to a higher frequency for the modified plate. The numerical model showed a lack of refinement especially in the PP-PE layer, which is represented only by one element layer.

5. Summary

Acoustic properties become more and more important in product design, therefore are different design rules available. There are different methods for the realisation in production. A new advanced one is the local change of material properties by bake hardening. A very effective way to quantify the acoustic influence is the measurement with scanning Laser Doppler Vibrometry, also for the quality assurance.

IOP Conf. Series: Journal of Physics: Conf. Series **1149** (2018) 012004 doi:10.1088/1742-6596/1149/1/012004

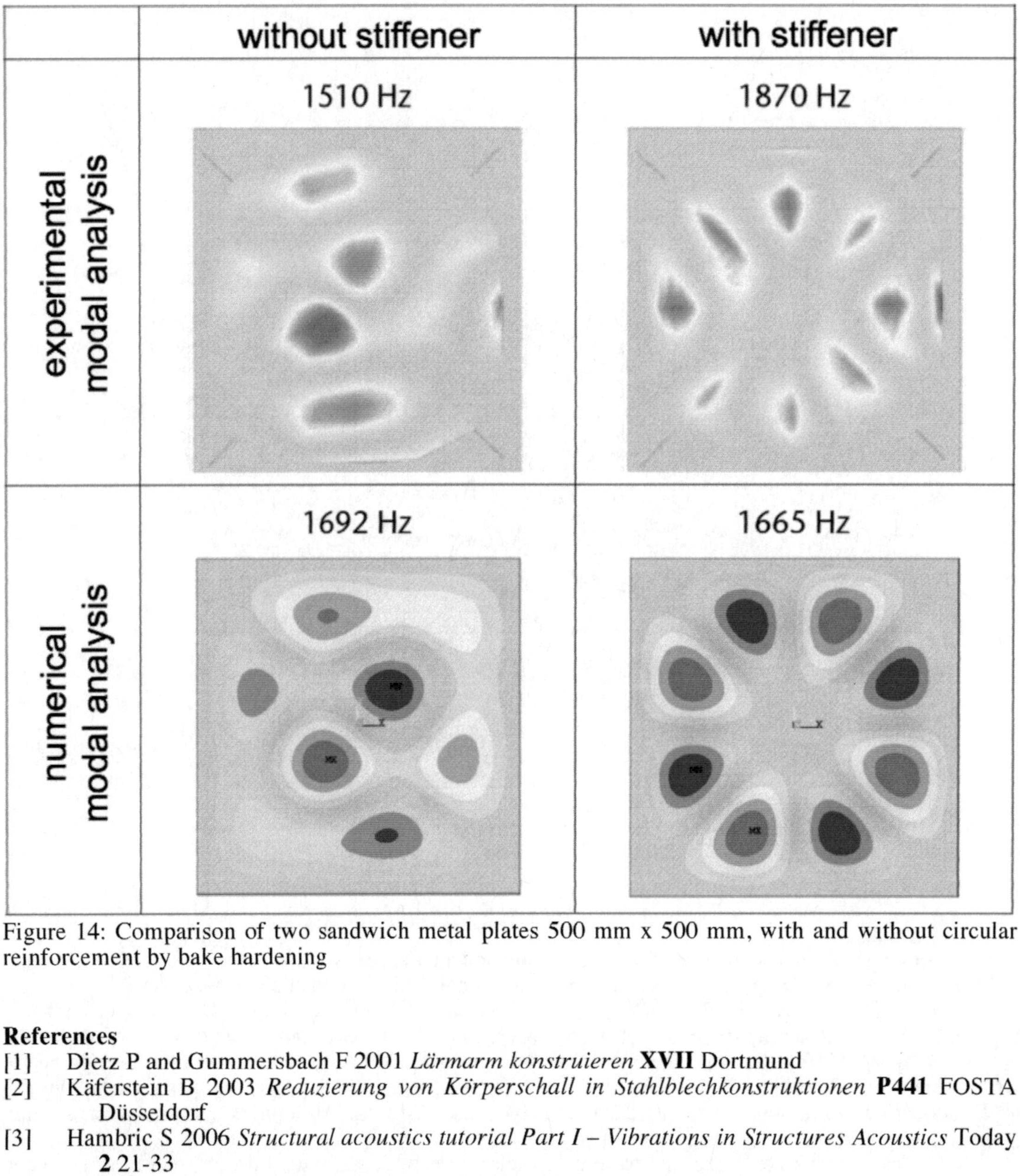

Figure 14: Comparison of two sandwich metal plates 500 mm x 500 mm, with and without circular reinforcement by bake hardening

References

[1] Dietz P and Gummersbach F 2001 *Lärmarm konstruieren* **XVII** Dortmund
[2] Käferstein B 2003 *Reduzierung von Körperschall in Stahlblechkonstruktionen* **P441** FOSTA Düsseldorf
[3] Hambric S 2006 *Structural acoustics tutorial Part I – Vibrations in Structures Acoustics* Today **2** 21-33
[4] Chaigne A 2014 *Structural Acoustics and Vibrations* (Handbook of acoustics) ed T D Rossing (Heidelberg Springer Science+Business Media) chapter 22 pp 960-965
[5] Lohrengel A, Rolshofen W and Thoden, D 2009 *Institutsmitteilung* **34** 107-114

IOP Conf. Series: Journal of Physics: Conf. Series **1149** (2018) 012005 doi:10.1088/1742-6596/1149/1/012005

Experimental modal analysis of a utility-scale wind turbine blade using a multi-camera approach

A. Sabato*, P. Poozesh, P. Avitabile, and C. Niezrecki

Department of Mechanical Engineering, University of Massachusetts Lowell, 1 University Avenue, Lowell, MA 01854, USA

*Corresponding author e-mail: Alessandro_Sabato@uml.edu

Abstract. Utility-scale wind turbine blades (WTBs) require proper certification to ensure that they meet the design specifications and for safe operation throughout their lifetime. Instrumentation during testing involves using wire-based sensors such as strain gauges and accelerometers. These systems are expensive, time-consuming to install, difficult to deploy on large-sized structures, require additional instrumentations (e.g., power amplifiers and data acquisitions), and produce results at only a discrete number of points. In this study, a multi-camera based measurement system is implemented and experimentally evaluated to estimate the modal parameters of a ~60-meter long utility-scale WTB. The measurement areas of the stereo-vision systems are limited based on the field of view (FOV) of the cameras. Therefore, two sets of cameras were used to increase the measurement area. The performance of the optical sensing system in identifying the modal parameters of the WTB was validated by using the multi-camera approach to measure the dynamic response of the blade subjected to impulse inputs (i.e., impacts). The multi-camera stereo vision system in conjunction with a data stitching algorithm enables the photogrammetry technique to cover larger areas for expanded measurements. From an analysis of processed data, it was observed that the multi-camera system allows for detecting the first five operating mode shapes with an error below 2% as compared with traditionally wired accelerometers. For higher order modes, an estimation is possible, but the accuracy of the system decreases and miscorrelation in the mode shapes up to ~10% resulted.

1. Introduction

Global wind power capacity grew from under 5 GW to nearly 485 GW in the last twenty years. This upsurge of almost two orders of magnitude is expected to continue in the next few decades [1]. The size of wind turbine blades (WTBs) has become physically larger and blades having a length in the range of 90-120 m are expected to be routinely manufactured in the future [2]. WTBs play a crucial role in the energy production process, as they are the elements used to convert wind kinetic energy into electrical energy. Therefore, they require to be certified through static and fatigue testing at large-scale facilities such as the Wind Technology Testing Center (WTTC) in Charlestown, MA. Testing aims to guarantee the blades' conformity to the design. Usually, these tests are conducted using wire-based sensors such as strain gages, accelerometers, and string potentiometers. These systems are time-consuming to install, difficult to deploy on large-sized structures, require additional instrumentations (e.g., power amplifiers and data acquisition systems), produce results only at a discrete number of points, and are expensive. For instance, the deployment of 200 contact-type sensors on a ~55-meter long WTB requires nearly three weeks. In addition, workforce costs range from \$35k to \$50k for the strain gauge system and up to \$1000 for a single high-sensitivity accelerometer.

In the last decade, non-contact measurements have become extremely appealing tools for SHM and condition monitoring. Techniques such as Global Positioning Systems (GPS), Laser Doppler Vibrometry (LDV), radar interferometry, and photogrammetry made it possible to extract full-field response and dynamic characteristics of structures while reducing the time associated with the inspection procedure. GPS proved to be an easy-to-deploy system for measuring dynamic characteristics and static displacements in real time. However, its reduced measurement precision is a severe drawback to the applicability of this technique for modal analysis purposes [3, 4]. LDV sensors are routinely used to measure structural velocity. Nevertheless, their applications in the monitoring of real world, utility-scale blades is a challenge due to the need of high-intensity laser beam for monitoring structures far away, the large displacements and low frequencies of these structures, and because of the duration of the measurement that requires sequential scanning of all the points of interests [5, 6]. In recent years, radar interferometry (also known as microwave interferometry) has emerged as an innovative technology for performing non-contact vibration monitoring of large structures. The system is based on continuous-wave step-frequency radar generating, transmitting and receiving the electromagnetic signals backscattered from the surface of the test structure [7 - 9]. This technique has been successfully employed for monitoring the points of a wind turbine tower lower than the blade edges [10]. However, when it comes to detecting large displacements occurring at lower frequencies, similar to those happening during WTB certification testing, radar interferometry becomes less efficient, and its applicability in real-world scenarios decreases. Recent advances in camera technology, optical sensors, and image-processing algorithms have made computer vision techniques a viable option for streamlining the blade testing protocol. This approach may not only expedite the certification process of WTB, but it may also help to reduce the levelized cost of energy. Three-dimensional (3D) Digital Image Correlation (DIC) and 3D Point Tracking (3DPT) systems have proven their efficiency in being able to perform non-contact, full-field inspections [11, 12]. They provide accurate quantitative information of structural deformations, full-field strain maps, geometry profiles, and modal parameters over a wide range of civil and mechanical engineering systems such as sections of WTs, rotating machinery, bridges, railroad tracks, and composite materials [13 - 17]. However, monitoring large-scale structures such as utility-scale WTBs in a testing facility is limited by the cameras separation, the maximum available working distance, and the narrow field of view (FOV) of the cameras.

In this research, a multi-camera measurement system, employing a stereo-photogrammetry technique in conjunction with dynamic spatial data stitching, is used to evaluate the dynamic characteristics of a utility-scale WTB subjected to dynamic loading. In particular, this study summarizes the results of optical measurements performed using two coupled stereo-vision systems over a ~20-meter long section of a ~60-meter utility-scale blade during an impact test. Advantages of the multi-camera system were demonstrated by using the proposed optical sensing technique, in conjunction with an output only system identification approach, to detect the blade's operating mode shapes and resonant frequencies. This approach has the merit to eliminate the need for the accelerometer measurement system and the drawbacks mentioned above associated with wire-based contact sensors.

2. Experimental setup

Two different tests were performed to evaluate the performance of the proposed optical system in identifying the modal parameters of the blade. The first set of tests was conducted using conventional accelerometers to estimate the modal reference parameters from a well-established and standardized measurement technique. Afterward, the multi-camera system was implemented within the test facility to identify the dynamic characteristics of the blade using 3DPT technique and an output-only system identification method. This section describes the test setup and modal survey performed on the blade

IOP Conf. Series: Journal of Physics: Conf. Series **1149** (2018) 012005 doi:10.1088/1742-6596/1149/1/012005

using both traditionally wired accelerometers and optical techniques through 3DPT to obtain the WTB's natural frequencies and mode shapes.

The WTB was mounted to a large concrete anchor which was used to cantilever the structure. As observed in Figure 1, the high-pressure side of the WTB was facing the laboratory ceiling, the root center of the blade from the floor was 6.5 meter, and the whole structure was oriented at an angle 6-degrees relative to the laboratory floor. A total of ten biaxial MEMS-based accelerometers were mounted at specific locations of the WTB to obtain dynamic parameters of interest (i.e., natural frequencies and mode shapes).

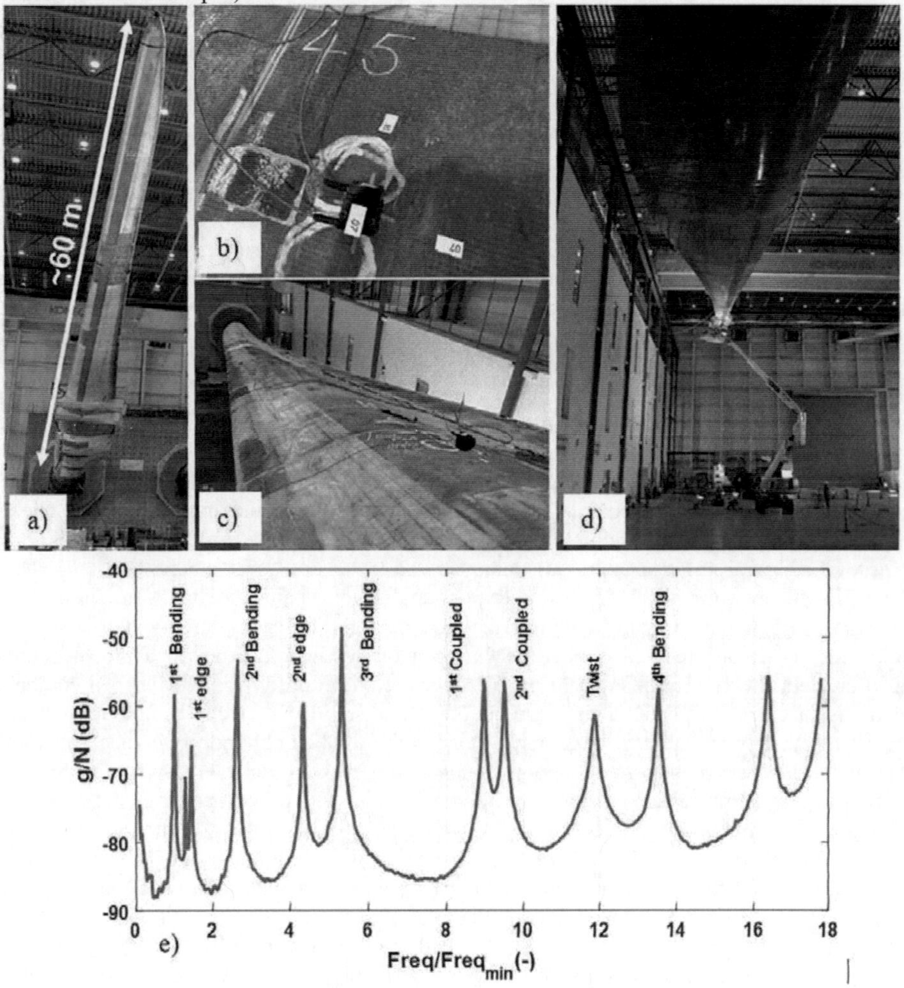

Figure 1 – Experimental modal test: a) ~60-meter utility-scale WT blade being tested; b) and c) accelerometers mounted on the high-pressure side of the blade; d) aerial lift equipment used for accessing the impact locations on the blade; e) sum of the cross spectrum and normalized resonant frequencies of the ~60-meter WTB computed using the accelerometers data.

The placement of the accelerometers was informed by the predicted mode shapes of the blade calculated using a FEM of the structure. In particular, eight accelerometers were used for measuring both in the flap and edge directions (i.e., X and Y direction), while two of them were used only for the flap direction measurements (i.e., X direction). During the test, the blade was excited at seven different locations using a modal hammer. In particular, five locations were excited both in edgewise and flapwise directions, while the remaining two were only impacted in the X direction. Because a single impact was not able to generate enough mechanical response to be measured by the optical system, the blade was excited by randomly impacting it several times before initiating the measurement. After 60 seconds of excitation, the response of the structure was obtained.

The accelerometers were installed on the WTB, and all cables were routed to a central hub from which the data acquisition hardware was connected using BNC cables. A 20-channel LMS data acquisition system was used to record the response of the blade as it was continuously impacted using the hammer. Fundamental frequencies (normalized to the first natural frequency, $Freq_{min}$) were computed from acceleration data and are shown in Figure 1e. Modal data recorded using the MEMS-based accelerometers were then used for performing a back-to-back comparison with those calculated using data obtained by using the optical system.

2.1. Experimental modal test using the optical system

To validate the performance of the optical sensing system in identifying the modal parameters of a utility-scale WTB, the multi-camera measurement system was used to measure the dynamic response of the blade.

When the size of the structure is beyond the FOV of the cameras, the working distance between the optical measurement system and the test structure needs to be increased. However, limitations on the maximum available working distance occur when working indoors. This increase in the FOV will also reduce the spatial measurement resolution (i.e., related to the number of pixels). Wider FOVs could be achieved by using (i) smaller focal length lenses, or (ii) higher working distances; however, these two options are not always available during the tests. In the first case image distortion may increase in the peripheral areas representing an issue for the measurements, while in the latter, wider FOVs would lower the resolution of the calculated data. For a utility-scale WTB within a test facility, the use of a standard stereo-photogrammetry to measure the entire blade may be limited by the camera separation or the required working distance to capture the area of interest. Therefore, larger FOVs can be obtained by stitching together, in a global coordinate system, the images recorded using the two adjacent sets of cameras as demonstrated in several prior studies [16, 18]. Thus, to perform a 3D-DIC measurement, it is desirable to couple several pairs of cameras to simultaneously monitor the distributed deformations within the line of sight of each stereo-vision system. Data obtained with individual standard optical systems can be stitched together in a global coordinate system to create a full-field displacement profile over the entire utility-scale WTB. To combine the individual displacement fields into a universal one, the recorded FOVs have to be laid out with several centimeters of overlap, and at least three optical targets are used as common reference points to find the rotation matrix that can align the two-point clouds obtained using each stereo-vision system.

The two DIC systems selected for this study consist of a pair of 4 Megapixel FASTCAM SA2 series high-speed cameras (i.e., Stereo-Vision A) and a pair of 12 Megapixel monochrome Toshiba Teli cameras (i.e., Stereo-Vision B). The first set of cameras was manufactured by Photron, employs a complementary metal–oxide–semiconductor (CMOS) image sensor with a 10 μm square pixel, has a resolution of 2048 × 2048 pixels when operated at 1080 fps, and was fitted with 24 mm focal length lenses manufactured by Nikkor. The second set is made by Toshiba, employs a CMOS image sensor with a 6 μm square pixel, has a resolution of 4096 × 3072 pixels, and was fitted with 24 mm focal length lenses. As shown in Figure 2, The Stereo-Vision A was positioned to have an approximate working distance of eight meters from the low-pressure side of the blade, a base distance of 3 meters and a 20° degree separation angle between the cameras' central line of sight. The Stereo-Vision B was positioned closer to the blades' tip section with an approximate working distance of 11 meters, a base distance of three meters, and camera angle of 16° degrees. Four stadium lights with 1500 W metal Halide bulbs were used to illuminate the measurement surface. Figure 3 shows the experimental setup used for performing the data acquisition during the modal test.

IOP Conf. Series: Journal of Physics: Conf. Series **1149** (2018) 012005 doi:10.1088/1742-6596/1149/1/012005

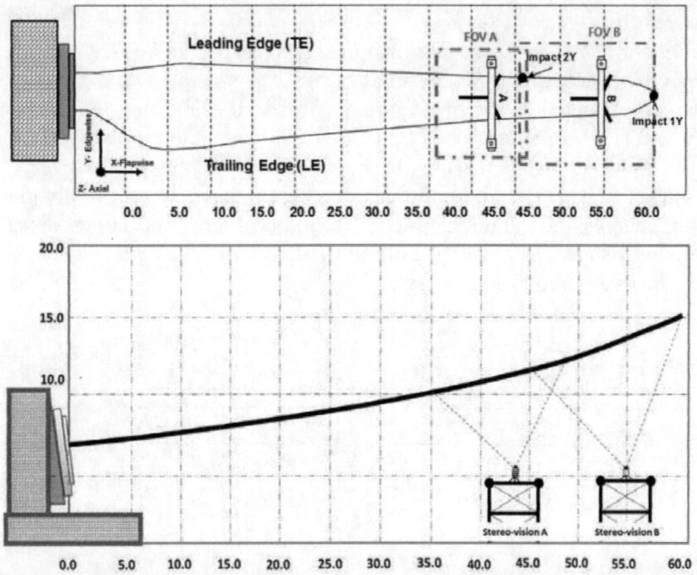

Figure 2 – Conceptual setup of the stereo-vision systems deployment and FOV of each individual system (not to scale).

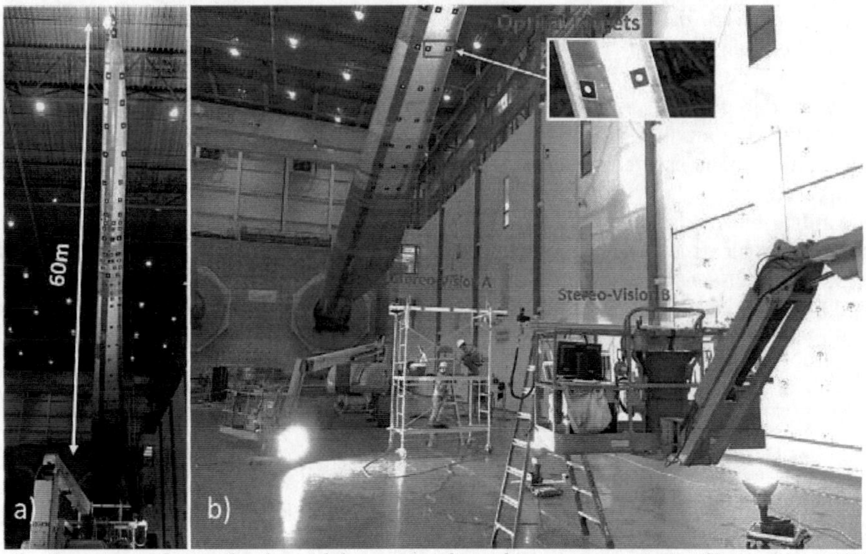

Figure 3 – Experimental setup for the modal test using the multi-camera system: (a) 60-meter long WTB with optical targets attached to the low-pressure side of the blade; (b) laboratory setup with optical targets detail.

All four cameras were synchronized to capture the motion of the cantilevered blade at the same time, by recording the position of the 56 optical targets shown in Figure 3 to measure their in-plane and out of plane displacements. Images were recorded at a 52 Hz sampling rate for 35 seconds. Before performing the measurements, the camera systems were calibrated using a large-area calibration procedure (not shown here for the sake of brevity) to obtain the extrinsic and intrinsic parameters of each vision system together with their radial distortion coefficients.

3. Analysis of the results

The multi-camera measurement was used to obtain the response of the WTB at locations of the optical targets as the structure was excited using a modal hammer at points 1Y and 2Y (see Figure 2). The position of the markers mounted on the blade was quantified at each time step using a commercially available 3DPT system (PONTOS developed by GOM). The point clouds obtained from stereo-vision A and B were then stitched together using the corresponding points in the overlapping areas for achieving a ~ 20-meter large FOV. The multi-camera stereo vision system in conjunction with data stitching algorithm enables the photogrammetry techniques to cover larger areas for expanded measurements. Figure 4 shows an example of the time history of the same targets, common to both FOVs, recorded by the two stereo vision systems.

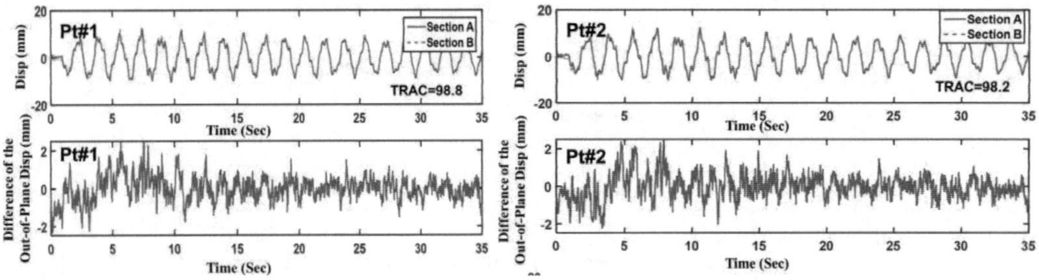

Figure 4 – Comparison of the displacements measured by stereo-vision A and B at optical targets in the overlapping area.

As can be observed, the displacement error of the corresponding points in the overlap area is generally smaller than ± 2 mm (i.e., on the same order of magnitude of the expected noise floor). Moreover, to validate the accuracy of the two sets of data, the time response assurance criterion (TRAC) was used to quantify the correlation between the two-time traces. TRAC values close to 100 indicate remarkable similarities, where values close to 0 indicate minimal or no similarity between the data sets. For the results shown in Figure 4, the TRAC values indicate the two-time traces for the different targets in the field of view A and B are almost identical (TRAC = 98.4).

To streamline the modal parameters of the WTB, impact testing without controlled force (i.e., output-only system identification) methods were used for processing the data recorded by the cameras. Using the acquired time data, spectral processing was performed to identify modal parameters of the WTB. The time domain signals were transformed in the frequency domain using the FFT algorithm and used for extracting the mode shapes of the WTB in flapwise and edgewise directions (see Figure 5).

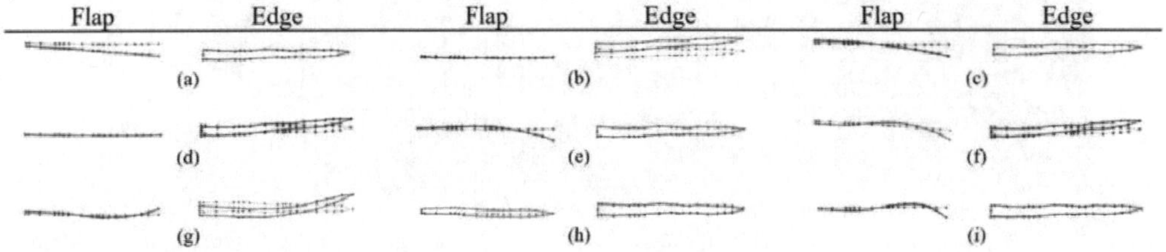

Figure 5 – Mode shapes of the ~60-meter WTB calculated from the data recorded using the multi-camera system: (a) 1st bending; (b) 1st edgewise; (c) 2nd bending; (d) 2nd edgewise; (e) 3rd bending; (f) 1st coupled; (g) 2nd coupled; (h) twist; (i) 4th bending.

Finally, Table 1 shows the accuracy of the proposed measurements technique (i.e., Optical) compared to the experiment performed using the input-output system identification method and accelerometers. The frequencies extracted from the optical data were normalized to the first natural frequency of the blade identified from the experimental modal analysis (i.e., F_{MIN}). From data summarized in Table 1 and an analysis of the Modal Assurance Criterion (MAC), it was observed that the multi-camera system allows for detecting the first five operating mode shapes with an error below 2% as compared with traditionally wire-based accelerometers.

Table 1: Comparison between frequencies measured using the accelerometers and those calculated using the data recorded by the multi-camera system.

Mode	Accelerometers	Optical	Difference	MAC
(-)	F_{ACC}/F_{MIN}	F_{OPT}/F_{MIN}	(Hz)	(%)
1st bending	1.000	0.990	0.010	99.2
1st edgewise	1.438	1.436	0.002	98.4
2nd bending	2.640	2.630	0.010	99.3
2nd edgewise	4.320	4.160	0.160	98.2
3rd bending	5.310	5.190	0.120	98.1
1st coupled	8.970	8.710	0.260	81.5
2nd coupled	9.500	9.210	0.290	73.7
Twist	11.850	11.440	0.410	86.1
4th bending	14.390	13.050	1.340	85.7

For higher modes, a measurement is still possible, but the accuracy of the system decreases and miscorrelation in the mode shapes up to ~10% resulted.

4. Conclusions

In this study, a novel measurement technique based on a multi-cameras 3D digital image correlation system is implemented and experimentally evaluated to estimate the modal parameters of a ~60-meter long utility-scale WTB. The measurement areas of the stereo-vision systems are limited based on the field of view (FOV) of the cameras. Therefore two synchronized sets of cameras were used to increase the FOV to increase the measurement area and cover an area having dimensions of approximately 4 x 20 m.

The performance of the optical sensing system was verified by using the displacement measured using the multi-camera system to determine resonant frequencies and mode shapes of a utility-scale WTB subjected to an impact test and compared to results obtained with traditional wire-based accelerometers. Analysis of the results showed that the proposed multi-camera approach is useful in measuring the first five operational natural frequencies of the blade with an error below 2% when a comparison with traditional contact sensors is performed. Higher order modes (i.e., sixth to ninth) can still be detected, but errors in the identification of the associated frequency rise to ~10%.

This research contributes to the overarching goal of developing a dynamic testing system that can accurately monitor the condition of utility-scale blades during the static, dynamic, and fatigue tests throughout the blade certification process. Unlike most of the conventional SHM sensing systems, which have limitation due to costs, wires, data transmission, power requirements, and only being able to provide discrete sensing, the proposed imaging system allows for displacement measurements over the entire surface of the utility-scale blade.

Acknowledgement
The authors would like to acknowledge the contribution of the personnel at the Wind Technology Testing Center (WTTC) in Charlestown, MA for their involvement in this research and for providing assistance and access to the utility-scale wind turbine blade used during the tests.

References

[1] Global Wind Energy Council (GWEC). *Global wind report. Annual market update 2015*, http://www.gwec.net/wp-content/uploads/vip/GWEC-Global-Wind-2015-Report_April-2016_22_04.pdf. (2016, accessed May 2018).
[2] International Energy Agency (IEA). *Technology roadmap. Wind energy*,

https://www.ieawind.org/wnd_info/IEA_Paris/Wind_Roadmap.pdf. (2009, accessed May 2018).

[3] Im SB, Hurlebaus S and Kang YJ 2011 Summary review of GPS technology for structural health monitoring. *J. of Struct Eng* **139.10** 1653-64.

[4] Casciati F and Fuggini C 2011 Monitoring a steel building using GPS sensors *Smart Struct Syst* **7.5** 349-63.

[5] Nassif HH, Gindy M and Davis J 2005 Comparison of laser Doppler vibrometer with contact sensors for monitoring bridge deflection and vibration *Ndt&E Int* **38.3** 213-18.

[6] Johansmann M, Siegmund G, and Pineda M 2005 Targeting the limits of laser Doppler vibrometry. *Proc. IDEMA* p 1-12.

[7] Gentile C 2010 Deflection measurement on vibrating stay cables by non-contact microwave interferometer *Ndt&E Int* **43.3** 231-40.

[8] Gentile C and Bernardini G 2010 An interferometric radar for non-contact measurement of deflections on civil engineering structures: laboratory and full-scale tests *Struct Infrastruct E* **6.5** 521-34.

[9] Pieraccini M 2013 Monitoring of civil infrastructures by interferometric radar: A review *Sci World J* http://dx.doi.org/10.1155/2013/786961

[10] Pieraccini M, Parrini F, Fratini M, Atzeni C and Spinelli P 2008. In-service testing of wind turbine towers using a microwave sensor *Renew Energ* **33.1** 13-21.

[11] Schmidt T, Tyson J and Galanulis K 2003 Full-field dynamic displacement and strain measurement using advanced 3d image correlation photogrammetry: part 1 *Exp Tech* **27.3** 47-50.

[12] Schmidt T, Tyson J and Galanulis K 2003 Full-field dynamic displacement and strain measurement-specific examples using advanced 3D image correlation. *Exp Tech* **27.4** 22-26.

[13] Poozesh P, Baqersad J, Niezrecki C, Avitabile P, Harvey E, and Yarala R 2017 Large-area photogrammetry based testing of wind turbine blades Mech Syst Signal Pr **86** 98-115.

[14] Lundstrom T, Baqersad J, and Niezrecki C 2016 Monitoring the dynamics of a helicopter main rotor with high-speed stereophotogrammetry *Exp Tech* **40.3** 907-19.

[15] Nonis C, Niezrecki C, Yu TY, Ahmed S, Su CF, and Schmidt T 2013 Structural health monitoring of bridges using digital image correlation. *Proc. SPIE - Health Monitoring of Structural and Biological Systems* (San Diego, CA) 869507.

[16] LeBlanc B, Niezrecki C, Avitabile P, Chen J, and Sherwood J 2013 Damage detection and full surface characterization of a wind turbine blade using three-dimensional digital image correlation *Struct Health Monit* **12.5-6** 430-39.

[17] Sabato A and Niezrecki C 2017 Feasibility of Digital Image Correlation for railroad tie inspection and ballast support assessment *Measurement* **103** 93-105.

[18] Sabato A, Beale C H and Niezrecki C 2017 A novel optical investigation technique for railroad track inspection and assessment *Proc. SPIE - Nondestructive Characterization and Monitoring of Advanced Materials, Aerospace, and Civil Infrastructure* (Portland, OR) 101692C.

IOP Conf. Series: Journal of Physics: Conf. Series **1149** (2018) 012006 doi:10.1088/1742-6596/1149/1/012006

Effect of Increased Damping in Subordinate Oscillator Arrays

John Sterling[1], Joseph Vignola, Ph.D[2], Jenna Gietl[2], Teresa Ryan[3], Ph.D[3], Noah Sonne[3], Sai Tej Paruchuri[4]

[1]Naval Surface Warfare Center, Carderock Division
9500 Macarthur Blvd, Bethesda, MD 20817
[2]Center for Acoustics, Vibrations and Structures
Department of Mechanical Engineering
The Catholic University of America
Washington, D.C. 20064
[3]East Carolina University, Acoustics and Vibrations Laboratory
Greenville, NC 27858
[4]Department of Mechanical Engineering
Virginia Polytechnic Institute and State University
Blacksburg, VA 24061

E-mail: `jsterling@gmail.com`

Abstract. Previous literature has shown that large structures with a significant number of smaller attached structures exhibit much higher apparent damping than was predicted by models. Further research into this effect led to the discovery that arrays of such small attachments can be designed to alter the response of the overall system. That work also showed that small variation in the distribution of attachment mass or stiffness degrade the performance of the system. Additive manufacturing has now matured such that several methods provide the capability to realize the tolerances required to effectively test these designs. This paper discusses the use of StereoLithography Apparatus (SLA) optical fabrication to test the error sensitivity of different materials using laser vibrometry to characterize the mechanical behavior of the system.

1. Introduction

Previous works have examined the vibration response of structures with several smaller structures attached and found that the primary structure would often appear more damped than classical theory would have predicted. This phenomenon of apparent damping has been a topic of focused research for some time, such that alteration of a system's resonant response with smaller resonant arrays has been well characterized[1, 2, 3, 4]. Such sets of smaller structures have been termed a subordinate oscillator array (SOA). An SOA can be designed to effect specific response modifications in time or frequency domain[5]. This work examines the application in which the SOA is designed to remove energy in a particular frequency band from the primary structure and move the energy into the subordinate oscillators to dissipate within the subsystem. It has also been shown that small levels of disorder can drastically impact the response of the system. However, because metals have been the only feasible manufacturing option, they were the only materials considered in simulations thus far. Since the 3D printed materials considered

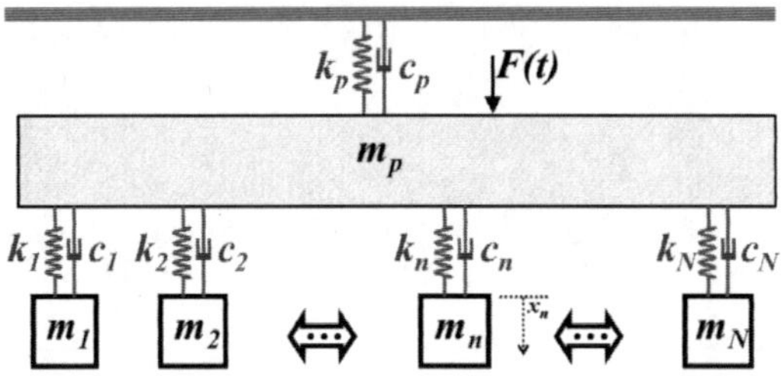

Figure 1: $N + 1$ degree of freedom model of a mechanical system with N subordinate elements attached to the primary element. Each element has a distinct mass m_n, stiffness k_n, and damping c_n.

in this work are plastic, the quality factor(Q), which is an indicator of damping, has been measured to be approximately an order of magnitude less than that of typical metals. The lower Q results in a more disorder-tolerant system. In this work, experiment and simulation will be put forth showing how the lower Q can overcome the sensitivity to disorder, such as the disorder resulting from manufacturing variation. This result allows consideration of more manufacturing techniques for the production of the subordinate oscillator arrays.

The lumped element model is shown in Figure 1. This paper considers only the frequency response of the system. Figure 2 shows several configurations of the system (Figure 2a) and one configuration with increasing amounts of added disorder (Figure 2b). For this paper, disorder is defined as a statistical average of deviation errors in individual oscillators.

Figure 2a includes four different distributions of the masses, m_n, stiffnesses, k_n with Q of 62.5. Selection of the property distributions allows for tuning of system bandwidth, as illustrated here with bandwidths ranging from 1.25% to 10%. The 10% disorder case from Figure 2a is used as an example in Figure 2b. Statistically inducing increasing levels of disorder (from 0.01% to 10%) into the frequency distribution of the oscillators shows how little disorder is required to degrade the response of the system. Figure 2b illustrates this progressively increasing effect of fabrication error on a bandpass response. This work examines results for the 3D printed resin with a Q of 14.1 in both analytical and experimental situations.

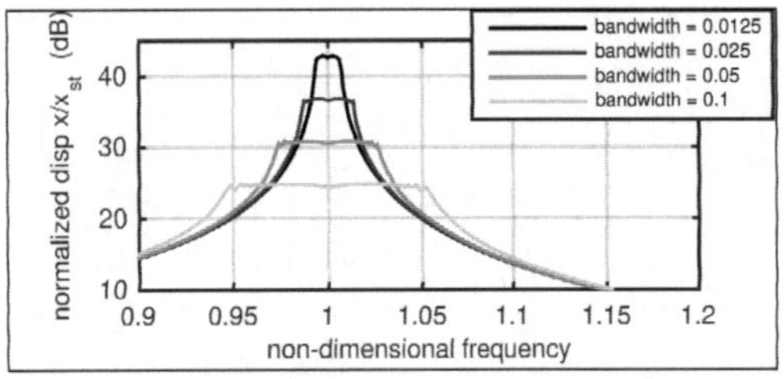

(a) SOA bandpass configurations

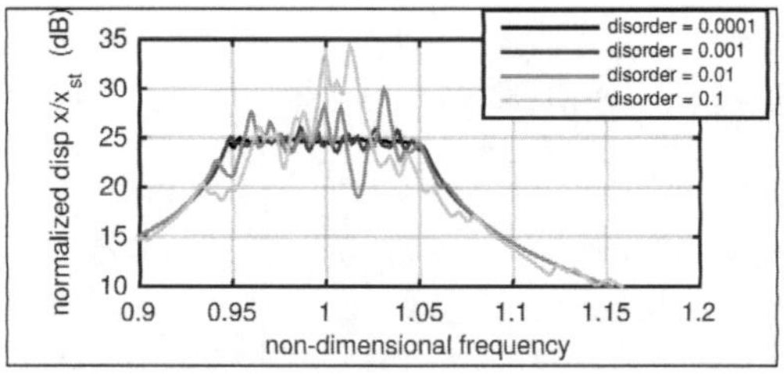

(b) SOA bandpass performance with disorder

Figure 2: (2a) The array of attachments can be designed to make the response of the primary element behave as a bandpass filter (2b) A bandpass response was designed to have a 10% bandwidth. The performance degrades as error in the property distribution increases from one part in 10^4, where the bandpass performance is unaffected, to 10% where the bandpass is response is not apparent.

IOP Conf. Series: Journal of Physics: Conf. Series **1149** (2018) 012006 doi:10.1088/1742-6596/1149/1/012006

2. Design Parameters of a Subordinate Oscillator Array

Previous work[5] has focused on simplifying the variables required to describe the SOA analytically. Instead of describing specific length, width, and thickness dimensions for each element, the array is specified according to distributions of frequencies and masses across N oscillators. In this method, design of effective bandpass arrays is simplified as the frequency distribution combines length and thickness and the mass ratio (defined as a ratio of the sum of masses of the individual elements to that of the primary mass). The mass ratio can be adjusted by changing subordinate cantilever width without an effect on the frequency, as long as the aspect ratio of the subordinate cantilever is maintained to be beam-like and not move to a more plate-like regime. The frequency distribution minimums, maximums, and frequency spacing define the band and shape of the effect. The mass distribution contributes to the magnitude of the bandpass response as well as edge effects. The comprehensive effects of the mass distribution are still under investigation.

Additional work [6] puts forward a simple equation relating array parameters that can predict the minimum number of elements required to produce a flat bandpass response in the primary. The minimum number of oscillators is determined by

$$N = \eta Q \Delta$$

where η is the modal overlap, Q is quality factor and Δ is the desired fractional bandwidth. The fractional bandwidth can be easily defined as the width of the filter effect. Modal overlap is a measure of how densely packed the modes are, calculated by half-power bandwidth divided by the Δf between peak frequencies. Modal overlap, η, must be greater or equal to 2 for the filter to function correctly[6]. With modal overlap $\eta > 2$, it is predicted to increase the robustness of the design and decrease disorder sensitivity.

3. Effects of Quality Factor on Bandpass Performance

When considering the requirements to design a bandpass response, it follows easily that a design using subordinate cantilevers with lower Q would require a lower N (fewer cantilevers) due to the modal overlap requirement. With wider individual peaks, a given band can be sufficiently covered with fewer elements in the the subordinate array. What had not been predicted was how lower Q materials would significantly reduce the disorder sensitivity of the overall response because the response of the oscillators were no longer so highly localized in the frequency domain. Figure 2b shows that with even moderate disorder, errors in individual elements can cause undesirable local behavior in the response of the system, disrupting the desired flat band response. In that higher Q situation, slight geometric error in even a single cantilever can cause the overall dB reduction in the band to be greatly reduced.

The desire to decrease the quality factor and also the cost of production leads 3D printed plastics to become an attractive option. Metal waterjet machining costs were in the hundreds of dollars each for previous prototypes and these 3D printed prototypes cost dollars each. In addition, in many applications, the use of 3D printed material still requires some element of compromise on some preferred property of the printed component. Those compromises are made in deference to the advantage of speed or design flexibility that 3D printing offers. In the bandpass application of the SOA, given the demonstration that a lower Q material has a lower sensitivity to disorder, the 3D printed component is actually the preferred option.

4. Specimen Production

Prior simulation work focused on metals, and with Q values of 100 or more, the metal designs required numerous oscillators. As such, error in any individual oscillator could have a significant impact on the overall response. In the era of rapidly developing 3D printing technology,

IOP Conf. Series: Journal of Physics: Conf. Series **1149** (2018) 012006 doi:10.1088/1742-6596/1149/1/012006

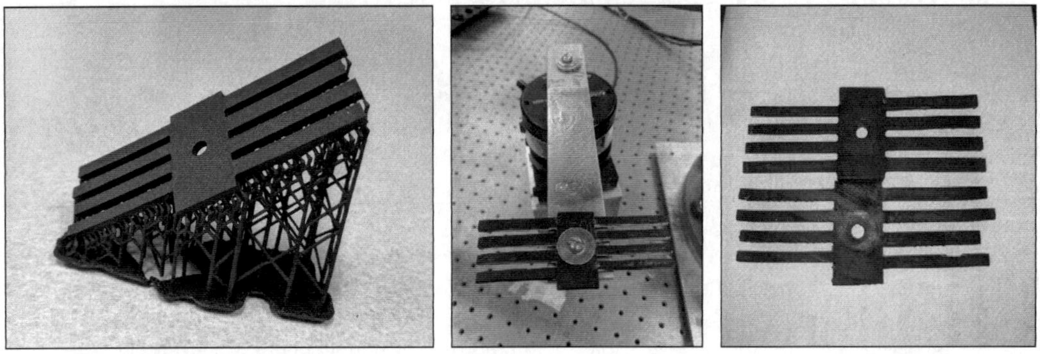

(a) SOA attached to scaffolding (b) SOA attached to beam (c) SOA comparison

Figure 3: (3a)A depiction of the SOA as printed by the Form 2 printer (3b) Shows the 10% percent SOA as attached to the beam and shaker. (3c)This is a comparative photo of the 10% SOA and the as-designed SOA one.

stereolithography apparatus printers, using lasers to harden resin, are able to construct parts meeting the tolerances that were formerly only available in a machine shop. The Form 2 SLA printer from FormLabs used for these prints has a smallest repeatable feature size of 150 microns. This is comparable to the performance of professional machine shops and makes disorder levels of 1/500 achievable even with these small parts. Figure 3a shows the final product of the 3D print process. The FormLabs GPBK-03 resin has a measured Q of 14.1 after a 30 minute IPA wash in the Form Wash machine and a 60 minute cure at 50 C in the Form Cure machine.

5. Laser Vibrometry Setup

A custom scanning laser Doppler vibrometrer system, developed at Catholic University of America and thoroughly described in [7], is used to characterize the printed arrays. The system consists of a Polytec OFV-5000 vibrometer controller and a Polytec OFV-501 remote head. A LABVIEW generated chirp excitiation is routed through a Krohn-Hite 7400 amplifier and into a Pasco 9324 shaker.

The simplest implementation of the SOA system is exhibited here as a primary beam on the shaker with an SOA mounted. Figure 3b shows the SOA system affixed to the primary beam and the laser measurement spot on the primary beam.

6. Experimental Results of Increased Damping

With the increased damping of SLA resin, the requirement of precise frequency distribution ordering is greatly relaxed without the possibility of a single, high-Q oscillator causing local disturbances. While the tolerances for the metal SOAs were almost impossibly tight, Figure 3c and Figure 5a show that even though there is a higher minimum disorder in the 3D printed parts, the results still shows a 17 dB peak reduction in this idealized situation. There is an offset in the center frequency response of the SDOF system from the SOA response systems due to the lack of accounting for the plate of material that connects the oscillators. Future work will account for that in order to properly match the center frequency of the SOA to that of the primary mass.

Prior results with low disorder metal samples in Figure 4a demonstrated less in-band ripple as shown in Figure 4b. Also evident is the fact that built in randomized disorder at one percent destroyed the broadband nature of the suppression. This is visible in how the 10% disorder case

IOP Conf. Series: Journal of Physics: Conf. Series **1149** (2018) 012006 doi:10.1088/1742-6596/1149/1/012006

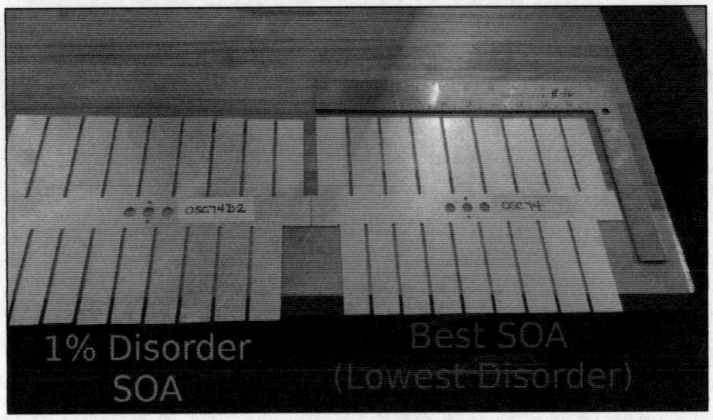

(a) 2 Steel SOAs

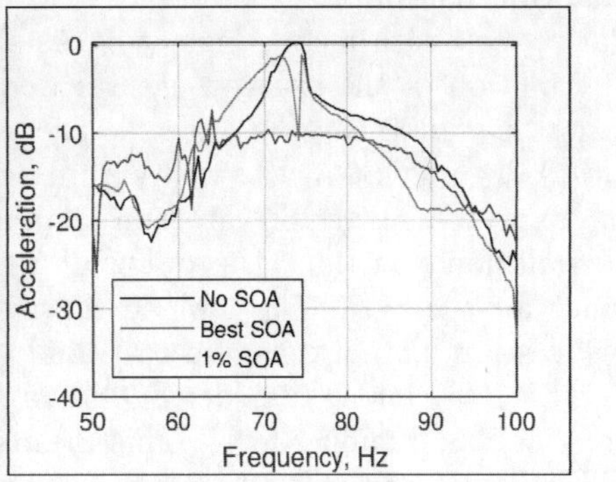

(b) Steel Results

Figure 4: (4a) This shows the steel 1% disorder and as-designed SOA side by side. Note how much closer in length the individual cantilevers are compared to the plastic ones. There is a small gap between the two pieces in the center with an arrow pointing at it for clarity. This shows how important the lower Q is for disorder tolerance (4b)This shows the initial disorder test response with steel SOAs. Note the low disorder curve works in a nearly ideal fashion, while the notch-filter curve has this response from a built-in 1% disorder

(orange curve) of Figure 5b provides a better reduction than the 1% disorder case (red curve) of Figure 4b.

These results show that SOAs with this high level of damping are effective and likely preferable in most situations as the low Q system with 10% disorder mitigates the hypersensitivity to disorder better than the 1% disorder displayed by the metals. This hypersensitivity was the limiting factor in the ability to implement these in real-world situations. These results also show a strong agreement between the MATLAB model and the experimental result (Figure 5).

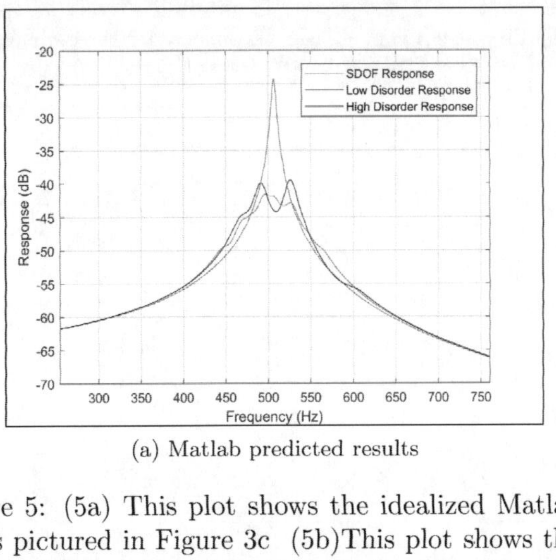

(a) Matlab predicted results

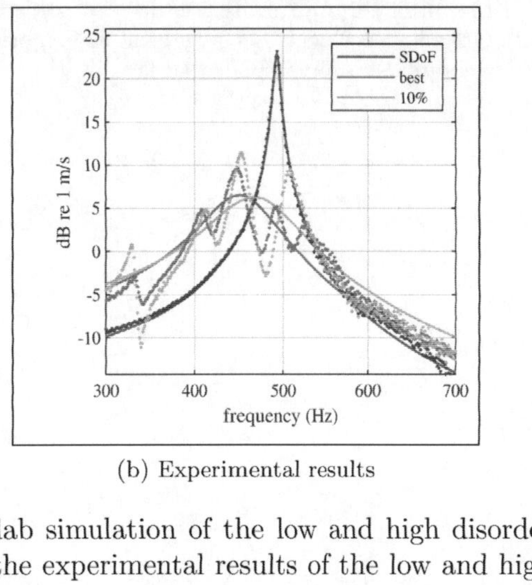

(b) Experimental results

Figure 5: (5a) This plot shows the idealized Matlab simulation of the low and high disorder SOAs pictured in Figure 3c (5b)This plot shows the experimental results of the low and high disorder SOAs using 3d-printed plastic. The dots represent the experimental data and the solid lines are a curve fit using half-power bandwidth, center frequency and Q of the data.

7. Conclusions

Early work with metal, high-Q SOAs showed an extreme sensitivity to error in the frequency distribution of the isolated natural frequencies of the oscillators. In such high-Q cases, error ratios of 1/1000 have a noticeable effect and 1/100 have significant effect. In contrast, the 10% low-Q disorder SOA was well within acceptable margins compared to the as-built SOA given the expected ranges of variation that may occur due to manufacturing variation, fatigue, or degradation over time. Use of low Q materials from 3D printers significantly expand the design space for the SOA. The lowered sensitivity of the arrays manufactured with plastic materials reduces production time and cost, and the arrays perform with equivalent or better metrics.

The 10%, low-Q disorder SOA had a nearly equivalent response to the as-designed low Q SOA at this damping level. This means an as-designed SOA has a high probability of success in the field in keeping its designed damping capabilities over time. This has prompted a future investigation into the feasible design space for the SOA. It also prompted investigation of a number of materials and manufacturing methods that were not previously considered as feasible alternatives. This is especially helpful for other, more traditional 3D printing methods which use nozzles instead of laser-hardened resin. In future work, the capabilities of different 3D printing methods and materials will be investigated and a design space will be created to inform future SOA manufacturing methods.

References

[1] L. Kitis, B.P. Wang, and W.D. Pilkey. Vibration reduction over a frequency range. *Journal of Sound and Vibration*, 89(4):559–569, August 1983.

[2] M. Strasberg and D. Feit. Vibration damping of large structures induced by attached small resonant structures. *J. Acoust. Soc. Am.*, 99(1):335–344, 1996.

[3] G. Maidanik. Induced damping by a nearly continuous distribution of nearly undamped oscillators: linear analysis. *Journal of Sound and Vibration*, 240:717–731, 2001.

[4] A. Akay, Z. Xu, A. Carcaterra, and I. Murat Ko. Experiments on vibration absorption using energy sinks. *J. Acoust. Soc. Am.*, 118:3043–3049, 2005.

[5] J.F. Vignola, J.A. Judge, and A.J. Kurdila. Shaping of a system's frequency response using an array of subordinate oscillators. *Journal of the Acoustical Society of America*, 126:129–139, 2009.

[6] Joseph Vignola, Aldo Glean, John Judge, and Teresa Ryan. Optimal apparent damping as a function of the bandwidth of an array of vibration absorbers. *The Journal of the Acoustical Society of America*, 134(2):1067–1070, August 2013.

[7] P. O'Malley, T. Woods, J. Judge, and J. Vignola. Five-axis scanning laser vibrometry for three-dimensional measurements of non-planar surfaces. *Meas. Sci. Technol.*, 20(11):115901, 2009.

IOP Conf. Series: Journal of Physics: Conf. Series **1149** (2018) 012007 doi:10.1088/1742-6596/1149/1/012007

Full-field modal analysis using high-speed 3D digital image correlation

E Di Lorenzo[1], P Lava[2], R Balcaen[2], S Manzato[1], B Peeters[1]

[1] Siemens Industry Software NV, Interleuvenlaan 68, 3001, Leuven, Belgium
[2] MatchID, Deinsesteenweg 94A, 9031, Gent, Belgium

E-mail: emilio.dilorenzo@siemens.com

Abstract. Digital Image Correlation (DIC) is a non-contact full-field image analysis technique which allows to retrieve strains and displacements in three dimensions at the surface of any type of material and under arbitrary loading. In recent years, high-speed and high-resolution cameras have been developed for static as well as for dynamic applications. As consequence, the application fields for DIC have broadened and it has proven to be a flexible and very accurate measurement solution for deformation analysis and material characterization. Nevertheless, nowadays DIC is often used in a qualitative manner rather than as a metrological tool. This is especially due to the time-consuming task related to the post-processing of the images. When compared to other vibration testing techniques, full-field approaches (such as DIC) allow a greater flexibility by providing a very dense number of experimental data over a single measurement. Another advantage is related to the fact that the geometry is automatically extracted from the images. In this paper, the possibility to combine global acceleration measurements on a small component with local full-field standard machine vision quasi-static camera measurements is investigated. In particular, the regularization properties of DIC and their impact on modal analysis will be studied in detail. Strains and displacements could be used in a second stage for modal analysis purpose in order to characterize the dynamic behaviour of the specimen in a certain frequency range. Different approaches could be used for combining together the data obtained during the tests. The most obvious approach would be the alignment of the time histories based on reference signals for Frequency Response Functions (FRFs) calculation prior to perform any further processing. Unfortunately this is not always possible because of synchronization issues. An alternative possibility, in case broadband random excitation is used, requires to process time data into auto and crosspowers and identify the modal parameters by using Operational Modal Analysis (OMA).

1. Introduction

Nowadays the typical setup when testing and validating numerical models of a mechanical structure comprises a certain number of transducers (e.g. accelerometers, strain gauges, etc.). By only using a limited number of transducers it is not easy to comprehensively and quickly measure the dynamic response, especially when dealing with large-size structures such as utility-scale wind turbines or helicopter blades.

Furthermore, placing transducers is a labor intensive and time consuming task and it could introduce electrical noise to the measured signals due to the extensive and unavoidable wiring. This is one of the reasons behind the development of image processing techniques to perform modal analysis of mechanical structures without contact and without having to instrument the specimen. Another reason for advancing in the research in this field is connected to the damage detection and evaluation in lightweight structures. In fact, for this type of systems such as composite materials and 3D printed structures it is mandatory to avoid adding weight by placing sensors on their surface. This allowed the development of embedded fiber optics within the material layers and the development of noncontact measurement techniques. Among the others, a scanning Laser Doppler vibrometer (SLDV) has seen an impressive increase of use in the last years, but it is limited to sequential measurements (i.e. one point at a time) and in order to measure at several locations it needs to scan the entire surface while the excitation signal is repeated over time. This could lead to an inconsistency in the measurements since the physical excitation or the system itself can change over time during a single measurements. An alternative technique which is further discussed in this paper is the Digital Image Correlation (DIC) which allows to measure the full-field response of a structure. These approaches allow to measure the dynamic behavior in a very accurate manner since they are able to measure at many more points if compared to conventional testing techniques using strain gauges, accelerometers or displacement sensors. Various applications in the field of experimental mechanics require a full-field data set for the analysis of a system, e.g. VFM and FEMU for material identification ([1], [2], [3]), the detection of structural damage on components, etc. Digital image correlation (DIC, [4]) is one of the many available methods on the market (besides Moiré interferometry, speckle pattern interferometry, the grid method [5], etc.) that provides the ability of producing those full-field datasets and especially DIC has gained more interest over the last years due to its simpler setup and lower hardware cost. DIC only requires a set of cameras and a specimen which has been speckled (i.e. a gradient must be present in the photographed specimen, which can be created with spray paint, a marker, stickers, etc.), which allows tracking of points on a sub-pixel level if one assumes the conservation of optical flow. Images are thus compared between cameras and over time in order to find correspondences, which leads to a set of displacement data.

2. Digital Image Correlation: working principle

Figure 1 shows the DIC working principle: the first image of camera 1 is used as a reference image to which the image of the second camera (taken at the same time) is compared. The corresponding points can be triangulated into a three-dimensional point in space (denoted as [X,Y,Z]) if the orientation of the cameras is known (by performing a pre-test calibration). The same matching of points can be done over time so each measurement point can be tracked over time, leading to a total deformation tracking of the surface area of a specimen (in which the deformation is indicated as [U,V,W]).

This process holds some intrinsic requirements towards the images: a gradient, as well as contrast, must be present so features can be tracked. The size of those features lead to the minimal size of the so-called subset (also called interrogation window – indicated in yellow in figure 1), which encompasses several features and which is tracked throughout the set of images. This is needed since a single pixel is not a unique item due to the limited amount of grey-scales which are stored in an image (usually 8 bit cameras are used, so a maximum of 256 grey scales are available). A cluster of pixels is however a unique set of data which can be tracked, while it will also reduce the measurement noise at the same time. This subset is restricted in the way it can deform by assigning a so-called "shape function" to it, leading to an even higher amount of robustness against noise (more information on the shape function can be found in [5] and [6]). This, together with image interpolation for sub-pixel accuracy [7], leads to high-quality and precise measurements of shape and displacements. From here on other metrics can be derived; strains can be identified by looking at the relative displacement between neighboring points [8], velocities, accelerations and strain rates can be derived if the timestamp of each image is known, stresses can be calculated if the correct material model is known [9], etc., which leads to the versatility of this optical-numerical technique [10].

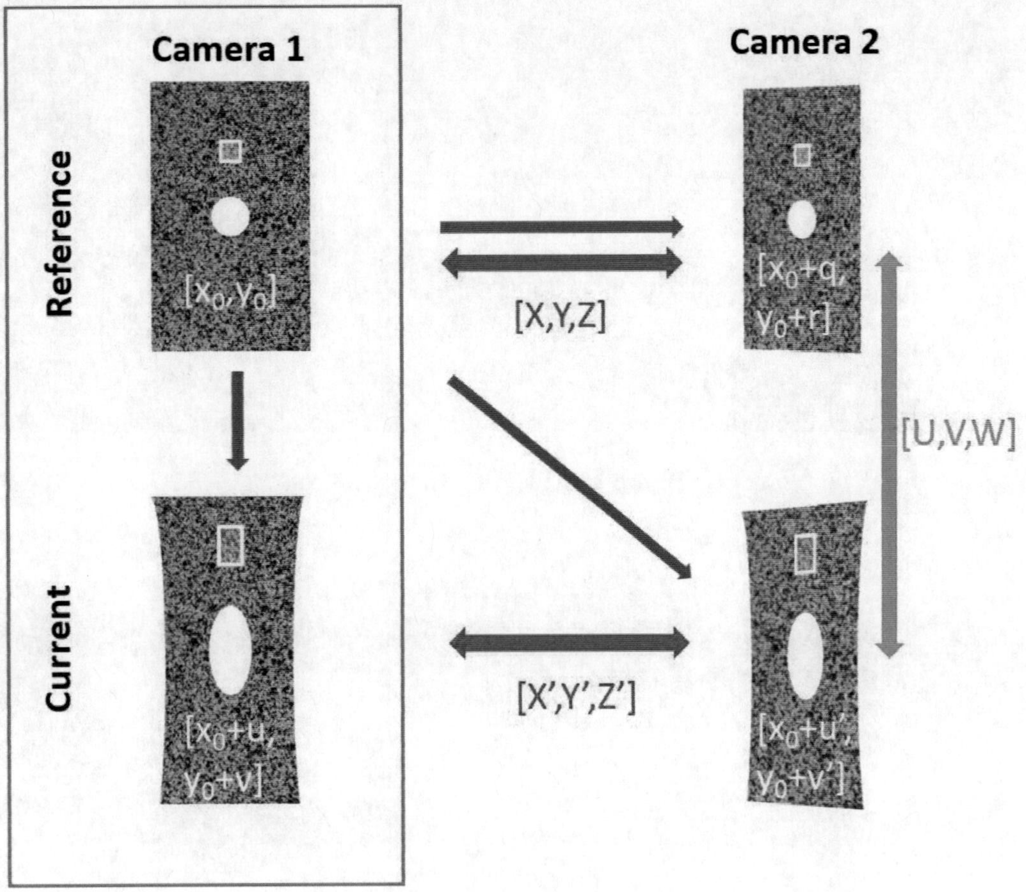

Figure 1. Principle of DIC.

3. Case study: helicopter blade

The possibilities of DIC as a measuring tool are further explored by using it as a data-source for performing a full-field modal analysis by measuring a tail blade of a helicopter, which is dynamically excited by a shaker. The high data-density of DIC leans itself as a proper tool for this since there is little surface preparation needed, while also having the benefit that the specimen is not locally loaded with accelerometers, which could alter the behavior of the structure in the case of fragile or small structures.

Figure 2 indicates the specimen on which a white base coat has been applied, which is needed to create enough contrast after applying the speckles (i.e. features). It also indicates the clamping on the right hand side, as well as the free end of the structure. Figure 3 indicates the setup after attaching all accelerometers (indicated as a red star), while the shaker is indicated in a blue dashed square. This state, as imaged by the camera, can eventually be seen in figure 4 (as imaged from the top).

Throughout the vibration test the specimen was imaged from the top with two AVT Mako U-130B cameras, of which the specifications can be seen in table 1. Each camera was equipped with a 12mm lens, leading to a standoff distance of about 2500mm. The applied speckle pattern was optimized for this specific setup with a pattern generator in order to achieve the maximum available spatial resolution, leading to a sampling of 6 pixels per speckle, a black/white ratio of 99.98% and a usage of the dynamic range of 55%. After performing a calibration the intrinsic and extrinsic parameters are determined (indicated in respectively in table 2 and table 3).

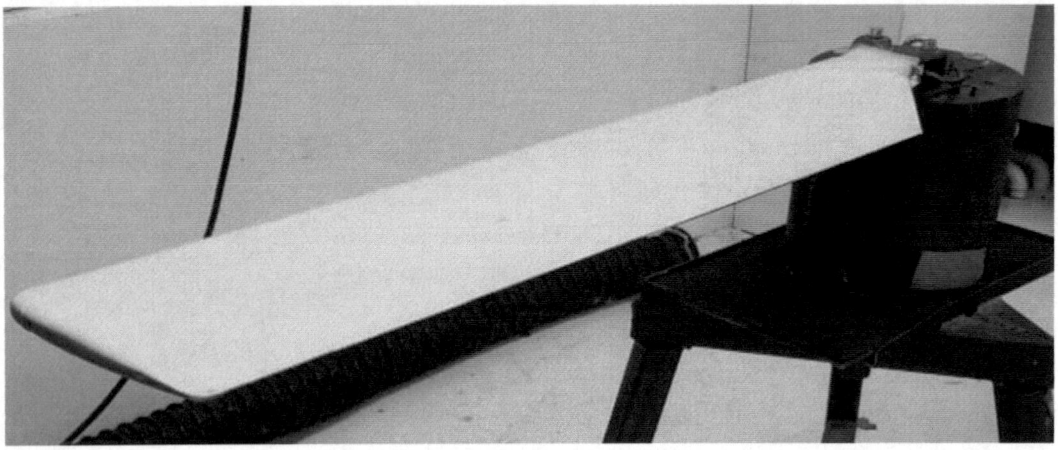

Figure 2. Blade with white base coat.

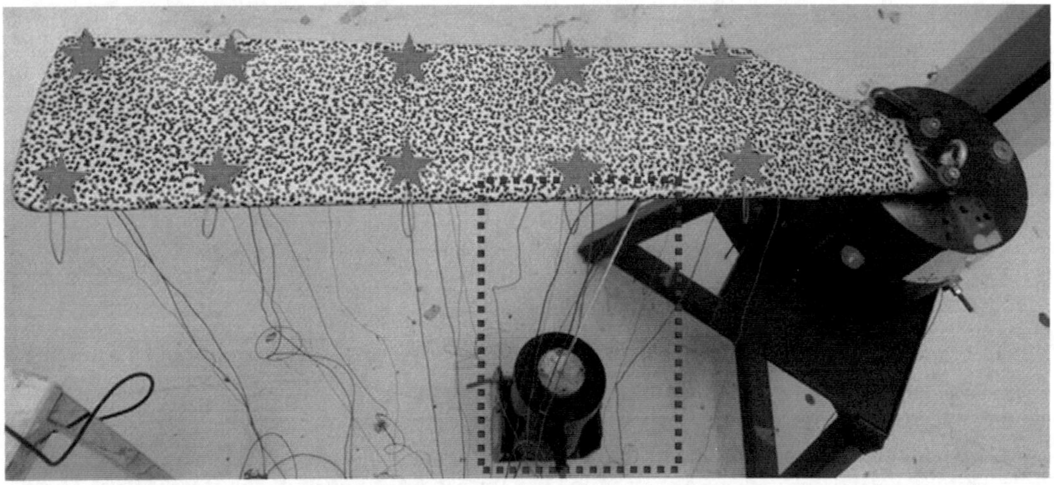

Figure 3. Speckled specimen with attached accelerometers indicates as a red star.

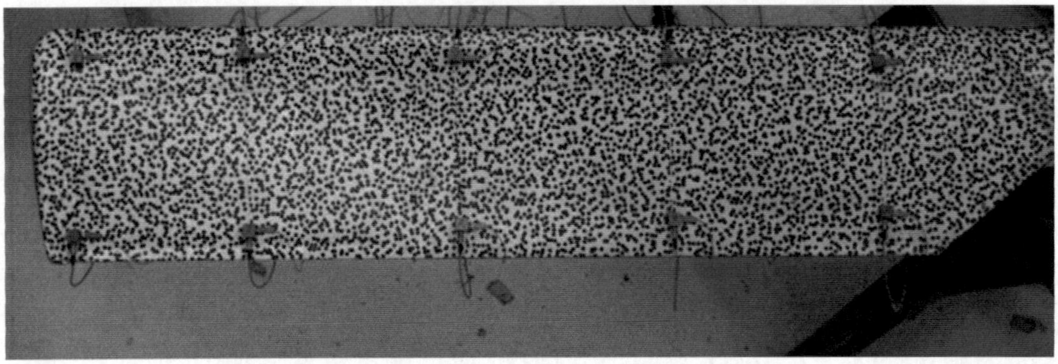

Figure 4. Specimen as imaged by the left camera.

Table 1. Mako U-130B specifications.

Parameter	Value
Resolution [pixels]	1280 (H) x 1024 (V)
Pixel size [μm]	4.8 x 4.8
Theoretical max. frame rate at full resolution [fps]	168
Mass [g]	60
Dimensions [mm]	49.5 x 29 x 29

Table 2. Camera intrinsic parameters.

Parameter	Camera 0	Camera 1
Fx, Fy [pixels]	[2649.28 ; 2649.35]	[2606.88 ; 2606.88]
Cx, Cy [pixels]	[665.40 ; 271.68]	[625.72 ; 292.30]
κ	[-0.375 ; 0.256 ; -0.375]	[-0.124 ; -0.003 ; 1.209]
P	[0.0033 ; 0.0019]	[0.0033 ; 0.0019]

Table 3. Camera extrinsic parameters.

Parameter	Translation camera 0 - 1	Parameter	Rotation camera 0 – 1
Tx [mm]	415.03	Θ [°]	2.73
Ty [mm]	-2.92	Φ [°]	-10.34
Tz [mm]	61.87	Ψ [°]	-0.06

4. Results: conventional modal analysis

The helicopter blade has been instrumented by means of ten accelerometers, as shown in Figure 3. One of the main objectives of this paper concerns the comparison between conventional modal analysis and full-field DIC-based modal analysis. On one hand the blade has been fixed to a rigid structure; on the other hand the blade is free to move. Several typical excitation signals have been sent to the shaker in order to dynamically excite the blade (random and sine testing). Hammer testing has also been performed and results of all tests are very similar in terms of modal parameters. Figure 5 shows the shaker signal during a sine sweep and two of the measured acceleration signals. Frequency Response Functions (FRFs) can then be calculated and a modal parameters estimator, such as Polymax [11], can be used in order to determine the natural frequencies, damping ratios and mode shapes of the helicopter blade.

IOP Conf. Series: Journal of Physics: Conf. Series **1149** (2018) 012007 doi:10.1088/1742-6596/1149/1/012007

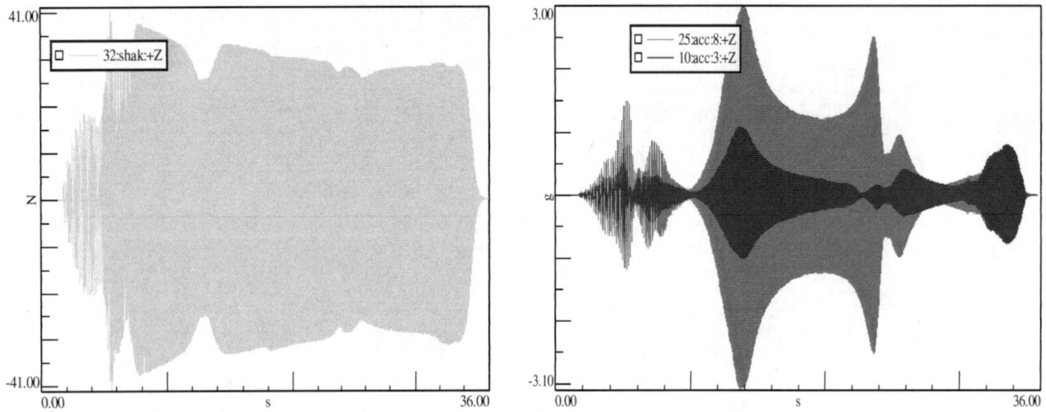

Figure 5. Shaker force signal during sine sweep (left); Accelerometers signals during the same sweep (right).

Table 4 shows the list of natural frequencies obtained with the conventional modal analysis in the frequency range up to 400Hz. The first four mode shapes are shown in figure 6.

Table 4. Helicopter blade natural frequencies and damping ratios in the range [0 - 400] Hz.

# mode	Natural frequency [Hz]	Description
1	20.1	1st bending mode
2	104.7	2nd bending mode
3	152.1	1st torsion mode
4	252.5	3rd bending mode

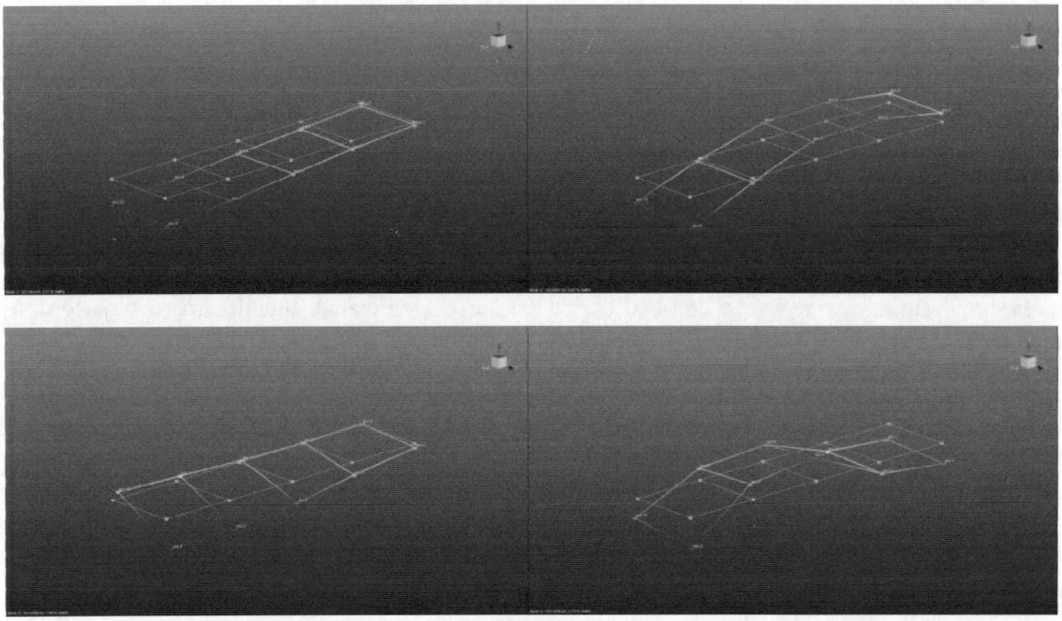

Figure 6. Helicopter blade mode shapes obtained by using conventional modal analysis based on ten 3D acceleration signals (marked points).

A discrete number of sensors is enough to characterize the global modes of a structure. On the other hand a full-field measurement technique allows to better identify local modes and to obtain enriched information which could be used for damage identification. The Modal Assurance Criterion (MAC) is used to validate the estimated modal model.

5. Results: full-field modal analysis

Various tests were performed (including hammer testing, shaker random testing and shaker sine testing), however only the results of one shaker sine test will be indicated further here for the sake of simplicity. During this test the shaker amplitude was fixed at 0.1V and the frequency spectrum ranged from 6 to 50Hz during a 30 seconds test. Throughout this test the blade was imaged by both cameras at their maximum synchronized recording speed and these images were post-processed with the settings mentioned in table 5 so a full-field displacement and strain map was obtained, with the respective noise levels indicated in table 6.

Table 5. DIC processing parameters.

Parameter	Value
Subset [pixels]	25
Step [pixels]	5
Interpolation	B-spline
Shape function	Affine
Strain window [pixels]	7
Strain interpolation	Q8

Table 6. Noise level displacements & strains.

Parameter	Value	Parameter	Value
Horizontal in-plane displacement U [μm]	4.5	Exx []	2.98·〖10〗^(-4)
Vertical in-plane displacement V [μm]	4.5	Eyy []	3.58·〖10〗^(-4)
Out-of-plane displacement W [μm]	26.6	Exy []	2.04·〖10〗^(-4)

Figure 7 indicates the full-field displacement map of the blade during loading, while figure 8 indicates an extraction of the blade tip during the first resonance frequency (indicating a peak displacement at image 300, which is around 20Hz). An enlarged side view of the blade can indicate the mode shape during this loading stage (as seen in figure 9). After deriving the time-displacement correspondence the full-field velocity and acceleration maps can be identified by a derivation of the identified displacement function (e.g. the velocity map in figure 10).

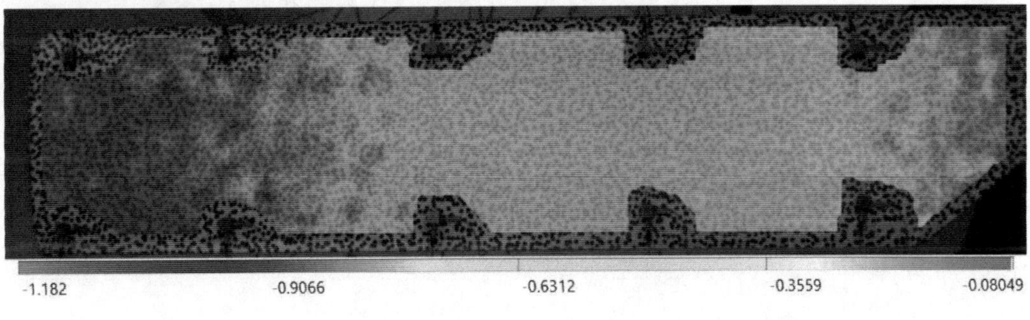

Figure 7. Out-of-plane displacement during loading [mm].

IOP Conf. Series: Journal of Physics: Conf. Series **1149** (2018) 012007 doi:10.1088/1742-6596/1149/1/012007

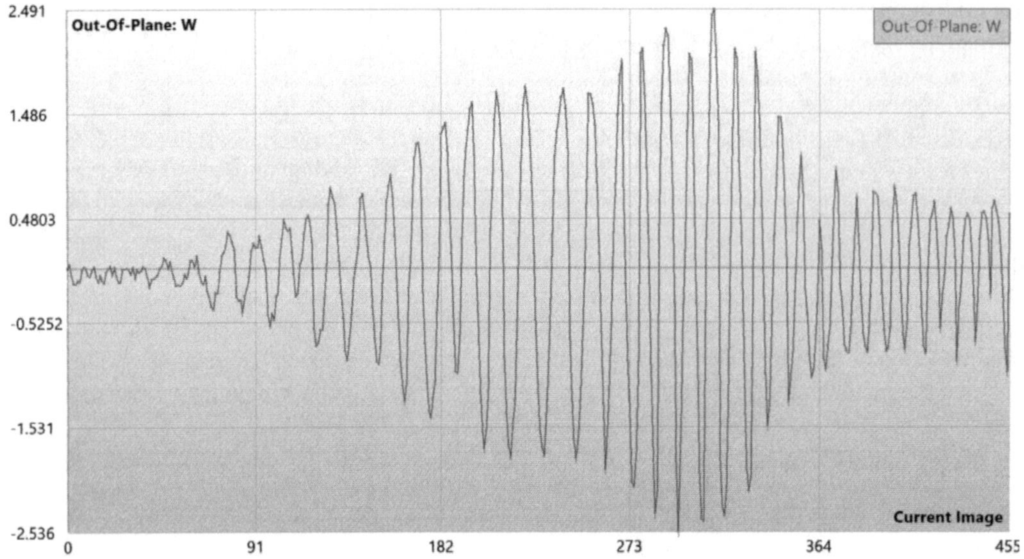

Figure 8. Blade tip displacement [mm].

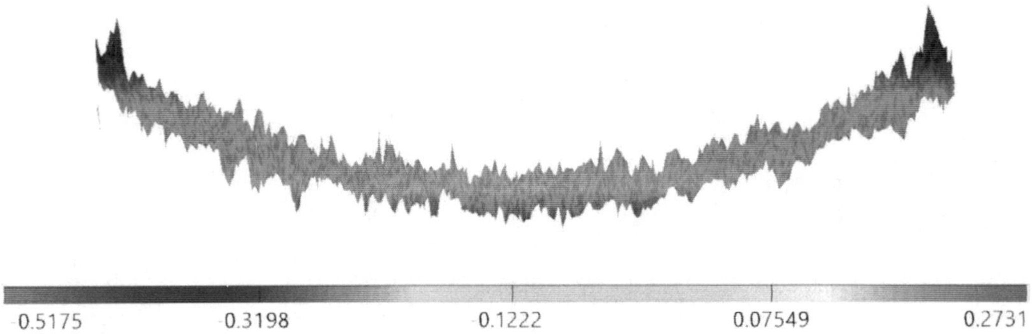

Figure 9. Blade deflection shape[mm].

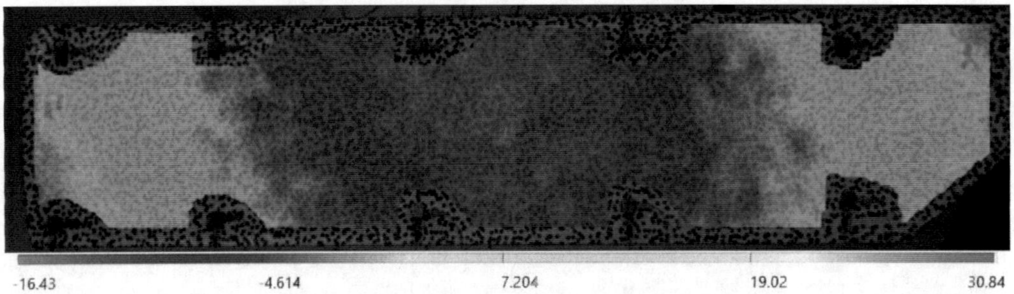

Figure 10. Velocity [mm/s].

IOP Conf. Series: Journal of Physics: Conf. Series **1149** (2018) 012007 doi:10.1088/1742-6596/1149/1/012007

One of the objectives of the work was the identification of the most practical ways to combine the data and results from the cameras and from the accelerometers. The most obvious approach would be to align the time histories based on reference signals prior to perform any further processing. This would have the advantage of allowing to calculate the Frequency Response Functions (FRFs) from the DIC displacements, but it would also require a perfect synchronization [12].

An alternative possibility is to process time data into half auto and crosspowers, identify the modal parameters using Operational Modal Analysis (OMA) and then combine the mode sets using a common reference [13]. During the performed sine tests the excitation profile in the frequency range of interest is equivalent to a white noise and OMA can then be reliably applied.

The steps going from the data acquisition to the modal analysis results for the DIC data are listed here:

1. Export displacement field of all images from MatchID software;
2. Manipulation and data reduction in Matlab to build the displacement time histories of all points measured during the entire measurements;
3. Import time histories in Simcenter Testlab for each measured Degree-Of-Freedom (DOF);
4. Import the geometry from MatchID software;
5. Calculate auto- and crosspowers and perform OMA

One of the main difficulties is related to the huge amount of data to be post-processed. In this case more than 10000 points were exported from the camera measurements and the geometry is presented in figure 11. Figure 12 shows the displacement time history measured by the cameras of a point randomly selected on the blade surface during the sine sweep test. The amplification of the oscillation around the resonance frequency is quite evident.

Figure 11. DIC measurement points and accelerometers (red points).

Finally, the mode shapes can be calculated. Unfortunately the cameras frame rate did not allow to go higher than 50Hz, which only allows the 1st mode identification (around 20Hz). This is explicitly shown in figure 13, where the 1st mode shape obtained by using the DIC modal analysis is compared to the one already shown in figure 6.

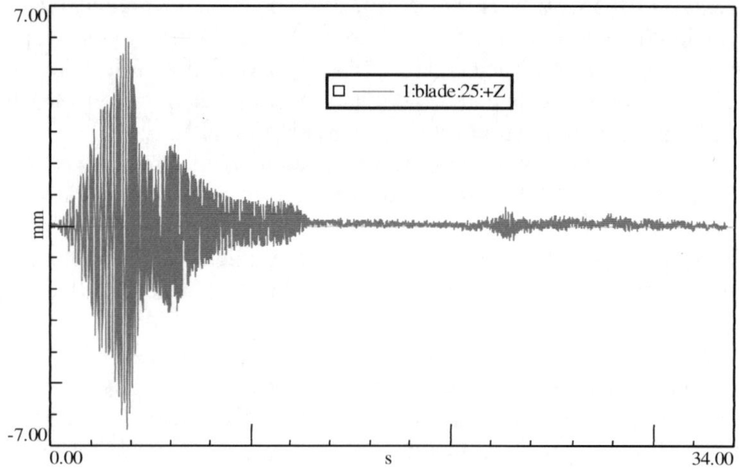

Figure 12. Displacement time history of a point on the helicopter blade surface measured by using the cameras.

Figure 13. 1st bending mode shape: Digital Image Correlation modal analysis (left) vs conventional accelerometers based modal analysis (right).

Conclusions

In this paper, an application which combines conventional modal analysis with standard machine vision quasi-static cameras and DIC analysis is presented. The quality of the measured DIC data is very good and its application in the modal analysis field in order to enrich the conventional modal analysis results with high density data is very promising. However some limitations need to be taken into account:

- Data reduction techniques should be implemented to allow faster and accurate processing with current processors capabilities
- Higher performance cameras need to be used for having a higher frame rate which allows to increase the maximum frequency to be analyzed.

A very interesting field of application is the combination of Digital Image Correlation and conventional modal analysis to perform a very accurate validation of Finite Element (FE) models due to the very high spatial resolution. This is one of the main advantages of DIC which shows also potential for identifying defects in lightweight strictures where the weight of sensors could be an issue when instrumenting the specimen for testing.

References

[1] Pierron F and Grediac M 2011 The virtual fields method *Springer Verlag New York*

[2] Denys K, Coppieters S, Seefeldt M and Debruyne D 2016 Multi-DIC setup for the identification of a 3D anisotropic yield surface of thick high strength steel using a double perforated specimen *Mech. of Mat.* **100** 96-108

[3] Tlemat H, Pilakoutas K and Neocleous K 2006 Modelling of SFRC using inverse finite element analysis *Mat. and Struct.* **39** 221-233

[4] Sutton MA 2008 Digital image correlation for shape and deformation measurements *Springer Handbook of Exp. Solid Mech.* 565-600.

[5] Grediac M and Sur F 2014 50th anniversary article: effect of sensor noise on the resolution and spatial resolution of displacement and strain maps estimated with the grid method *Strain* **50** 1-27

[6] Yu L and Pan B 2015 The errors in digital image correlation due to overmatched shape functions *Meas. Sci. and Tech.* **26** 045202

[7] Schreier WH and Sutton MA 2002 Systematic errors in digital image correlation due to undermatched subset shape functions *Exp. Mech.* **42** 303-310

[8] Schreier HW, Braasch JR and Sutton MA 2000 Systematic errors in digital image correlation caused by intensity interpolation *Opt. Eng.* **39** 2915-21

[9] Wattrisse B, Chrysochoos A, Muracciole JM and Némoz-Gaillard M 2001 Analysis of strain localization during tensile tests by digital image correlation *Exp. Mech.* **41** 29-39

[10] Lava P, Coppieters S, Van Hecke R, Van Houtte P and Debruyne D 2012 Digital image correlation in the classroom: determining stress concentration factors with webcams *Exp. Tech.* **37** 72-80

[11] Peeters B, Guillaume P, Van der Auweraer H, Cauberghe B, Verboven P and Leuridan J 2004 The PolyMAX frequency-domain method: a new standard for modal parameter estimation? *Shock and Vibr.* **11** 395-409

[12] Manzato S, Di Lorenzo E and Mackel P 2018 Enhancing standard GVT measurements with digital image correlation *Proc. of 36th Int. Modal Analysis Conf.* (Orlando FL)

[13] Peeters B, Van der Auweraer H, Vanhollebeke F and Guillaume P 2004 Operational modal analysis for estimating the dynamic properties of a stadium during a football game *Shock and Vibr.* **11** 395–409

IOP Conf. Series: Journal of Physics: Conf. Series **1149** (2018) 012008 doi:10.1088/1742-6596/1149/1/012008

3D vision method applied to measure the vibrations of non-flat items with a two-mirror adapter

T. Durand-Texte[1,2], M. Melon[1], E. Simonetto[2], S. Durand[2] and M.-H. Moulet[3]

1 LAUM UMR CNRS 6613, avenue Olivier Messiaen, F-72085 Le Mans cedex 9, France
2 Laboratoire Géomatique et Foncier, CNAM, 72000 Le Mans, France
3 Centre de Transfert de technologie du Mans, 72000 Le Mans, France

E-mail: `manuel.melon@univ-lemans.fr`

Abstract. In order to measure low frequency vibrations with a stereo sensor, it is interesting to increase the angles between the cameras and the measured surface, as the out-of-plane displacements are then more visible on the images. Even if Digital Image Correlation has proven to be a valid tool to measure vibrations, the initial pairing process remains difficult, in this context, because of the large pan angle, all the more so if the measured object presents significant variations in depth (for example a loudspeaker). This conference paper thus presents a new method specifically designed to rectify images (referred to as the IRIs method), which successfully allows initializing the vibration measurement with a high number of measurement points. In the same way, the conventional single-camera pseudo stereo system with a four-mirror adapter, which is largely used to perform displacement measurement, remains rather complex to operate. This paper thus proposes a single-camera simplified system, with a two-mirror adapter only. The ensuing global protocol is more user-friendly, and even if the results obtained for vibration measurement are a little less accurate with the two-mirror adapter, the operational modal shapes have been successfully retrieved and match very well those obtained with the conventional set-up.

1. Introduction

The measurement of structural acoustic sources can be achieved by various means, that may be gathered together into two different branches: on the one hand, vibration measurement methods which focus on the vibration of the source; on the other hand, acoustic field measurement approaches which use microphones. The former are very popular because they usually do not require an anechoic chamber and thus can be performed in normal rooms. Two classical measurement devices are generally used in this context: accelerometers [1] and laser vibrometers [2]. However, these single-point techniques are not appropriate for sources that exhibit non-stationary or non-linear behaviours. In that case, full-field methods, for which all the points of interest are measured simultaneously, become particularly interesting. As a consequence, over the past decades, several optical full-field methods have been adapted to vibration measurements: deflectometry [3], digital holography [4] and dynamic photogrammetry [5] for example. While allowing performing full-field measurement, they present additional assets, such as non-intrusiveness or low sensitivity to the ambient acoustic noise. In a context of shape or vibration measurement, dynamic photogrammetry and more precisely 3D vision methods are at the core of the approach presented in this paper.

In order to perform shape and deformation measurement, Digital Image Correlation (DIC) is largely used [6]. Along with the development of high-speed cameras, its relevance to measure vibrations has been increasingly studied over the past decade [5, 7–13]. These cameras are also used by other vibration measurement techniques, such as videography [14, 15], stereophotogrammetry [16], digital holography [4] or fringe projection [17].

For the 3D vision methods, the protocol usually designed requires two views of the studied object to perform triangulation and obtain the data needed to retrieve the 3D shape and displacement of the object. Originally, two high-speed cameras were used [7, 8]. Because of the cost of these devices and the associated problems to synchronise them neatly, another set-up, originating from robotics [18–20] is very often used. It requires a single high-speed camera and a four-mirror adapter to generate two virtual cameras, and thus views, from a single real one. This pseudo stereo system with a four-mirror adapter has been validated several times to measure vibrations [21–25].

In this context, the use of large angles between the measured surface of the object and the cameras has proven to provide better precision for out-of-plane displacement [6] and thus to lead to better results for vibration measurement [25]. However, the initial pairing process, using DIC tools, remains difficult if the pan angle is large and if the measured object presents variations in depth (typically a loudspeaker). This conference paper thus presents a method designed for this specific context and a simplified set-up, in order to make the whole protocol more effective, user-friendly, while keeping a high degree of precision in the vibration measurement.

2. 3D iterative rectification of images (referred to as the IRIs method)

The usual image rectification methods, such as the epipolar image rectification [26, 27], are no longer sufficient to pair a high number of measurement points automatically, notably if a large pan angle is used (from 60° to 90° approximately) and if the studied object presents significant variations in depth (cf. figure 1). A new protocol has thus been designed in order to quickly and easily perform shape measurement with these specific parameters. The resulting synoptic diagram is shown figure 2.

An initial coarse shape measurement is performed with rectified images [26–28], with few measurement points, paired either automatically or even manually (cf. figure 3). From this initial shape, two ortho-images (one per view) are calculated by defining a pixel matrix in the 3D space of the object. The 3D positions of the matrix pixels are obtained by using the reference shape and are projected onto the initial images to perform image interpolation (cf. figure 4). As the ortho-images obtained display a higher degree of similarity, more points may be automatically paired. They are then projected onto the initial images in order to calculate a new shape, which can be used, in turn, to calculate new ortho-images, and so on...

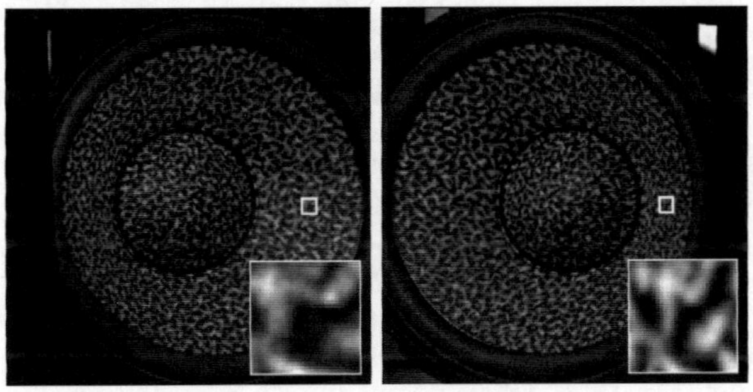

Figure 1. Rectified images: (*left*) from the left camera, (*right*) from the right camera.

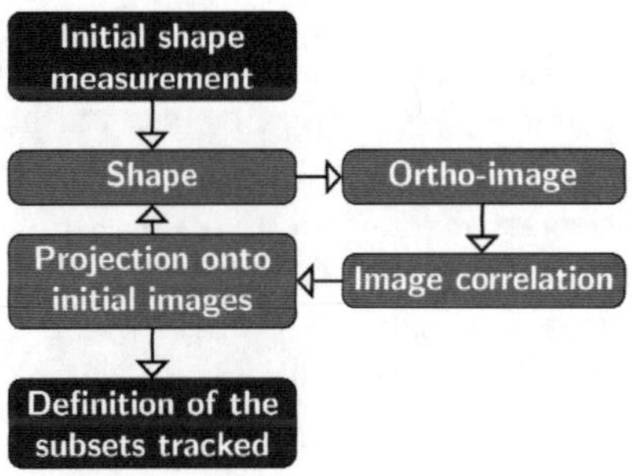

Figure 2. Synoptic diagram.

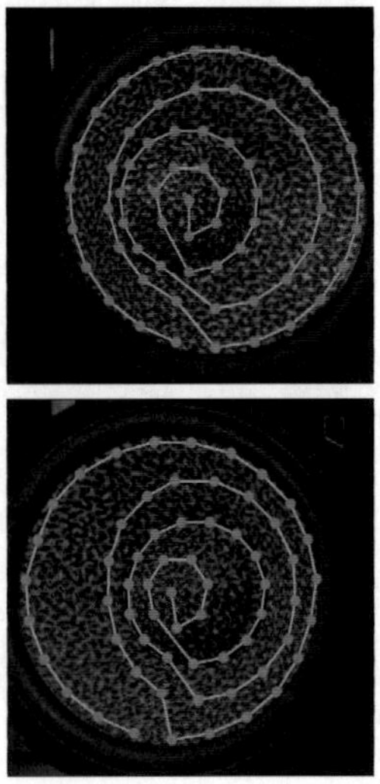

Figure 3. Initial paired points used to measure a coarse shape.

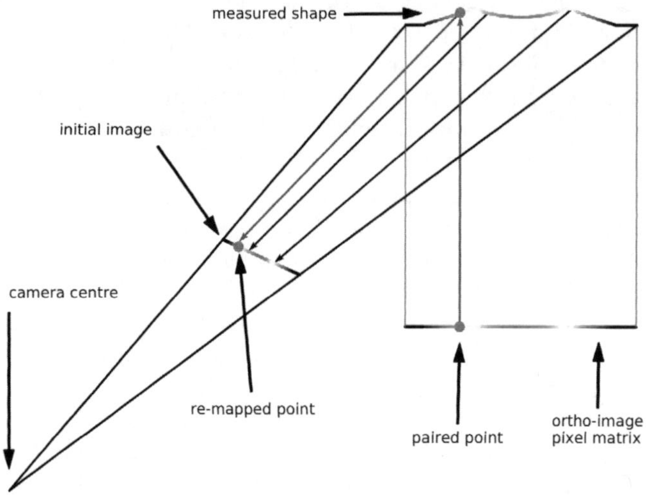

Figure 4. Sketch of the protocol used to calculate ortho-images.

After a few iterations, a large number of points can be paired (cf. figure 5) with an increasingly high precision as the ortho-images are increasingly similar. The correlation coefficient of the ortho-images directly indicates the degree of convergence of the process. For this experiment, when the third ortho-images are superimposed, ≈92% of correlation is achieved, with most of the differences coming from specular reflections (cf. figure 6).

In order to measure vibrations, pixel areas may now be defined. For this work, a few points have been chosen on a circle around each measurement point in the ortho-images. These points have been projected onto the initial images so as to calculate ellipses, whose areas define subsets tracked in the video. This protocol allows triangulating the full-field vibration signals of the object.

Once this technique established, it was tested on a loudspeaker, using the pseudo stereo system with a four-mirror adapter. This set-up was noticeably complex to manipulate, particularly because of the central mirrors. Moreover, the large pan angle clearly hindered the initial pairing process of the protocol presented in this paper. An attempt at reorganising the various elements differently was thus made, resulting in a new set-up requiring only a two-mirror adapter.

3. Pseudo stereo system with a two-mirror adapter

As mentioned previously, the pseudo stereo system with a four-mirror adapter (cf. figure 7) is not user-friendly, especially because of the central mirrors. Its use is all the more complex since it presents a shadow area in the middle of the image. The new set-up presented in this paper offers an alternative (cf. figure 8) : it requires only two mirrors and is specifically adapted to measure vibrations. Indeed, the real high-speed camera is placed on the side of the object and generates a view that is sensitive to out-of-plane displacement. The other, virtual, view is generated by the two mirrors and positioned in front of the object: it is thus sensitive to in-plane displacements. It may be noted that the principle of having a view set in front of the object is also used in trinocular stereo systems to initialise and facilitate the pairing process [29]. As a consequence, the pan angle is reduced by two, which simplifies the initial pairing process, while keeping a view that is very sensitive to vibrations. The whole system is more user-friendly, as there are only two large mirrors, and the shadow area in the image is reduced. The non-symmetry of the views

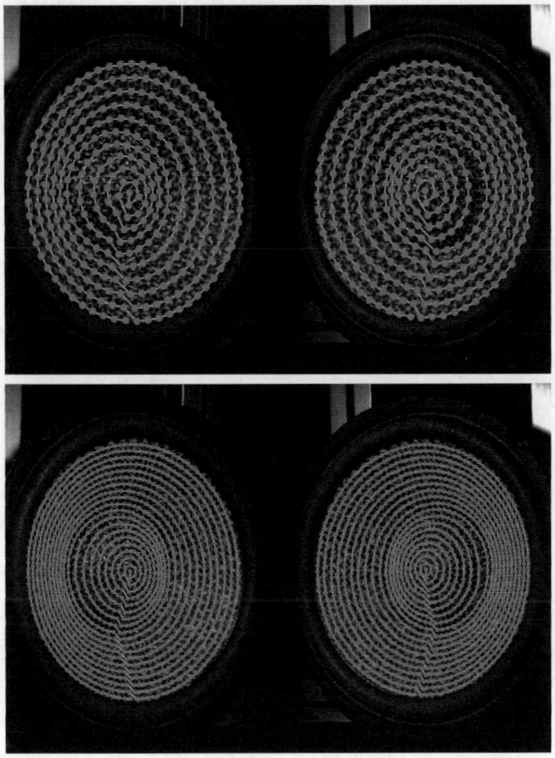

Figure 5. Measurement points re-projected onto the initial images: (*left*) first iteration, (*right*) second iteration.

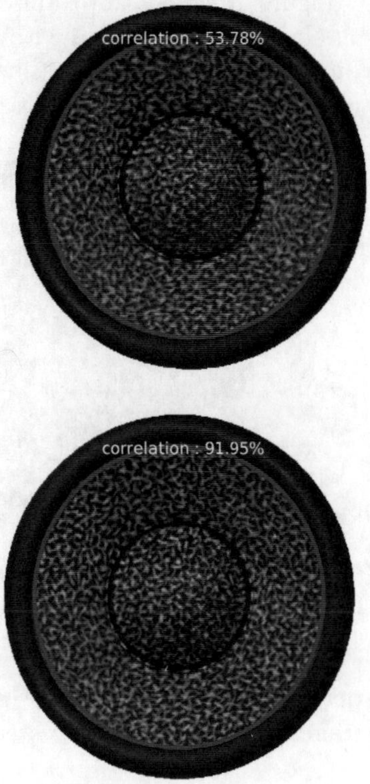

Figure 6. Ortho-images combined in two colors: (*top*) first iteration, (*bottom*) third iteration

induces biais in the shape measurement, but more pixels are available for the vibration-sensitive view.

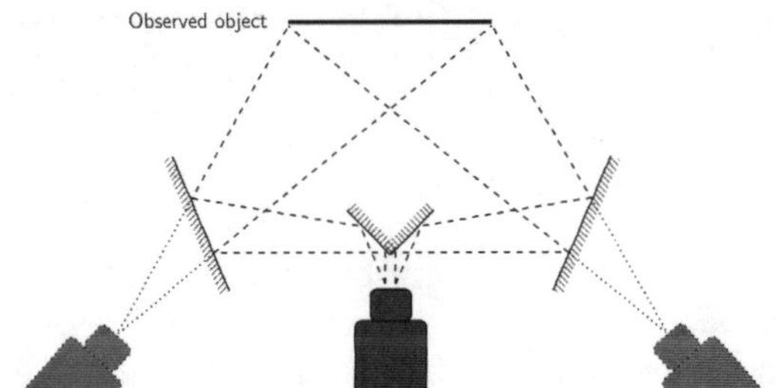

Figure 7. Sketch of the set-up with a four-mirror adapter.

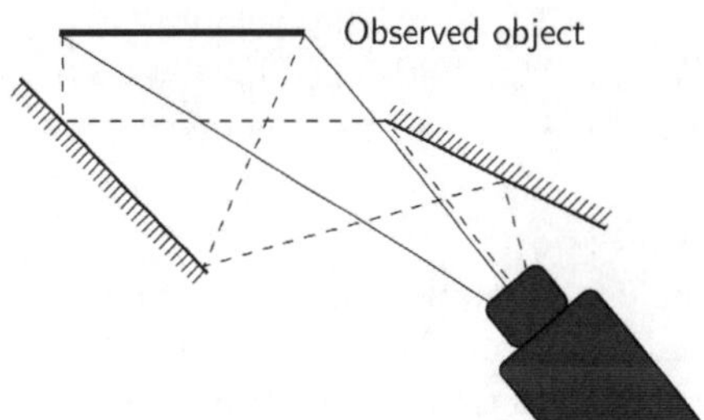

Figure 8. Sketch of the set-up with a two-mirror adapter.

A test on a loudspeaker has been carried out : the two-mirror adapter set-up has been used with the IRIs method to initialise the pairing process. The experimental set-up is presented figure 9. Figure 10 shows an example of image obtained with the new pseudo stereo system. The results have been compared with those obtained with the four-mirror adapter set-up for the full-field measurement, and with those obtained with a laser vibrometer (Polytec) for a single point measurement, in order to validate the whole protocol.

Figure 9. Picture of the experimental set-up.

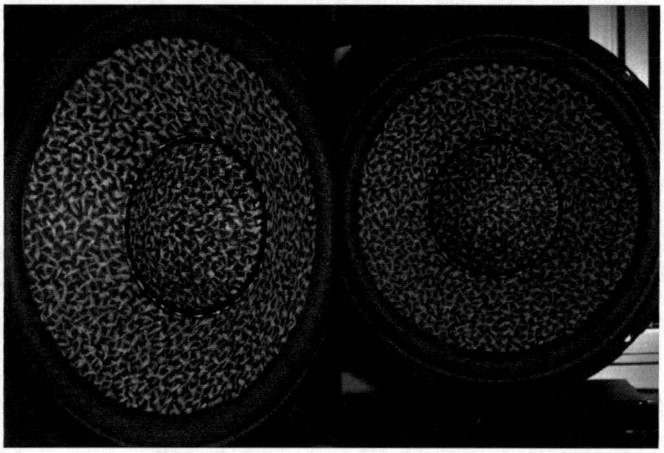

Figure 10. Example of image obtained with the pseudo stereo system with a two-mirror adapter.

The Frequency Response Functions (FRFs) of the central measurement point of the loudspeaker are calculated from the excitation signal at its terminals and from the resulting displacement signals measured with the two-mirror adapter and four-mirror adapter set-ups, and with the laser vibrometer. The results of the vision methods are shown figure 11 and match well those obtained with the reference technique, which validates the approach. The proposed set-up is a little less precise than the conventional one; yet the differences between the values obtained with the laser vibrometer and with the two-mirror adapter set-up are globally below 10%.

From the FRFs of all the points, the Operational Modal Shapes (OMSs) can be retrieved and displayed for both pseudo stereo systems (cf. figures 12 and 13). The images match neatly, which validates the technique once again.

4. Conclusion

In conclusion, firstly, large pan angles improve the precision of the vibration measurement but make it more difficult to pair initial measurement points if the object of study presents significant variations in depth. The IRIs method proposed here has been designed to perform this pairing process from a coarse shape measurement, while taking into consideration all the constraints mentioned above. It successfully allows initializing vibration measurement with a high number of measurement points. Secondly, if only one high-speed camera is at one's disposal, the conventional pseudo stereo system with a four-mirror adapter may be used to measure vibrations at the expense of the number of pixels available. Nonetheless, this set-up is not particularly user-friendly. Hence the attempt at designing a new set-up, which involves a different, simpler adapter with two mirrors. With this new system, the results seem to be a little less precise but the whole protocol allows retrieving similar operational modal shapes, which validates the technique presented in this paper.

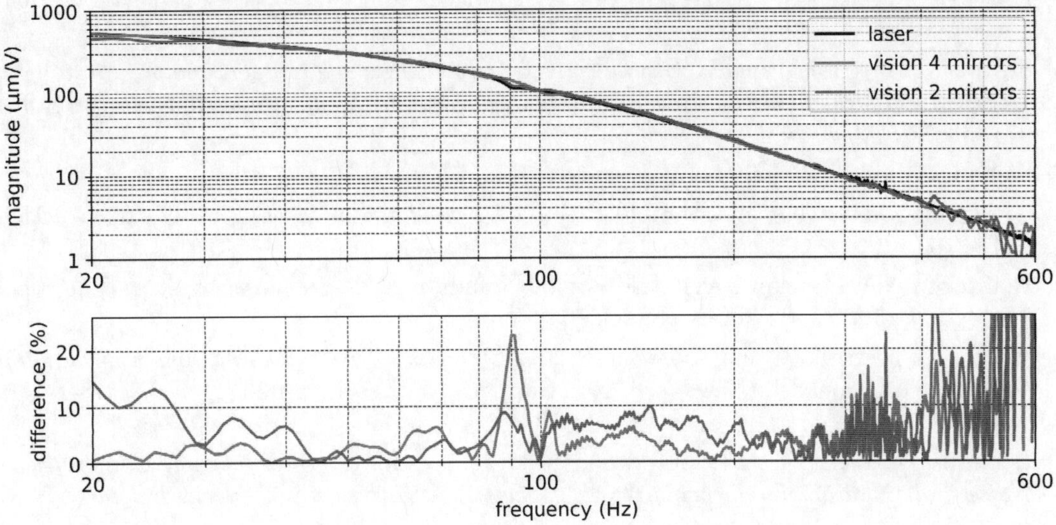

Figure 11. FRFs measured with the laser vibrometer and the vision set-ups

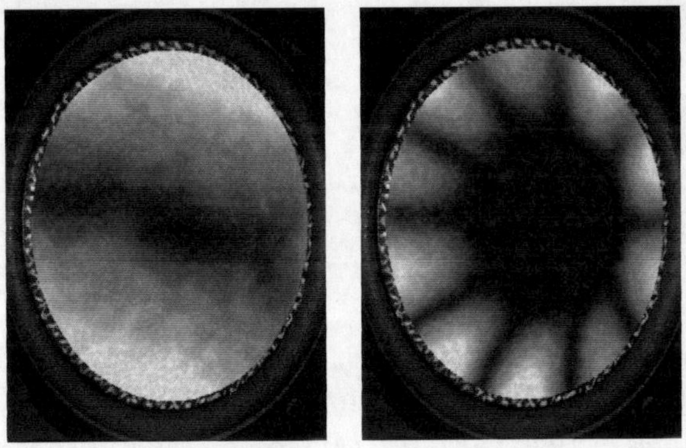

187 Hz 321 Hz

Figure 12. OMSs obtained with the two-mirror adapter set-up.

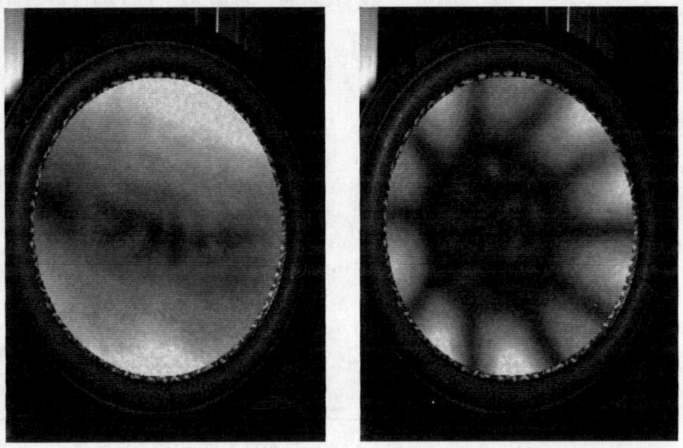

189 Hz 323 Hz

Figure 13. OMSs obtained with the four-mirror adapter set-up.

Acknowledgments
The work presented here is supported by the institute "Le Mans Acoustique" and funded by the "Pays de la Loire" region and the "European Regional Development Fund".

References

[1] Brandt A 2011 *Noise and Vibration Analysis: Signal Analysis and Experimental Procedures* (Wiley) ISBN 978-0-470-74644-8

[2] Fahy F and Walker J 2004 *Advanced Applications in Acoustics, Noise and Vibration* (CRC Press) ISBN 9780415237291

[3] Berry A and Robin O 2016 *CFA 2016* (Le Mans, France)

[4] Poittevin J, Picart P, Faure C, Gautier F and Pézerat C 2015 *Appl. Opt.* **54** 3185–3196

[5] Baqersad J, Poozesh P, Niezrecki C and Avitabile P 2017 *Mechanical Systems and Signal Processing* **86** 17–34

[6] Sutton M, Orteu J J and W Schreier H 2009 *Image Correlation for Shape, Motion and Deformation Measurements. Basic Concepts, Theory and Applications* (Springer US) ISBN 978-0-387-78746-6

[7] Siebert T, Becker T, Spiltthof K, Neumann I and Krupka R 2007 **46**

[8] Tiwari V, Sutton M A and McNeill S R 2007 *Experimental Mechanics* **47** 561–579 ISSN 1741-2765

[9] Helfrick M N, Niezrecki C, Avitabile P and Schmidt T 2011 *Mechanical Systems and Signal Processing* **25** 917 – 927 ISSN 0888-3270

[10] Warren C, Niezrecki C and Avitabile P 2011 *Structural Dynamics and Renewable Energy, Volume 1* ed Proulx T (New York, NY: Springer New York) pp 39–44 ISBN 978-1-4419-9716-6

[11] Beberniss T, Eason T and Spottswood S 2012 *Proceedings, International Conference on Noise and Vibration Engineering (ISMA), KatholiekeUniversiteit Leuven, Belgium*

[12] Zappa E, Mazzoleni P and Matinmanesh A 2014 *Optics and Lasers in Engineering* **56** 140 – 151 ISSN 0143-8166

[13] Cameron C, Sirohi J and Uehara D 2015 *56th AIAA/ASCE/AHS/ASC Structures, Structural Dynamics, and Materials Conference*

[14] Karsen C V, Bouman T and Gwaltney G 2014 *Topics in Modal Analysis, Volume 7* Conference Proceedings of the Society for Experimental Mechanics Series ed Allemang R, Clerck J D, Niezrecki C and Wicks A (Springer New York) pp 637–644

[15] Chen J G, Wadhwa N, Cha Y J, Durand F, Freeman W T and Buyukozturk O 2015 *Journal of Sound and Vibration* **345** 58–71

[16] Lundstrom T, Baqersad J and Niezrecki C 2016 *Experimental Techniques* **40** 907–919 ISSN 1747-1567

[17] Jian W, Yong L, Zhang Z and Gao P 2014 *Proc. SPIE 9276, Optical Metrology and Inspection for Industrial Applications III* vol 9276 pp 9276 – 9276 – 8

[18] Inaba M, Hara T and Inoue H 1993 *Proceedings of 1993 IEEE/RSJ International Conference on Intelligent Robots and Systems, IROS 1993, Tokyo, Japan, July 26 - 30, 1993* pp 1857–1865

[19] Wang R, Li X and Zhang Y 2008 *Journal of the European Optical Society - Rapid publications* **3** ISSN 1990-2573

[20] Li J, Dan X, Xu W, Wang Y, Yang G and Yang L 2017 *Optics & Laser Technology* **95** 1 – 7

[21] Castellini P, Chiariotti P, Martarelli M, Zappa E and Lavatelli A 2017 *Shock & Vibration, Aircraft/Aerospace, Energy Harvesting, Acoustics & Optics, Volume 9* ed Harvie J M and Baqersad J (Cham: Springer International Publishing) pp 285–291 ISBN 978-3-319-54735-0

[22] White T G, Patten J R W, Wan K H, Pullen A D, Chapman D J and Eakins D E 2017 *Strain* **53**

[23] Yu L and Pan B 2017 *Mechanical Systems and Signal Processing* **94** 374–383

[24] Yu L and Pan B 2017 *Optics and Lasers in Engineering* **95** 17 – 25 ISSN 0143-8166

[25] Durand-Texte T, Simonetto E, Durand S, Melon M and Moulet M H 2018 *Submitted to Mechanical Systems and Signal Processing*

[26] Faugeras O 1993 *Three-dimensional computer vision: a geometric viewpoint* (MIT press) ISBN 9780262061582

[27] Szeliski R 2011 *Computer Vision: Algorithms and Applications* vol 5

[28] Hartley R I and Zisserman A 2004 *Multiple View Geometry in Computer Vision* 2nd ed (Cambridge University Press) ISBN 0521540518

[29] Wanek J M and Wu C H 2006 *Ocean Engineering* **33** 723 – 747 ISSN 0029-8018

Estimation and compensation of motion blur for the reduction of uncertainty in DIC measurements of flexible bodies

E Zappa, S Turrisi and A Lavatelli

Department of Mechanical Engineering, Politecnico di Milano – Via La Masa 1, 20156 Milan, Italy

E-mail: emanuele.zappa@polimi.it

Abstract. Digital image correlation (DIC) is a useful technique to measure displacement/strain fields both for static and dynamic problems in experimental mechanics. When monitoring moving objects with digital cameras, motion blur may occur if the shutter time reaches the time scale of the motion of the measurand. Consequently, motion blur is one of the most relevant problems in those dynamic DIC applications where shutter time cannot be set arbitrarily. This work deals with the problem of compensating motion blur effects on a generic DIC image. The problem of motion blur compensation to reduce DIC uncertainty is discussed in literature in the case of rigid target, where the amount of motion blur is the same for the whole image area. Deformable targets, instead, pose the additional problem when motion blur is variable within a single frame. In this paper a subset-based technique is proposed to estimate and compensate the motion blur for each image region. The approach is tested on synthetically deformed and blurred images of a notched beam specimen.

1. Introduction

This paper proposes a new approach to mitigate the effects of motion blur on 2D Digital Image Correlation (DIC) measurements. DIC is an optical method that employs tracking and image registration techniques for accurate optical flow measurements [1]. This method is often used to measure full-field displacement and strains, being indeed widely applied in many areas of experimental mechanics. Whenever possible the motion blur is reduced during image acquisition by reducing the exposure time and increasing the illumination level. However, this is not always possible since lighting conditions cannot be easily controlled, such as in outdoor conditions or in the case of a camera mounted on a flying drone [2]. The uncertainty of DIC measurements is composed by a correlation bias (namely bias error) [3, 4] and a random error [5] that generates dispersion of measured displacements. When the images are affected by motion blur, random error is the main uncertainty component [6, 7]. Conversely, the contribution of the bias error is less relevant in presence of motion blur [8, 9, 10]. In previous works [10, 11], the authors modeled the physical nature of motion blur phenomenon and how this results into uncertainty. Given that knowledge, a further step is to try to compensate the undesired effects of motion blur, through image deconvolution.

To remove motion blur, it is of primary importance to estimate its intensity in a reliable and appropriate way. Inaccurate blur estimation leads to the generation of artifacts in compensated

IOP Conf. Series: Journal of Physics: Conf. Series **1149** (2018) 012009 doi:10.1088/1742-6596/1149/1/012009

images. In this paper, the estimation of motion blur transformation is carried in the frequency domain. In brief, DIC always require an image of the measurand in reference position, hence it is possible to estimate the Optical Transfer Function (OTF) between the reference texture and the deformed one. In presence of motion blur, the OTF is represented by a bi-dimensional *sinc* kernel. In principle, it is possible to recover the blur transformation by fitting the OTF data with a *sinc* kernel model. This approach is developed at subset level, to deal with variable motion blur.

Once the motion blur is estimated in a robust way, the paper discusses motion blur removal, using the well known Wiener filter method. This choice is justified by the presence of noise in digital images. Noise itself should be modeled correctly, hence this issue is discussed within the body of the paper.

Eventually, the whole approach is tested and validated against a synthetic experiment on a notched cantilever beam. Data shows that the algorithm is able to remove motion blur from DIC images, also in those situations where it is highly variable inside the acquired frame. At the same time, the compensation of motion blur is able to reduce the uncertainty in DIC measurements of about 80% in cases where motion blur exceeds 7 pixels.

2. State of the art

The phenomenology of motion blur on digital images has been described first by Potmesil et al. [12]. Their work dates back to the 80s and they solved the problem of describing motion blur in analytic way. Nonetheless it enabled the research of methods to remove motion blur in digital images.

The problem of motion blur compensation is relevant in many fields: from general photography [13], to image stabilization in smartphones [14], coming up to the niche of scientific imaging [15]. In any case motion deblurring involves two separate problems:

- determining the motion blur transformation parameters
- removing the motion blur transformation from the acquired image

The two problems can be solved in different ways and each of them can be formulated in separate domains [16], depending on the approach of the proposed solution. In the next paragraphs it is highlighted a review of the literature.

2.1. Estimating motion blur

A sound review of the state of the art of motion blur estimation techniques can be found in [16] and [17]. To focus better the work further presented this paper, it is necessary to highlight the two possible ways which the motion blur estimation problem can be formulated:

- referenced estimation
- blind estimation

Referenced estimation is the problem of estimating motion blur in an acquired image, given that an unblurred image of the scene is available. Blind estimation, instead, is the estimation of motion blur transformation when only the motion blurred scene is available. Even if they deal with the same phenomenon, the two problems are very different from a mathematical point of view. For what concerns referenced deconvolution it is always possible to formulate estimator for the Point Spread Function (PSF) (or OTF in the frequency domain) given that the original scene is available. Conversely, the blind estimation problem is ill-posed since there are not data describing what the scene would appear without blur. Hence solving a blind problem [18] will always involve statistical tools [19, 20] and a-priori hypothesis on what the scene would appear like [21].

If the reader considers the subject of this paper, the problem is indeed a referenced one: for a vibrating target it is usually possible to have a still image of it, as long as mechanical excitation is off. The main idea of referenced estimation lies on the estimation of the OTF starting from reference and blurred images. Example of uses of referenced deconvolution procedure can be found in [22, 23, 24].

The authors tried to focus on an approach dedicated to vision based measurement systems. The main idea of the work is to fit the OTF estimator with a frequency domain model of motion blur, trying to preserve all metrologic information nested in the image (similarly to what documented in [24]). Another goal is to estimate a motion blur which is variable inside a given image. A preliminary study on the feasibility of a monodimensional motion blur estimator for DIC images has been presented in [25]. This work will extend those concepts to the generic bi-dimensional problem.

2.2. Removing motion blur

The task of removing motion blur is instead a well defined problem in all cases where the motion blur PSF (or the OTF) is retrieved. As described in [12], the motion blur is a convolutive operator. Consequently its removal is a problem of signal deconvolution once the convolution kernel is known. Signal deconvolution is a trivial task when working with deterministic signals. Unfortunately, digital images do not fall in the category of deterministic signals, since a moderate level of noise is unavoidable during actual image acquisition. Noise introduces spectral components in the blurred images which are uncorrelated with the reference one. Consequently, the deconvolution problem becomes ill-posed.

For this class of problems, the Wiener filter [26] is considered the best solution. In fact, both the Wiener causal filter and its finite impulse response (FIR) counterpart are optimal filters for the signal deconvolution problem in presence of additive noise.A full description about modern digital implementation of Wiener filters can be found in [27]. To apply the Wiener filter it is necessary to estimate the Optical Transfer Function due to motion blur and the noise-to-signal ratio (NSR) to provide a model of the uncorrelated noise. This paper will also provide a detailed description of the NSR estimation procedure.

3. Motion blur estimation procedure

It is not possible to formulate a universal model able to describe the generation of motion blur in an image of a vibrating target. In fact, even in the ideal case of noise-less image formation, there are phenomenon such as optical distortions or sensor non-linearity, that limit the development of universal models. Stated the impossibility to retrieve a generic universal formulation, it is necessary to limit the field of investigation with a suitable set of hypothesis. In particular:

- existence of a reference image $R(m, n)$ of the vibrating target standing still in equilibrium position
- the non-blurred image $Q(m, n)$ of the vibrating target for a given position is supposed to be equal to the reference one rigidly translated of a quantity A_0 in the direction ϕ, so that $Q(m, n) = R\left(m - A_0 \cos\phi,\ n - A_0 \sin\phi\right)$
- exposure time is a fraction of vibration period and small enough to describe motion during exposure as a uniform rectilinear one.

When the fore mentioned hypothesis are verified, it is possible to state the following corollaries:

(i) the motion blurred image $P(m, n)$ is generated from $Q(m, n)$ after the convolution with a simple 2D rectangular window of length W_0 pixels

(ii) motion blur appears into the image in the form of rectilinear swipe trails as shown in figure, orientated in the motion direction

IOP Conf. Series: Journal of Physics: Conf. Series **1149** (2018) 012009 doi:10.1088/1742-6596/1149/1/012009

(iii) in the direction of swipe trails the problem is mono dimensional, since each line of the acquired frame is generated from its equivalent line in the reference frame after rigid translation and convolution

As a consequence the blur phenomenon exemplified in figure 1 is described by the length of convolution window W_0 and the direction of swipe trails ϕ. In particular the length (in pixels) of convolution window is linearly dependent to instantaneous speed of marker during exposure, as described in [11].

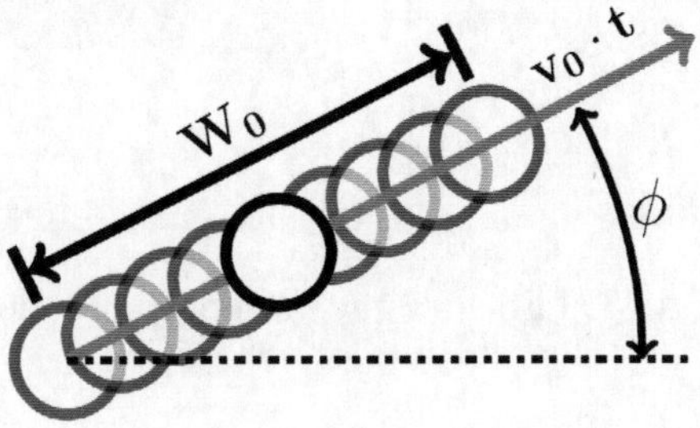

Figure 1: Swipe trails descriptions as for linear blur model

Given all the hypothesis and all the theoretical background highlighted previously, it is possible to express the transfer function, in the frequency domain (k, l), between the reference image and the acquired one, in equation (1) (where n is the indext of the harmonic component, while k and l are the vertical and horizontal spectral component respectively). This result is called Optical Transfer Function (OTF) [28] and takes into account both rigid translation and motion blurring.

$$\begin{cases} n = \sin(\phi) \cdot k + \cos(\phi) \cdot l \\ \mathcal{H}(k,l) = W_0 \cdot sinc(W_0 \cdot n) \cdot e^{-j \cdot 2\pi \cdot A_0 \cdot n} \end{cases} \tag{1}$$

It is possible to highlight that the model is a 2D sinc surface kernel as shown in figure 2a. It requires to identify 3 parameters in order to describe the OTF:

- motion blur direction ϕ
- rigid translation between reference image and acquired one A_0
- blurring window size W_0

To sum up, a multi-parametric model that is able to describe the transformation between a reference image and a generic one has been formulated. Now it is necessary to describe how it is possible to estimate those parameters, starting from a reference image and a motion blurred one.

If the reference image $R(m, n)$ and the blurred one $P(m, n)$ are available, it is possible to estimate the OTF $\mathcal{H}_{\mathcal{E}}(k, l)$ between the two as in eq. (2), where the $\mathcal{F}$ operator indicates the 2D Fourier transform:

$$\mathcal{H}_{\mathcal{E}}(k,l) = \frac{\mathcal{F}(P)}{\mathcal{F}(R)} \tag{2}$$

Given that, the transformation parameters can be retrieved by minimizing the error between the estimated OTF $\mathcal{H}_{\mathcal{E}}(k, l)$ and the theoretical one $\mathcal{H}(k, l)$, as in equation (3):

$$\begin{Bmatrix} \phi \\ W_0 \\ A_0 \end{Bmatrix} = \operatorname{argmin} \left\{ \mathcal{H}_{\mathcal{E}} - W_0 \cdot \operatorname{sinc}\left[W_0 \cdot (\sin\phi \cdot k + \cos\phi \cdot l)\right] \cdot e^{-j \cdot 2\pi A_0 \cdot (\sin\phi \cdot k + \cos\phi \cdot l)} \right\}^2 \tag{3}$$

IOP Conf. Series: Journal of Physics: Conf. Series **1149** (2018) 012009 doi:10.1088/1742-6596/1149/1/012009

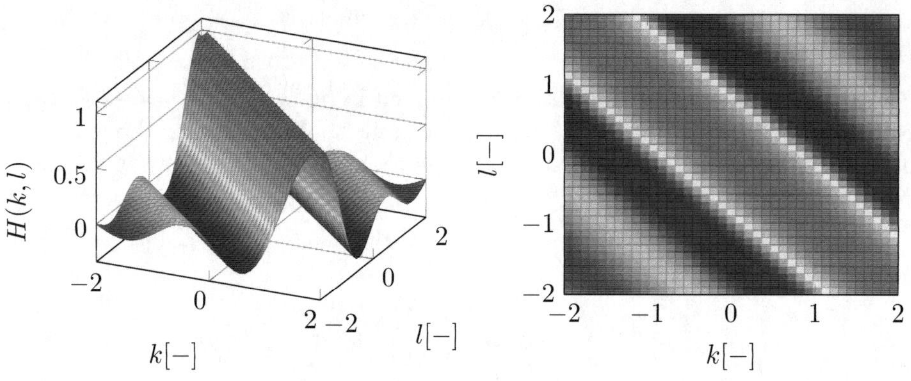

(a) 3D surface view (b) Plane view

Figure 2: Shape of the OTF as for the parametric model of equation (1)

Unfortunately, the fitting procedure cannot be performed directly in one step. In fact a set of problems limits the robustness of a direct fitting approach:

- image digitalization deals to relevant spectrum leakage problem [28]
- as a consequence of the previous point, the estimation of OTF $\mathcal{H}_{\mathcal{E}}$ should be performed with adequate processing
- noise introduces uncorrelated components in the spectral representation
- background effect: even without blur and noise it's not in general true that the acquired image is a simple translation of the previous one, as long as there may be elements on the scene background that does not move or vibrate.
- motion blur may be variable inside an image if the measured target is deforming
- fitting the sinc kernel is itself a badly conditioned problem, since the fitting error gradient is much higher for the amount of motion blur than for the direction angle

As a consequence, a dedicated procedure to retrieve a good and robust estimation of motion blur has been developed. The main idea is to focus on the morphology of the OTF to retrieve a first estimate of the motion blur parameters. Then the fitting procedure becomes a better defined bounded minimization problem.

The last step has been to deal with the problem of having variable motion blur inside an acquired image. The solution has been found in mocking the DIC behavior. In this sense, the reference image is divided into subsets. Then a template matching algorithm is run to find the position of the subsets in the generic deformed and blurred image. Then OTF estimation is run at subset level, letting us retrieve the mean value of motion blur inside the subset. For sure this procedure has some limitation as the subset size becomes smaller: in fact the smaller the subset, the less the frequency resolution of OTF is. Therefore the uncertainty in the parameter estimation becomes larger. To validate the approach, the algorithm has been tested in 5.1, to understand what are the limit in shrinking subset size.

4. Motion blur restoration with Wiener filter

As explained in section 2.2, the knowledge of both the Point Spread Function PSF due to motion blur and the noise-to-signal ratio NSR is necessary to implement Wiener filter. Even though the PSF is completely identified once the blurring window size W_0 and direction ϕ are estimated, an accurate definition of the NSR is key to achieve a good and robust deconvolution. During the image grabbing process various noises (e.g. heteroscedastic noise, shot noise, thermal noise, cut-off noise) and illumination lighting fluctuations are unavoidably present and most of them are characterized by a peculiar spectral distribution. The authors decided to use as support for Wiener deconvolution the function `deconvwnr.m`, which is already implemented in *Matlab*. It considers the noise as an additive and stochastic process, allowing to express the NSR as a scalar constant which approximates the NSR in every point of frequency coordinates.

4.1. NSR estimation

The algorithm adopted to estimate the noise-to-signal ratio to be used for Wiener deconvolution is discussed and presented in this section. It is based on the analysis of a sequence of static images of the target. In detail:

- The first image of the series (I_{REF}) is chosen as reference and its Root Mean Square (σ_S) is calculated, according to eq.(4), where $\bar{p}$ is the mean value of the image, p_{ij} is the pixel intensity and m, n are the number of rows and columns of the image, respectively.

$$\sigma_S = \sqrt{\frac{\sum_{j=1}^{n}\sum_{i=1}^{m}(\bar{p}-p_{ij})^2}{m \cdot n}} \tag{4}$$

- The mean value $\bar{p}$ is subtracted from all the still images to compensate any type of variations or fluctuations in lighting conditions.
- The reference image is subtracted from each one of the remaining images (I_{NOISE}), obtaining a matrix difference called I_{DIFF}. The standard deviation of I_{DIFF} is computed and the results are averaged for the number of involved images, obtaining σ_N.

In this way it is possible to estimate the NSR as the ratio between the averaged standard deviation of the image series σ_N, which represents an index for the noise level, and the RMS of the reference image σ_S, which represents an index for the signal level (5):

$$NSR_E = \frac{\sigma_N}{\sigma_S} \tag{5}$$

5. Validation on synthetic images

5.1. Estimation results by changing the image size

To address the problem of motion blur at subset level, it is of utmost importance to understand how the estimation results are influenced by the image size. Since blur may change inside the image of a moving deformable body, it is necessary to divide the frame itself into small regions where motion blur is almost homogeneous. This assures the hypothesis of applicability of the deconvolution algorithms. Therefore, a simulation activity is developed involving a synthetic pattern. Four squared images of different size are considered and motion blur, having a length of 6 px and orientation of 45°, is numerically simulated in each of them according to the sub-pixel accurate technique described in [6, 29]. To resemble operating conditions, a zero mean Gaussian noise with standard deviation equal to 1% of the full dynamic range of the image (i.e. 1% of 255) is also added to the images. This is because modern cameras usually operate with a noise level between 0.5% and 1% of the dynamic range [30]. Then, motion blur estimation algorithm is applied on the four images, bringing to the results reported in Tab.1. It can be confirmed the goodness of the estimation technique even in the case of the smallest images. However, it was not possible to decrease the image size below 70 x 70 px. This physical limit is dictated by the lacking of points necessary to perform a robust fitting of the sinc function.

Table 1: Motion blur estimation tests on images having different size. Imposed blur is 6 px on a 45 degrees direction

Image size [px]	$w_{est}[px]$	$\theta_{est}[^o]$	**Blur error [px]**	**Angle error [deg]**
70x70	5.80	44.6	0.20	0.4
100x100	5.97	45.1	0.03	0.1
200x200	5.98	44.9	0.02	0.1
400x400	5.98	45.0	0.02	0.0

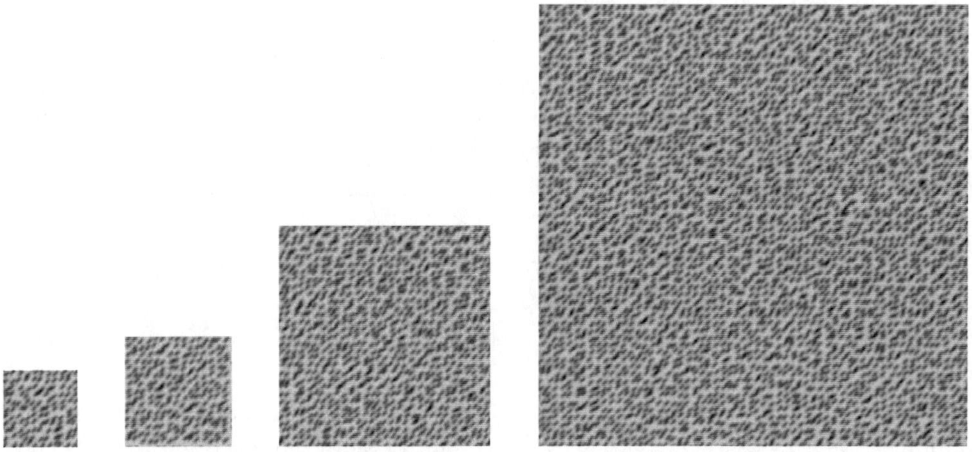

(a) 70x70 (b) 100x100 (c) 200x200 (d) 400x400

Figure 3: Motion blur simulation ($w_{nom} = 6px$, $\theta_{nom} = 45^o$) on synthetic images having different size

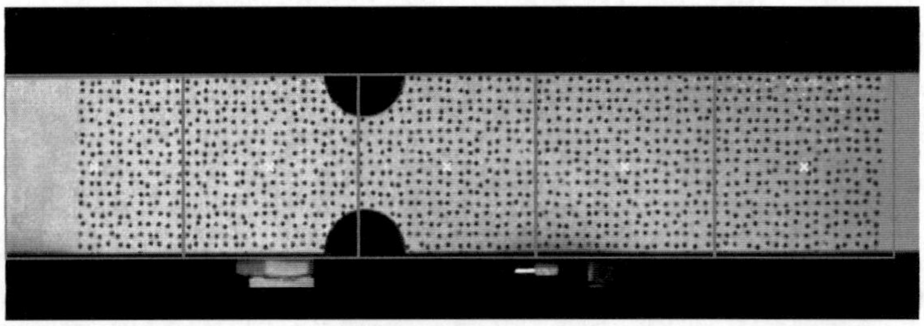

(a) Reference image

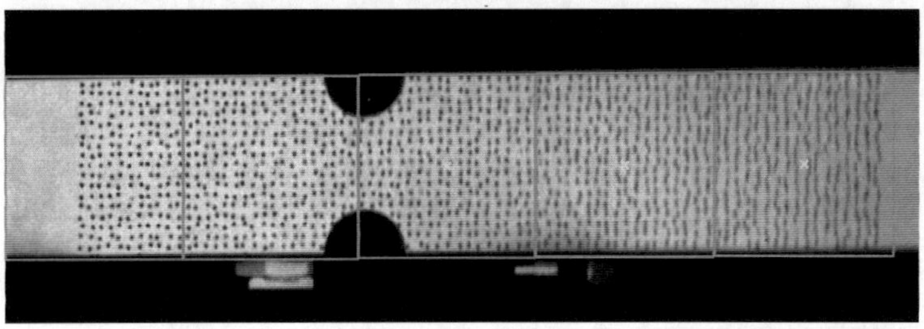

(b) Deformed image

Figure 4: Synthetic experiment of a dynamic beam shear: the images are labelled with the subsets used for motion blur compensation

5.2. Simulation of dynamic beam shear

The effectiveness of the proposed algorithms is tested on the synthetic experiment represented in Fig.4. To resemble an actual experimental activity, a static image of a notched beam is used as the reference one for DIC. Then, motion blur is numerically simulated in the vertical direction, linearly increasing from 0 px on the left side of the image (simulated clamped constraint) until reaching 10 px on the right hand side. The post-processing activity starts considering the reference image and dividing the beam surface into adjacent subsets along its longitudinal axis. Then, the template matching identifies the ith subset location in the respective blurred image. A subset size of 165 px guarantees homogeneous blurring conditions inside the same region, which is a requirement that must be satisfied for the correct working of the deconvolution algorithms. For each subset the data fitting with a 2D sinc allows to estimate both the intensity and the direction of motion blur (Fig.5). As soon as those parameters are known, it is possible to compensate the blur with a Wiener filter.

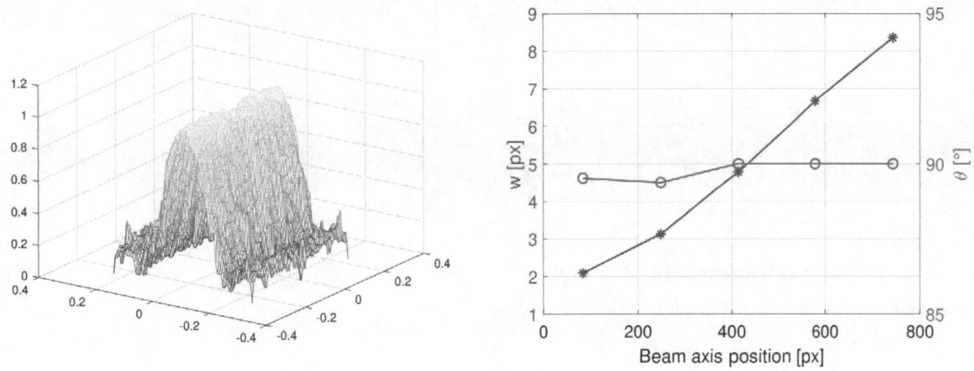

(a) OTF estimator (b) Estimation results

Figure 5: Estimation of blur parameters at subset level

5.3. DIC analysis after motion blur compensation

In the context of DIC, uncertainty analysis is often carried out by means of synthetic experiments [31, 32], where image transformations are numerically imposed. In the current study, Ncorr [33] software is used as support to perform DIC analysis on the images presented in section 5.2. DIC processing returns as output the displacement of the subsets in which the image has been divided by the software. Each subset has been selected to have a radius of 21 px and a spacing of 2 px. Since a vertical and variable motion along the longitudinal beam axis has been simulated, the displacement matrices obtained for each image are used to calculate both the mean displacement v_{mean} and the standard deviation of displacement σ_v of a column of subsets. The mean displacement returned by DIC estimates the motion of the beam axis. The standard deviation, instead, represents a viable index to express the uncertainty of the data. DIC results before and after blur compensation are reported in Fig.6. In agreement to what has been previously found in [25], standard deviation of displacement field σ_V(the random error) is the most relevant component of uncertainty when motion blur phenomena occur. After the compensation of motion blur (dashed line in Fig.6) standard deviation is strongly reduced. This demonstrates how deconvolution becomes a viable solution to improve DIC accuracy in motion blurred images. It is important to highlight the presence of a slight bias in the reconstruction of vertical displacement when motion blur exceeds 7 px. This issue will be investigated in future works, however it doesn't affect the overall judgment on the algorithm since the amount of motion blur is higher than the usual range of DIC experiments (where it is mostly below 4 px).

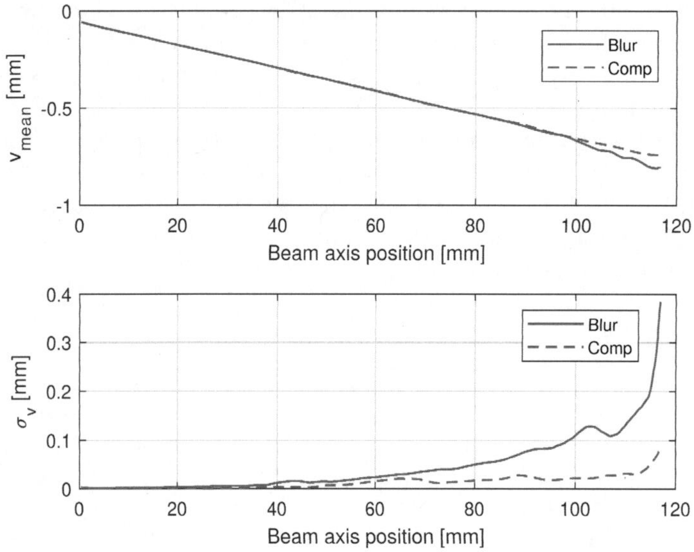

Figure 6: DIC results for the synthetic blur experiment before (solid line) and after (dashed line) motion blur compensation: the upper graph represents the mean value of vertical beam displacement, the lower graph its uncertainty

6. Conclusion

In this work the authors demonstrated a generic procedure to detect a generic amount of motion blur in a generic spatial direction. This procedure has been implemented at subset level, thus enabling to work with images having variable motion blur (such as the ones coming from vibrating deformable bodies). The validation of the method is performed by means of a synthetic experiment. However the authors already performed experimental tests which will be documented in future works. The results of the synthetic experiment show a sensible reduction of DIC random uncertainty due to the application of motion blur compensation. Data show that also a slight bias is introduced. This issue will be better investigated in future works.

References

[1] M. Sutton, J. Orteu, and H. Schreier, *Image Correlation for Shape, Motion and Deformation Measurements: Basic Concepts,Theory and Applications.* Springer US, 2009.

[2] D. Reagan, A. Sabato, and C. Niezrecki, "Feasibility of using digital image correlation for unmanned aerial vehicle structural health monitoring of bridges," *Structural Health Monitoring.* Published online on October 10, 2017. DOI:10.1177/1475921717735326.

[3] B. Pan, "Bias error reduction of digital image correlation using gaussian pre-filtering," *Optics and Lasers in Engineering*, vol. 51, no. 10, pp. 1161–1167, 2013.

[4] H. Schreier, J. Braasch, and M. Sutton, "Systematic errors in digital image correlation caused by intensity interpolation," *Optical Engineering*, vol. 39, no. 11, pp. 2915–2921, 2000.

[5] Y. Wang, P. Lava, S. Coppieters, M. De Strycker, P. Van Houtte, and D. Debruyne, "Investigation of the uncertainty of dic under heterogeneous strain states with numerical tests," *Strain*, vol. 48, no. 6, pp. 453–462, 2012.

[6] E. Zappa, A. Matinmanesh, and P. Mazzoleni, "Evaluation and improvement of digital image correlation uncertainty in dynamic conditions," *Optics and Lasers in Engineering*, vol. 59, pp. 82–92, 2014.

[7] E. Hack, X. Lin, E. Patterson, and C. Sebastian, "A reference material for establishing uncertainties in full-field displacement measurements," *Measurement Science and Technology*, vol. 26, no. 7, 2015.

[8] D. Wang, Y. Jiang, W. Wang, and Y. Wang, "Bias reduction in sub-pixel image registration based on the anti-symmetric feature," *Measurement Science and Technology*, vol. 27, no. 3, p. 035206, 2016.

[9] P. Mazzoleni, F. Matta, E. Zappa, M. Sutton, and A. Cigada, "Gaussian pre-filtering for uncertainty minimization in digital image correlation using numerically-designed speckle patterns," *Optics and Lasers in Engineering*, vol. 66, pp. 19–33, 2015.

[10] A. Lavatelli and E. Zappa, "A displacement uncertainty model for 2-d dic measurement under motion blur conditions," *IEEE Transactions on Instrumentation and Measurement*, vol. 66, pp. 451–459, March 2017.

[11] A. Lavatelli and E. Zappa, "Modeling uncertainty for a vision system applied to vibration measurements," *IEEE Transactions on Instrumentation and Measurement*, vol. 65, pp. 1818–1826, Aug 2016.

[12] M. Potmesil and I. Chakravarty, "Modeling motion blur in computer-generated images," *Computer graphics*, vol. 17, no. 3, 1983.

[13] D. Dansereau, A. Eriksson, and J. Leitner, "Richardson-lucy deblurring for moving light field cameras," in *IEEE Computer Society Conference on Computer Vision and Pattern Recognition Workshops*, vol. 2017-July, pp. 1783–1794, 2017.

[14] R. Fergus, B. Singh, A. Hertzmann, S. T. Roweis, and W. T. Freeman, "Removing camera shake from a single photograph," *ACM Trans. Graph.*, vol. 25, pp. 787–794, July 2006.

[15] M. Bertero, P. Boccacci, G. Desidera, and G. Vicidomini, "Image deblurring with poisson data: from cells to galaxies," *Inverse Problems*, vol. 25, no. 12, 2009.

[16] T. Chan, B. Amizic, R. Molina, S. D. Babacan, T. Bishop, and A. Katsaggelos, "Blind image deconvolution," in *Blind Image Deconvolution*, pp. 1–41, CRC Press, May 2007.

[17] S. Tiwari, V. Shukla, A. Singh, and S. Biradar, "Review of motion blur estimation techniques," *Journal of Image and Graphics*, vol. 1, no. 4, pp. 176–184, 2013.

[18] A. Levin, Y. Weiss, F. Durand, and W. Freeman, "Understanding blind deconvolution algorithms," *IEEE Transactions on Pattern Analysis and Machine Intelligence*, vol. 33, no. 12, pp. 2354–2367, 2011.

[19] R. Nelwin and S. Athira, "Adaptive blind deconvolution and denoising of motion blurred images," in *2016 IEEE International Conference on Recent Trends in Electronics, Information and Communication Technology, RTEICT 2016 - Proceedings*, pp. 1171–1175, 2017.

[20] T. Chan and C.-K. Wong, "Total variation blind deconvolution," *IEEE Transactions on Image Processing*, vol. 7, no. 3, pp. 370–375, 1998.

[21] Y. Liu and S. Hu, "Blind restoration based on detection and segmentation of motion blurred image," *Xi Tong Gong Cheng Yu Dian Zi Ji Shu/Systems Engineering and Electronics*, vol. 39, no. 3, pp. 662–667, 2017.

[22] H. Sun, M. Desvignes, and Y. Yan, "Motion blur adaptive identification from natural image model," in *2009 16th IEEE International Conference on Image Processing (ICIP)*, pp. 137–140, Nov 2009.

[23] X. Feng and B. Thierry, "Sure-based motion blur estimation," pp. 373–377, 2012.

[24] A. Stern and N. S. Kopeika, "Analytical method to calculate optical transfer functions for image motion and vibrations using moments," *J. Opt. Soc. Am. A*, vol. 14, pp. 388–396, Feb 1997.

[25] S. Turrisi, E. Zappa, and A. Lavatelli, "Image deconvolution techniques for motion blur compensation in DIC measurements," in *2018 IEEE International Instrumentation and Measurement Technology Conference (I2MTC) (I2MTC 2018)*, (Houston, USA), May 2018.

[26] N. Wiener, *Extrapolation, interpolation and smoothing of stationary time series.* The MIT Press, 1st M.I.T. paperback ed., 1964.

[27] T. Kailath, A. Sayed, and B. Hassibi, *Linear Estimation.* Prentice Hall Information and, Prentice Hall, 2000.

[28] J. Goodman, *Introduction to Fourier Optics.* Roberts & Company, 2005.

[29] M. Potmesil and I. Chakravarty, "Modeling motion blur in computer-generated images," in *Proceedings of the 10th Annual Conference on Computer Graphics and Interactive Techniques*, SIGGRAPH '83, (New York, NY, USA), pp. 389–399, ACM, 1983.

[30] P. Bing, Q. Kemao, X. Huimin, and A. Anand, "On errors of digital image correlation due to speckle patterns," vol. 7375, pp. 7375 – 7375 – 7, 2009.

[31] R. Balcaen, P. Reu, P. Lava, and D. Debruyne, "Stereo-dic uncertainty quantification based on simulated images," *Experimental Mechanics*, vol. 57, pp. 939–951, Jul 2017.

[32] M. Bornert, F. Brémand, P. Doumalin, J.-C. Dupré, M. Fazzini, M. Grédiac, F. Hild, S. Mistou, J. Molimard, J.-J. Orteu, L. Robert, Y. Surrel, P. Vacher, and B. Wattrisse, "Assessment of digital image correlation measurement errors: Methodology and results," *Experimental Mechanics*, vol. 49, no. 3, pp. 353–370, 2009.

[33] J. Blaber, B. Adair, and A. Antoniou, "Ncorr: Open-source 2d digital image correlation matlab software," *Experimental Mechanics*, vol. 55, pp. 1105–1122, Jul 2015.

3D Digital Image Correlation for vibration measurement on rolling tire: procedure development and comparison with Laser Doppler Vibrometer

S. Galeazzi[1,°], P. Chiariotti[1,*], M. Martarelli[2,~], E. P. Tomasini[1]

1Università Politecnica delle Marche, via Brecce Bianche, 60131 Ancona, Italy
2Università degli Studi e-Campus, via Isimbardi, Novedrate (CO), Italy

° s.galeazzi@pm.univpm.it, ste.galeazzi@gmail.com
* Tel. +390712204441, Fax +390712204813, e-mail address: p.chiariotti@univpm.it
~ milena.martarelli@uniecampus.it

Abstract. Noise generated from rolling tire is strictly related to vibrations caused by the impacts of tread blocks on the road surface. Performing vibration measurements on rolling tires is a quite complex task and non-contact methods should be exploited. Laser Doppler Vibrometer (LDV) is the technique that has been mainly used so far, but it has some disadvantages, among which the measurement duration represents the main issue. Digital Image Correlation (DIC) could represent an appealing alternative for this kind of measurement. Recent developments in Digital Image Correlation combined with the modern fast cameras could represent an innovative instrument for vibration measurement at high frequency (up to about 1000 Hz depending on rolling speed). This paper analyses the possibility of using 3D DIC for vibration measurement on rolling tire by a direct comparison with LDV. The advantages and disadvantages of both the techniques are highlighted along the paper by analysing data of an impact test (static case) and a rolling measurement.

1. Introduction

Nowadays, the reduction of tire noise is one of the biggest and most difficult challenges for tire manufacturers both for the new severe limits imposed by regulations and for NVH requirements coming from car manufacturers. In order to design silent tires, it is necessary to understand how tires generate noise.

Even though a lot of researchers have worked on this topic, the phenomenon is still not well understood because it involves several mechanisms which interact with each other. It must be also considered that tire structure is very complex because it contains rubber and other materials and their combination determines tire behavior: this makes the understanding of noise generation mechanisms quite tough and, therefore, the definition of general rules for tire design become a difficult task.

Anyway, it is well known [1] that tire noise is generated by two main groups of mechanisms: the air-related and the vibrational ones.

Tire noise is generally divided into two groups, depending on the frequency range:

- Structure-borne noise: low frequency vibration results into hub force which are responsible for the so called in-vehicle noise, i.e. noise transmitted into the vehicle cabin;
- Air-borne noise is the high frequency noise and it is related to aeroacoustic phenomena (pipe, air pumping…) but also to the high frequency vibration caused by the hitting of tread blocks on the road.

This work focuses on tire vibration as noise generating driver. Since tire vibrations involves low and high frequency phenomena, an experimental characterization of the tire require the identification of a measurement technique able to cope with this frequency range (10 – 800 Hz) and with highly dynamic vibration amplitudes.

Moreover it should be pointed out that measurements on rolling tire are quite complex to perform. In literature there are some works on this topic presenting measurement performed by Laser Doppler Vibrometer (LDV). LDV can be considered as the state of art for this kind of measurement because it is a non-contact technique and its sensibility makes it possible to measure a wide range of vibration conditions (both in terms of frequency and dynamics). However, when going to tire vibration measurement, it shows some limits, especially due the length of the test, which makes the hypothesis of stationary condition under discussion. Indeed, if the number of points to measure is very high, measurement duration is very long and there could be variation in materials due to the variation of temperature, as well as wear effects that might change the tire structural behavior from point to point during the test.

The recent development in Digital Image Correlation (DIC) technique combined with the modern fast cameras could represent an alternative technique to LDV to perform this kind of measurement. Several works proved how this technique can be used for vibration measurement [6,7,8], with a comparison with LDV but only in static condition and for small and simple specimen, so it is necessary to understand if DIC can be used also in rolling conditions and with an object of a complex shape and a particular material, like a tire is. According to the authors knowledge, at the moment in which this paper is written there is only a paper about this topic [9] in which the mode shapes in static condition are extracted using 3D DIC system.

The main issues related to DIC, when considering rolling tire application, are related to the cameras resolution and frame rates with respect to the phenomena that have to be investigated. In fact, global tire modes require a full framing of the tire by the cameras to be appreciated. However, since DIC measures tire displacement and since high frequency tire vibrations are characterized by very small displacement (in the nanometer range), the same optical set-up (meant as camera view with respect to the tire) cannot be exploited for both global modes and local tread vibrations analyses.

Given this practical considerations, it should be pointed out that there are other reasons that make DIC attractive for this kind of vibration measurement:

- from a single measurement it is possible to extract information for a big number of points, so it could be used as a validation tool for the FE (Finite Element) models [5, 10];
- it is less dependent to shape factors (e.g. it gives the possibility to measure on tire sidewall, crown, tread and grooves indistinctly), surface optical characteristics, presence of geometric irregularities (e.g. textures on the sidewall) as LDV can be;
- DIC is less affected by large displacement that can take place due the initial loading of the tire at the contact patch area;
- being a full-field technique (all data are measured simultaneously) there is no need of reference signal for phase realignment as it happens for LDV measurement, given its sequential nature.

In this paper the capabilities of 3D-DIC technique are studied through a comparison with the TLDV, in order to define if it can be an effective tool for tire vibration measurement in rolling conditions. The comparison is made considering both a static (impact test) and a rolling case.

2. Experimental setup

All the results reported in this work refer to measurement performed on a commercial 18" tire with a real pattern on tread. This aspect is very important because in rolling condition the excitation responsible of the high frequency vibrations comes from the pattern: the impacts of tread blocks on the road, combined with other effect such as stick-slip or stick-snap, generates their movement which vibrate with a frequency of these blocks which vibrate with a frequency related to number of blocks, their length and rolling speed.

The tire is fixed on the hub of a so called tire-stand: it is a big mass that can be moved up and down in order to press the tire on the underlying surface to control the tire load. Under the tire there is a drum driven by an electric engine and it is covered by a replica road surface to replicate the excitation coming from real roads. The inflation pressure is 2.5bar and the vertical load is 6.0kN both for the impact test and the rolling condition and rolling speed is 80kph.

2.1. Laser Doppler Vibrometer

The measurement of a moving surface by means of LDV is based on the Doppler Effect which generates a frequency shift in the light beam when it is reflected by the moving surface itself.

In the static case, a median circumference of 80 points is defined in order to scan the tire sidewall and the FRFs of each point is defined by the average of 5 hammer impulse in order to reduce the noise of LDV signal.

Both the signals of hammer and LDV are acquired by LMS SCADAS Mobile while the laser spot is moved by a LabView VI which is used to define the measurement circumference and to move the laser spot trough these points. All the data are processed in LMS Test.Lab and mode shape and the modal parameters are obtained by means of the Polymax algorithm.

In rolling condition, the measurement setup is almost the same; the only differences are the excitations and the reference sensors (hammer impact/load cell in static condition and rolling tire/accelerometer fixed on the tire stand for the dynamic case).

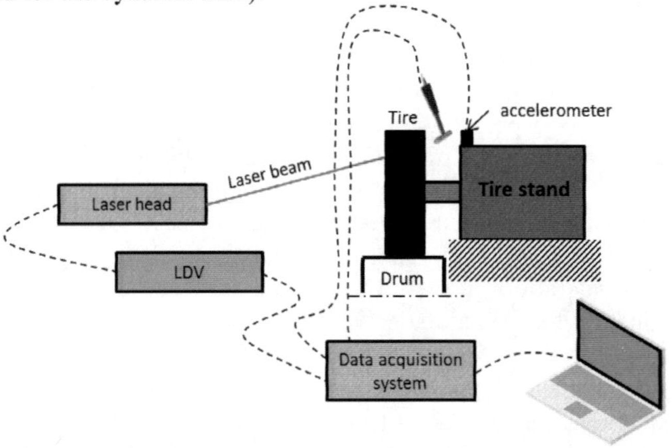

Figure 1 – LDV experimental setup.

2.2. 3D Digital Image Correlation

In this section the 3D-DIC setup is described with a theoretical introduction about the correlation algorithm used to perform this kind of measurements.

2.2.1. Theory. Digital Image Correlation (DIC) is a non-contact technique for full-field measurement of displacement and deformation based on the comparison between a reference image and an image in deformed condition: the tracking of measurement points defines the displacement and/or deformation map. These points are defined by painting a "speckle pattern" on the surface: it is a random pattern of points with high contrast with the surface (white dots on dark background for the case presented in this work). The points are organized in virtual areas, made by a certain number of pixels, called subset and for each of them the displacement is calculated with respect to the central point of the subset. Point P(x,y) moves to P'(x',y') where x' = x + u and y' = y + v are the new coordinates after the deformation, u and v are the displacement components in the reference system of the cameras.

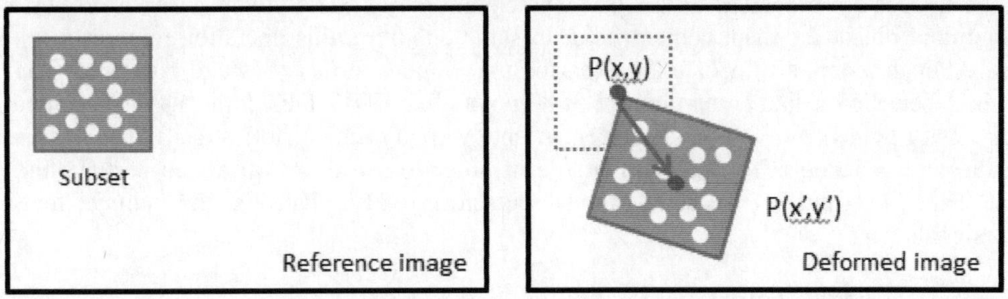

Figure 2 - DIC principle: the comparison between the reference and the deformed images defines the new coordinates of the measurement point P.

3D-DIC uses two cameras in order to measure displacement and deformation in three directions. The cameras must be placed at the same distance from the target object and they must have the same angle in order to have a correct 3D reconstruction. The distance between cameras and the object depends on the size of the object but also to resolution needed to measure the desired phenomenon which in turns depends on the excitation, as will be discuss later.

For the static case a classical Lagrangian approach is used, it means that the displacements are defined by following each measurement point and comparing the current image with the reference one; for the rolling condition an Eulerian approach is used (see section 4.3).

2.2.2. Experimental setup and procedure. The setup is made of two synchronized fast cameras characterized by a maximum frame rate of 4000fps, a maximum frame size of 1024 x 1024 pixels and an internal storage of 8GB which defines the number of acquired frames (5400) and accordingly the measurement time that is very short (1.35 seconds): the images are stored into the internal memory of the cameras and then transferred to the PC through a LAN connection. This operation requires less than 15 minutes. The illumination of the scene is provided by 2 LED lamps of 175 Watt each. A DAQ unit complete the system: it is used to acquire the load cell which will be used in post-processing for the calculation of the FRF and it is triggered with the master camera.

The cameras must be placed in order to respect the geometric relation presented in fig.4.

Finally the tire surface is prepared by means of a particular printer which generates a random pattern of white dots in contrast with the black surface of the tire.

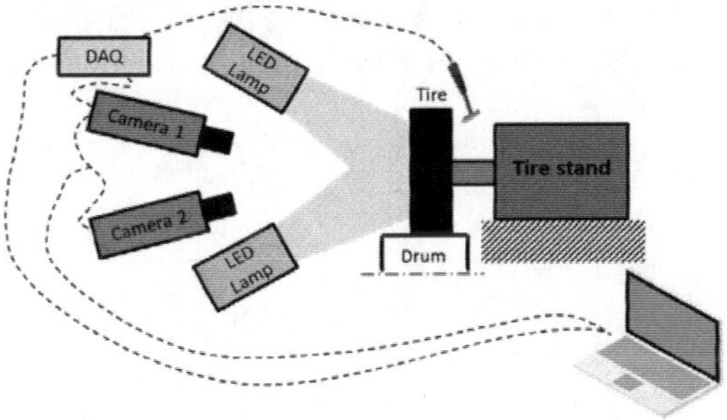

Figure 3 – 3D-DIC experimental setup.

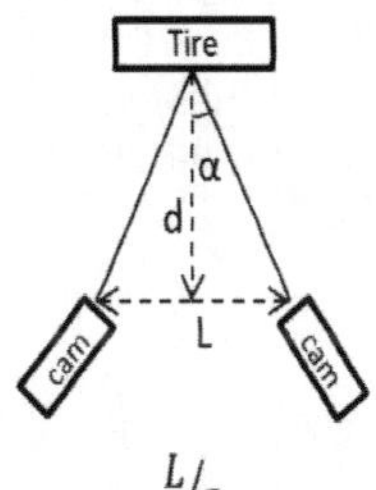

$$\sin \alpha = \frac{L/2}{d} \cong 0.25$$

Figure 4 – Cameras position.

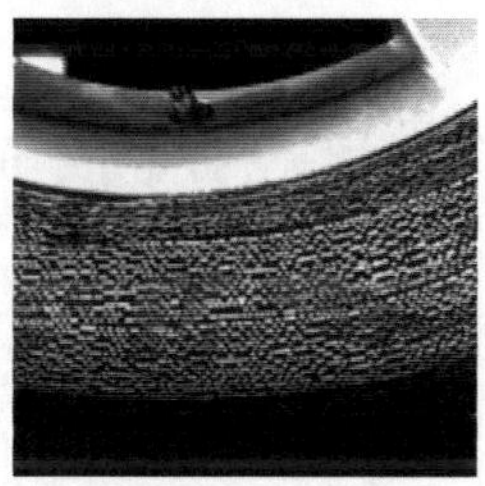

Figure 5 – Example of speckle pattern.

The acquisition of calibration and measurement images is performed by the VIC-Snap software which is used for the setting of all the acquisition parameters, such as frames size, fps, exposure and son on.

All the images are processed with Correlated Solution VIC-3D software in order to perform the correlation and obtain the displacement field. In static condition this operation requires less than 30 minutes. After this step, all the data are exported into Matlab format files and post-processed with a script which select a median circumference of 80 points like LDV (DIC defines a displacement fields of thousands of points), convert the displacement into velocity time histories (for the comparison with LDV data) and generate a Test.Lab format file in order to use this software to perform the modal analysis. This file simulates the synchronous acquisition of 81 channels: the hammer and the 80 velocity signals.

3. Case study 1: modal analysis

In this section the comparison between the two techniques in static condition is discussed.

In literature so many works can be found on this topic, e.g. [12, 13] and [9] are just two examples of modal analysis performed on a tire measuring vibrations with non-contact techniques.

3.1. Excitation

The first comparison is made in static condition with a classic impact test with hammer hitting the sidewall. Since the focus is the measurement of sidewall out-of-plane vibration, it is very important to define the optimal hammering conditions because the excitation in that direction must be high enough to generate appreciable velocity or displacement in the measurement direction. The optimal solution in terms of high excitation and low noise is hitting the so-called opposite serial side that is the sidewall opposite to the one measured. In this way the hammer does not cover the measurement surface and it is avoided the saturation of DIC system: the displacements in the impact area are much higher than the other points so the range to measure would be too wide and the smallest displacement could not be correctly measured.

3.2. Results

Figure 5 shows the FRFs obtained for the two systems in the same condition of load and inflation pressure. It can be seen that there is a good correlation between LDV and DIC both in terms of amplitude and frequency. There is a small frequency shift for some modes and a difference in terms of amplitude but this could be related to the number of averages: the FRF obtained by LDV is the results of the average of 5 impacts instead of the DIC's FRF which is the results of a single measurement. As expected the hammer is able to excite tire structure up to about 200Hz, where there is the biggest difference of the curves: due to its higher sensibility, the LDV is capable to detect the cavity mode (the peak at 200 Hz) instead of DIC where the peak is scarcely visible. However it must be noted that the cavity is well detected by DIC in the vertical direction, in fact the spectrum of displacement present a clear peak at 200 Hz, but it is not reported here since the comparison is made on the out-of-plane component. Moreover, the small frequency shift but it could be related to a higher noise level which characterizes the DIC technique if compared with the LDV one.

The results are summarized below.

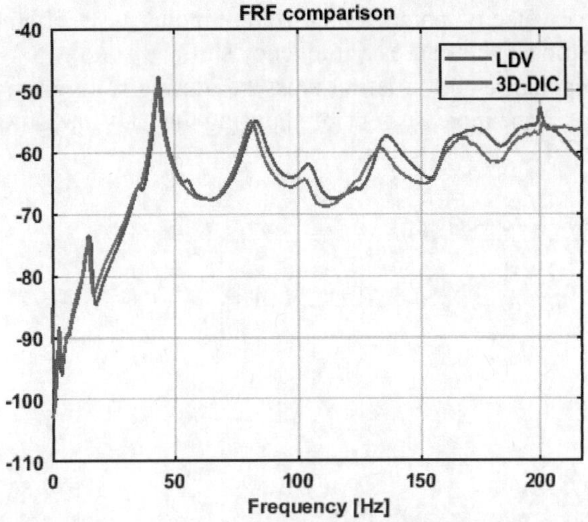

Figure 6 – FRF comparison.

	LDV	3D-DIC
	Frequency [Hz]	Frequency [Hz]
Mode 1	15.1	14.9
Mode 2	43.2	42.9
Mode 3	78.9	78.3
Mode 4	103.2	102.3
Mode 5	129.6	129.3
Mode 6	156.3	156.1

Table 1 – Frequency comparison for the detected modes.

4. Case study 2: Rolling tire

In this section the results obtained in rolling condition are discussed.

4.1. Scanning Laser Doppler Vibrometer – Approach and measurement grid definition

The measurement of tire vibration and the visualization of the operational deflection shapes is something known in literature: Castellini et al. performed this kind of measurement on a tire in free rolling condition by mean both of a Tracking LDV [3] and Scanning LDV [2]. In [4] the modes of a rolling tire are extracted considering the excitation coming from a cleat impact and measuring tire vibration with a single-point LDV.

SLDV measurement on rolling tire is performed using the Eulerian approach described in [2]: with TLDV a Lagrangian approach is used because the laser spot follows a single point along its trajectory, with the Eulerian approach the laser spot is fixed on a measurement point and the object is moving.

LDV measurements require stationary condition: the stationarity of the boundary conditions is guarantee by a continuous control of the load, the rolling speed of the drum and the temperature of the room. The only parameters that changes during the measurement are tire temperature and inflation pressure, so first of all it is necessary to evaluate if these parameters have an effect on the measurements. To do this, two circumferences of 80 points each have been considered, one closer to the rim and the second closer to external radius of the tire. In this way several aspects can be evaluated:

- effect of tangential velocity: points with different radius have different tangential velocities, so it must be understood if this causes a frequency shift;
- effect of tire temperature and inflation pressure: during rolling tire temperature and inflation pressure increase, so the increase generated during the LDV measurement could have an effect on the measurement.

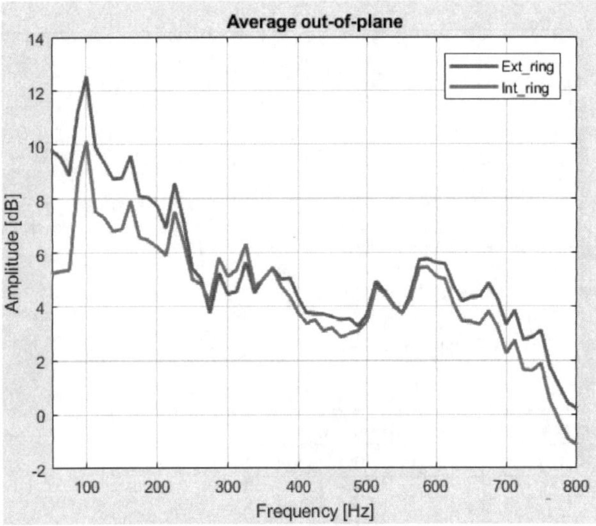

Figure 7 - Average AutoPower of the out-of-plane velocity for each circumference.

Figure 8 - Visualization of the measurement circumferences.

The figure shows the average AutoPower of the out-of-plane velocity for each circumference. This figure shows three main outcomes:

- the amplitude of vibrations depends on sidewall position: points closer to the rim have lower mobility if compared with external ones due to the rim constrain; there is no effect in terms of frequency so every circumference can be considered representative of tire behavior, the only difference is the amplitude level;
- tangential speed does not affect the frequency;
- the variation of tire temperature and inflation pressure during measurement does not affect the measurement; maybe the duration of the acquisitions is not so long to generate a significant increase of these parameters.

In conclusion, the measurement circumference can be chosen just considering the one which gives the highest level of signal.

Moreover, the sensibility of LDV allows a good measurement in the whole frequency range of interest, in fact it is able to measure both the low frequency vibration, responsible of the so called "Structureborne noise" (up to about 250Hz), and the high frequency vibration, responsible of the "Pattern noise" (the frequency range is related to the number of blocks, their dimension and the rolling speed; for the analyzed tire at 80 kph the range is 500 - 700 Hz).

It is interesting also to analyze the different contributions coming from different area of the rolling tire.

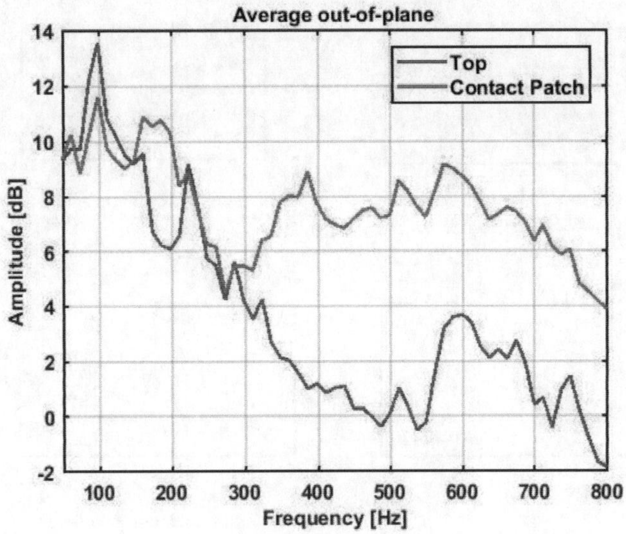

Figure 9 - Average AutoPower of the contact patch area and superior part of the tire to evaluate the contributions in the frequency range of interest.

Figure 10 - Visualization of contact patch area and top area of the tire.

From this figure, the main contribution in the high frequency range comes from the portion of sidewall that is closer to the contact patch, while the spectra in the low frequency range are comparable. The reason is the excitation: at low frequency the whole structure receives the same excitation, but at higher frequency the small displacement due to the contact with road generates waves propagating for a short stretch of sidewall both toward the Trailing and the Leading Edge. This effect will be more and more noticeable with 3D-DIC measurement.

Moreover, the correspondence of the spectra in the low frequency range suggests that LDV sensibility make possible to measure in the whole frequency range of interest.

The same effect appears considering the circumference closer to the rim, the only difference is a downward shift in terms of amplitude moving from tire treads towards the rim, so spectra are not reported just to avoid redundancy.

The analysis of LDV data proved that the high frequency is mainly caused by the contact patch area of the sidewall and this consideration will be helpful for DIC case (see paragraph 5).

4.2. 3D-DIC – Approach for dynamic measurement

In rolling condition the images are processed using an Eulerian approach instead of the classical Lagrangian approach used in DIC algorithms. Classic DIC uses a Lagrangian approach: the first frame is used to define the geometry of the object and the initial positions of the measurement points. The measurement area is divided in a certain number of interrogation areas called subset and the displacement field is defined by following each subset along its trajectory and the displacement is defined by the difference between the current position of the points and its position in the initial frames. The results are the absolute displacements of the points, frame by frame.

The approach used here is the so called incremental DIC [11]: incremental DIC was developed for those cases in which the deformation is too wide that it is not possible to find the correspondence of points in current frame and the reference one, so an intermediate reference frame is needed. With the Eulerian approach the reference image changes step by step, so frame by frame the measured displacement it is not an absolute displacement, but it is delta of displacement. The idea is to fix the observation area and analyse what happens behind that area, in the same way as the LDV measurement.

IOP Conf. Series: Journal of Physics: Conf. Series **1149** (2018) 012010 doi:10.1088/1742-6596/1149/1/012010

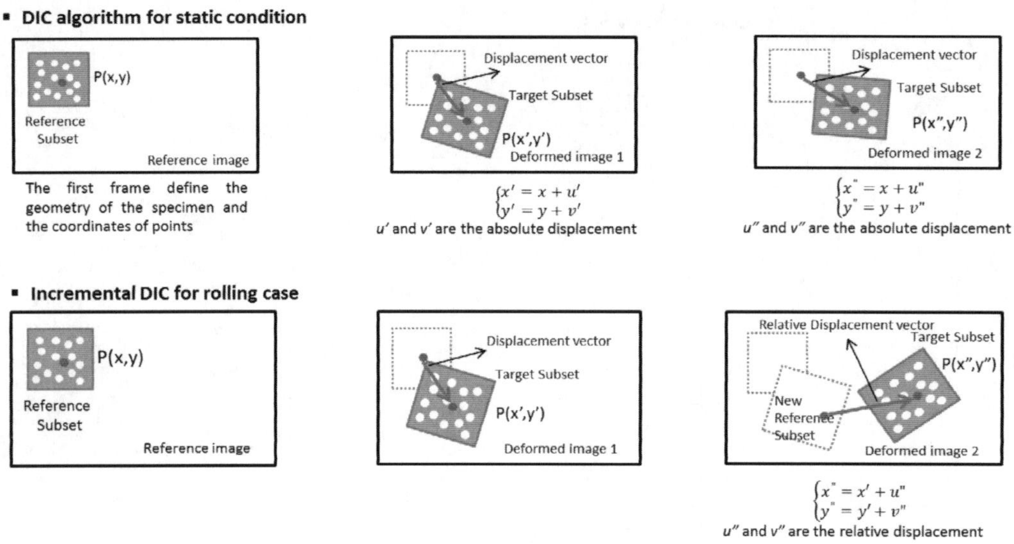

Figure 11 - Scheme of classic and incremental DIC algorithms.

To obtain good results with 3D-DIC, the frequency range of interest must be defined first, because this choice defines the optimal framing.

One of the major advantages of 3D-DIC is that it is a full-field, but the system resolution is defined by the ratio of the dimension of the framed area and the number of pixels of the camera, so the full-field framing could not be the optimal one. Tire generally has a diameter of about 600 mm and the sensor used has just 1 megapixels, so it is not possible to have a very high resolution: it means that bigger displacement will be measured but the smallest ones will not. The resolution can be increased using a sensor with a higher number of pixels or changing the framing as will be described below.

This kind of framing can be used if the idea is to measure the low frequency vibrations and it is useful to visualize how the contact with the road excite tire.

The following image is an instantaneous map of the out-of-plane displacement when the tire is rolling. The effect described analysing the LDV spectra is confirmed: the excitation coming from the contact with the road generates two small areas in the lower part of the tire characterized by higher displacement if compared with the rest of the structure.

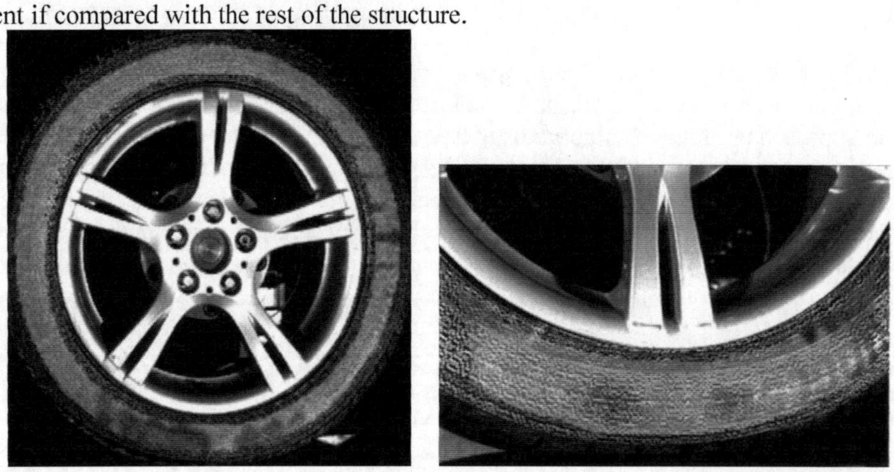

Figure 12 - Out-of-plane displacement field for full-field framing (left) and closer framing (right).

When the framed area is smaller, the resolution increases since the pixels are focused on a smaller area and smaller displacement can be correctly measured.

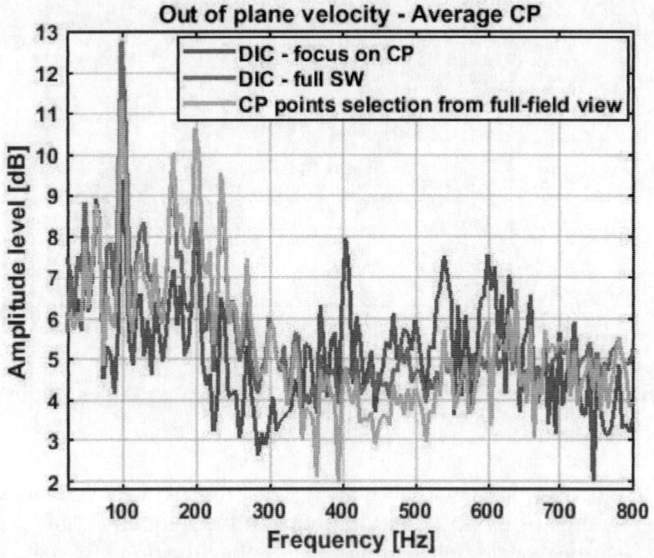

Figure 13 - Comparison of the average AutoPower of out-of-plane velocity for the two configurations of 3D-DIC system.

In figure 8 are reported the average AutoPower of out-of-plane velocity for the following case:

- an external circumference of 80 points, like LDV case (red curve);
- a group of 20 points in the contact patch zone starting from the full-field view (yellow curve);
- a group of 20 points for the closer view (blue curve).

Focusing the cameras on the contact patch area it is possible to measure the high frequency vibrations, with a loss of information at lower frequencies: main peaks are detected in terms of frequency, but the amplitude is lower if compared with the full-field view and the measurement in that region is noisier.

It must be remarked that, to measure the high frequency vibrations, it is not sufficient to measure the entire sidewall and perform the FFT just on the points localized in the contact patch area (yellow VS blue spectra). In fact, working with global framing the range of the displacement field is very wide and the DIC's sensibility it is not so high to measure it. This is probably the main disadvantage of DIC with respect to LDV.

5. SLDV VS 3D-DIC: techniques comparison in rolling condition

This paragraph presents the comparison of the measurement performed with the two techniques to evaluate the effective capabilities of DIC technique for this kind of measurement.

The figure below shows the comparison of the SLDV with the DIC measurement performed with a framing on the entire sidewall. The LDV spectrum is the average of the 80 points along the measurement circumference; the same for the DIC's one: this measure provides information about thousands of points, so a circumference of 80 points is selected.

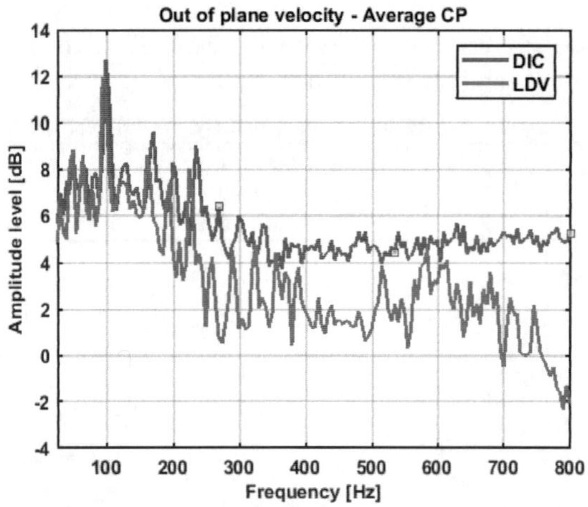

Figure 14 - Comparison of AutoPowers obtained with full framing DIC and the entire circumference of LDV.

As previously stated, this comparison confirms that the full field view measurement is not capable to measure in the whole range of interest: the vibrations in the frequency range of pattern noise (500 – 700 Hz in this case for the considered rolling speed), but the low frequency vibrations are well detected. There is an amplitude shift in the whole measurement range, but this could be related to number of averages: DIC measurement is very fast (1.35 sec) so it is intrinsically affected by an higher noise level with respect to LDV technique, so it could be reduced by performing several measurement and averaging them or using a bigger subset dimension during image processing.

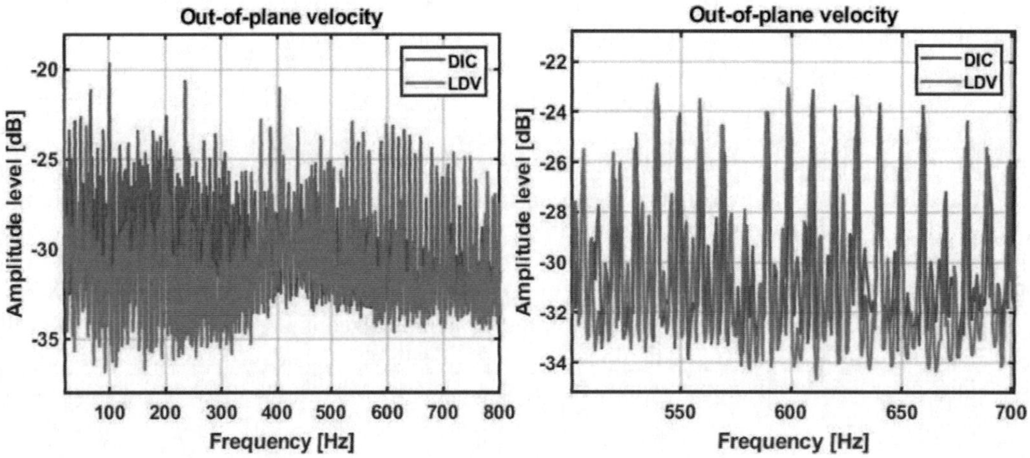

Figure 15- Comparison of AutoPowers obtained with closer framing DIC and the portion of circumference localized in the contact patch area of LDV (left); zoom in the frequency range of interest (right).

The figure above, instead, shows a good correlation between LDV and the 3D-DIC measurement performed with the cameras focused on the contact patch of the tire. The shape of the spectrum is well reproduced, the amplitude levels are more comparable in the high frequency region (above 500 Hz) and the noise level of the signal is significantly reduced if compared with the previous case. The resolution increase due to closer framing allows measuring the small displacements which characterise the high frequency vibration and a reduction of noise level.

6. Conclusion

The comparison between LDV and 3D-DIC measurement suggests the possibility to use this new technique as an effective tool for vibration measurement both for the low and the high frequency, the static and the dynamic case. In static condition there is a very good correlation in terms of amplitude and frequency, there is just a little difference between spectra, but it could be reduced increasing the number of averages in order to reduce noise (with LDV 5 impacts for each point are considered, while DIC measurement is performed with a single hammer impact) and with a better optimization of the speckle pattern. In this case the DIC technique seems to be better than LDV because the post-processing time is not so long so the very short acquisition time can be effectively can be considered an effective advantage because we can perform a stationary measure without variation in terms of boundary conditions.

For the rolling case further investigations are needed, especially the difference in terms of frequency must be understood. The main advantage of DIC technique (it is a full-field measurement) can't be exploited when measuring at high frequency due to the resolution of the cameras. It is necessary to consider a smaller area, which can be analyzed with a high spatial resolution since thousands of measurement points can be detected. In this case the long post-processing time is the biggest limitation if compared with LDV, but in order to obtain a comparable number of points with LDV the measurement should last about an hour with significant temperature variation which affect material behaviour.

In conclusion, a new procedure for tire vibration measurement using 3D-DIC has been developed and validated with a direct comparison with LDV, both in static and dynamic conditions, and two different setups have been described according to the frequency range of interest.

References

[1] Sandberg U and Ejsmont 2002 Tire/road noise reference book Informex Ejsmont & Sandberg Handelsbolag Harg, SE-59040 Kisa, Sweden

[2] Castellini P and Ballatore M 2010 How to measure actual tire shape in the rolling condition using Scanning LDV Mech. Sys. And Sign. Proc. 24 (2010) pp 736-45

[3] Castellini P 2002 Vibration measurements by tracking laser doppler vibrometer on automotive components Shock and Vib. 9 (2002) pp 67-89

[4] Kindt P, delli Carri A, Peeters B, Van der Auweraer H, Sas P and Desmet W 2011 Operational modal analysis of a rotating tire subject to cleat excitation Proc. IMAC (Jacksonville) 28 (2010)

[5] Moser R and Lightner III J G 2007 Using three-dimansional digital image correlation techniques to validate tire finite element model Exp. Tech. 31 (4) pp 29 -36

[6] Castellini P, Chiariotti P, Matrarelli M, Zappa E and Lavatelli A 2017 Experimental modal analysis on vibration data measured by digital image correlation Proc.The Soc. For Exp. Mech. Series (2017) pp 285-291

[7] Helfrick M N, Niezrecki C, Avitabile P and Schmidt T 2010 3D digital image correlation methods for full-field vibration measurement Mech. Sys. and Sig. Proc. 25 (2011) 917-27

[8] Beberniss T J and Ehrhardt D A (2015) High-speed 3D digital image correlation vibration measurement: recent advancements and noted limitations Mech. Sys. and Sig. Proc. 86 (2017) pp 35-48

[9] Patil K, Basqersad J and Sheidaei A (2017) A multi-view optical technique to obtain mode shapes of of a tire Shock & Vib, Aircraft/Aerospace,Ener. Harv, Aco. & Opt. vol 9 211-7

[10] Moser R, Sube H J, Turner J L and Zakelj P (2009) 3D digital image correlation: application to

IOP Conf. Series: Journal of Physics: Conf. Series **1149** (2018) 012010 doi:10.1088/1742-6596/1149/1/012010

tire testing TSTCA vol 38, no. 2 (2010) pp 100-18
[11] Zhou Y, Sun C and Chen J (2013) Adaptive subset offset for systematic error reduction in incremental digital image correlation Opt. and Laser Eng. 55 (2014) pp 5-1
[12] Abd_Elsalam A, Gohary M A and El-Gamal H A (2016) Modal analysis on tire with respect to diferent parameters AEJ vol 54, Issue 4 (2017) pp 345-57
[13] Pieters R S (2007) Experimental modal analysis of an automobile tire under static load DCT rapporten vol 2007.112

IOP Conf. Series: Journal of Physics: Conf. Series **1149** (2018) 012011 doi:10.1088/1742-6596/1149/1/012011

Material Defect Reconstruction by Non-Destructive Testing with Laser Induced Ultrasonics

H Selim[1], M Delgado[2], J Trull[1], R Picó[3] and C Cojocaru[1]

[1.] Universitat Politècnica de Catalunya, Physics Department, Terrassa, Barcelona, Spain
[2.] Universitat Politècnica de Catalunya, Electronic Engineering Department, Terrassa, Barcelona, Spain
[3.] Instituto de Investigación para la Gestión Integrada de Zonas Costeras, Universitat Politècnica de Valencia, Valencia, Spain

E-mail: hossam.eldin.mohamed.selim@upc.edu

Abstract. Aligned with current research efforts and industrial applications on non-destructive testing, in this work, a hybrid system combining remotely induced laser ultrasonics with conventional transducer detection is studied for defects detection in metallic parts. The processing of the induced acoustic emission waves is proposed to be approached by means of a high-resolution volumetric signal processing procedure based on the synthetic aperture focusing technique for the benefit of the final 2D visualization of the defects. The advantages of the hybrid non-destructive testing approach and the performance of the processing technique are experimentally validated.

1. Introduction

Nowadays, Non-Destructive Testing (NDT) methods are implemented in industrial applications for the detection of fractures and defects in manufactured assets and components. The most extended NDT techniques involve ultrasonic transducers for both the generation of ultrasound as well as for its detection after propagation through the material under study. The analysis is typically based on the measurement of the time of flight of ultrasonic pulses considering the distance from the emitter to the receiver, which allows the identification of discontinuities. Different configurations of this approach were extensively discussed in many research papers and applications, such as the measurement of the crack location, height and width [1–3], or of the crack penetration using bulk waves [4]. Although these methods have the advantages of low cost, and easy implementation and provide satisfactory results in many applications, the emitter transducers also show drawbacks. Some of them are the low output power that prevents such systems from being used remotely, low frequency bandwidth range that makes necessary the use of microphone arrays or ultrasonic scanners increasing the system's overall cost, small surface areas that prevent covering large object areas at once, and quite low spatial resolution in the excited volume.

As an alternative, photonic approaches based on laser induced ultrasonic and optical detection showed up as valuable competitors to the conventional ultrasonic NDT techniques, offering the possibility of remote excitation and detection at a much higher resolution [5,6]. Optical methods have important advantages such as the non-contact application, remote control and generation of broadband frequency waves (from kHz to GHz), high output power and the possibility to easily scan large object areas. The main drawbacks of optical detection methods are their high sensibility and critical stability in detection, as well as elevated costs that are acceptable only for high-resolution measurements in certain fields of application.

Considering the advantages and disadvantages of both NDT approaches, in this work, the study of a hybrid NDT scheme, where the ultrasounds are induced by laser pulses and their acquisition is done with conventional transducers, is proposed, thus, combining the main advantages of both optical and ultrasonic devices [7,8]. Laser induced ultrasound technique allows remote inspection and control, without the need of a direct contact. This technique features a broader frequency bandwidth compared with the conventional ultrasonic transducer, covering any ultrasonic bandwidth needed for various applications. It provides, then, an enhanced resolution in the excited volume and represents an easy tool for scanning the sample. On the other hand, the detection with ultrasonic transducers simplifies the operation of the whole NDT system keeping a good resolution in the range of millimeters. In this study, it is proposed to analyze the resulting ultrasonic waves by means of a high resolution volumetric signal processing procedure based on the synthetic aperture focusing technique. This technique has been validated in classical ultrasonic arrangement, and its performance is proposed to be analyzed in front of such hybrid NDT approach. The study is supported by experimental results with the aim of damage localization and defect reconstruction in a metallic part.

2. Two-dimensional scan and SAFT analysis

Synthetic Aperture Focusing Technique (SAFT) is based on the back-propagation theory which creates a virtual image of the object under study by focusing the recorded data of a 1-dimensional or 2-dimensional A-scan processes. The algorithm superimposes the relatively low resolution A-scan data at every point of the scanned area, creating a higher resolution focused image with a higher signal to noise ratio (SNR), that is referred to as the Synthetic Aperture (SA) [9,10]. Depending on the implemented setup, the analysis can be performed in the time domain or in the frequency domain. SAFT technique can be used to generate a 2-dimensional image of a particular plane of the object or can be generalized to the 3-dimensional mode to cover all planes in the volume of the object. In the 3-dimensional mode, the reconstructed image can project maps of the defect in any plane, regardless of the plane used for the scan. This is a very flexible property that allows the scanning of the sample at any angle or plane, making the method more appropriate from the applicability point of view, since in general, the position and size of the defect that wants to be reconstructed is not known. In the time domain analysis, the Time Of Flight (TOF) is calculated across all SA points, starting from the source position at a particular scan point to all geometrical target points in the object's volume and back to the signal receiver at each particular scan point. For each geometrical point, an integration of all received signals at a particular value of TOF is calculated, to produce the final pixel value of this particular scan point in the final image. This integration of the signal at each pixel results in a higher contrast, resolution and better SNR value. The basic description of the SAFT algorithm is shown in figure 1, assuming a 3D object under inspection by a transmitter-receiver set in a pitch-catch mode.

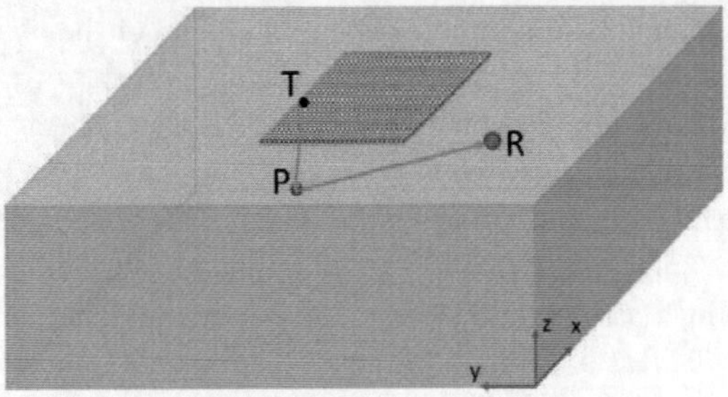

Figure 1. Geometry of the SAFT algorithm processing procedure, including the transmitter (T), the receiver (R), and the point of interest in the object's volume (P).

The TOF of the ultrasonic signal generated by a transmitted, *T*, and detected by a receiver, *R*, after it has been propagated to and reflected by an arbitrary point in the object volume, P, can be estimated following equation (1).

$$TOF_{(i,j,k)P} = \frac{|\bar{d}_{(i,j,k)P} - \bar{d}_{(i,j,z)T}| + |\bar{d}_{(i,j,k)P} - \bar{d}_{(i,j,z)R}|}{c} \quad (1)$$

where the *TOF* is the time of flight corresponding to the point *P,* and *d* is the displacement vector at positions of *T* ,*R* and *P*, while *i, j* and *k* represent the indexes of the volume image points in the *x*, *y*, *z* planes, respectively. The parameter *c* denotes the speed of the longitudinal waves in the material of the object [9–11]. This is done for every point in the A-scan measurements to produce a preliminary unfocused image y_R of the particular point *P*. The combination of all the individual images results into a high resolution focused image y_f. It is assumed that the scan is done at *M***N* points denoting a number of 2-dimensional transmitter steps, and also, only one receiver collects all the transmitted A-scan data from all transmitter scan points at an arbitrary point in the SA. Then, the final signal is given by equation (2).

$$y_f(P) = \sum_{j=1}^{N} \sum_{i=1}^{M} a(TOF_{(i,j)_P}, i, j) * y_R(TOF_{(i,j,k)_P}, i, j) \quad (2)$$

where y_R is the received signal at the $TOF_{(i,j,k)_P}$ at scan point indexes *i* and *j,* and y_f is the focused received signal from all the transmitter scan points at this particular receiver *R*. The $a(TOF_{(i,j)_P}, i, j)$ represents the weighting or apodization function [12]. For more than one receiver, we should modify equation (2) accordingly and consider the integration due to the scan of all receivers.

This derivation of the final image reconstruction using SAFT is only valid if the ultrasonic transmitter is considered as a point source which is completely applicable in the case of laser-generated ultrasonics. The receivers also need to be regarded as measuring data using a point-like transducer.

It should be noticed that, with regard to the part under inspection, the dimensions of the object should be large enough to avoid reflections on the object's boundaries, which could generate misleading data. The SAFT algorithm, the way it is described in the aforementioned analysis, can work with ultrasonic signals travelling at a single speed, if there is a composite structure that will result in different velocities of the sound inside each layer; the equation should be modified to consider each plane as a separate problem and combine all together at the end to generate the final image. Thus, the wave dispersion due to working in composite or inhomogeneous materials will result in violating the SAFT algorithm.

3. Experimental set-up

The experimental set-up is shown schematically in figure 2. A Nd:YAG laser is used emitting pulses of 8ns at a wavelength of 532nm and with 10mJ energy per pulse. The laser beam is directed towards a galvanometer that is programmed to scan a 2-dimensional area of the object's surface. The energy carried by the laser pulse is rapidly absorbed into a shallow volume of the material, and creates a localized thermo-elastic expansion inducing a stress wave that generates an acoustic pulse. For detection, we use two conventional Olympus Ultrasonic transducers, model V133-RM, with a 2.25 MHz central frequency, coupled to the surface of the object. Both sensors are placed on the laser beam incidence's surface, in a pitch-catch mode. The acquisition system is comprised of an Olympus preamplifier 5662, and a high-performance Gage A/D card model Oscar Express CSE4444, that allows a sampling frequency of 50 MHz at 16 bit of resolution. The recorded signals are filtered by a bandpass filter and added to an interpolation algorithm to remove the noise and low/high frequency components that are not of interest. We subtracted the background noise from the main signals and an averaging algorithm was performed to remove the DC components from the signal.

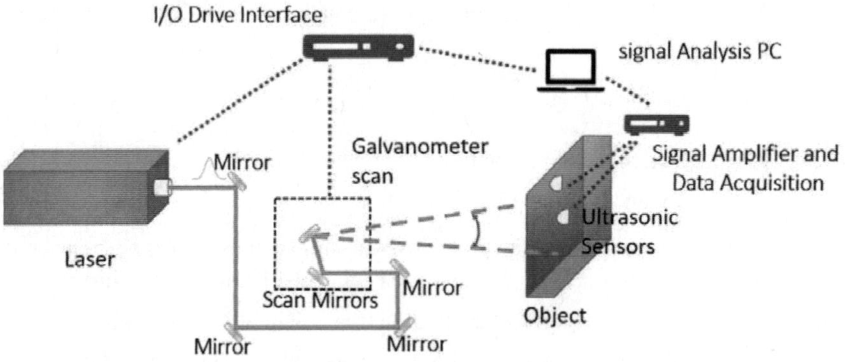

Figure 2. Schematic representation of the experimental set-up

Figure 3 shows the geometry of the 300mm x 200mm x 20mm aluminium sample considered in this study, having a cylindrical hole of 8mm diameter (the defect). Details about the scanning area and the sensors position are also shown in the figure.

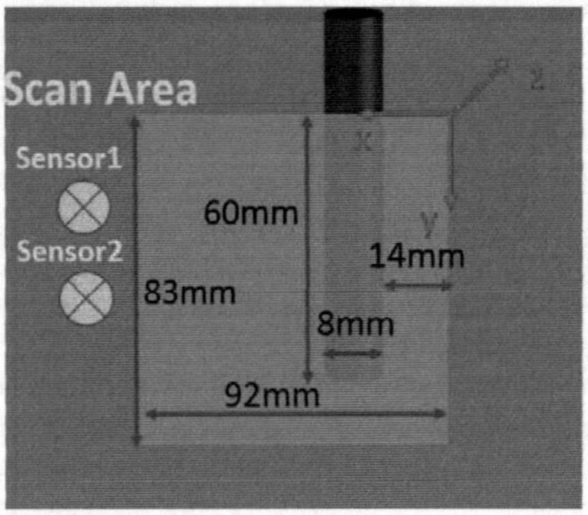

Figure 3. Aluminium sample with defect and the scan plane (in grey) for SAFT experiment.

IOP Conf. Series: Journal of Physics: Conf. Series **1149** (2018) 012011 doi:10.1088/1742-6596/1149/1/012011

The galvanometer shown in figure 2 is programmed to perform a 2-dimensional laser impact scan on the sample. As aforementioned, a 2-dimensional scan area of 92mm x 83mm is considered, resulting in 101 x 91 scan points corresponding to the *x* and *y* directions, respectively. The laser moves and impacts on each defined point of the object's surface, generating a broadband ultrasound signal. The experiment is done in a pitch-catch mode, where two receivers are placed at the positions indicated in figure 3. The programming and processing software for this experiment has been done using Matlab.

4. Experimental Results

Once the ultrasonic signals are collected, the SAFT algorithm is applied. The weighting function has been chosen to be a Gaussian shape. Thus, with the data obtained from the 2-dimensional laser scan on the sample and applying the signal pre-processing and the SAFT algorithm described above, the volumetric reconstruction of the defect under the scanned area is obtained as shown in figure 4. It must be noted that the volumetric representation considers only the reconstruction of the volume under the scan area. Also, it should be noted that commonly used SAFT described in the literature incorporates 1-dimensional line scanning instead of the 2-dimensional scanning area proposed in this study. In the classical case, only the reconstruction of one projection of the defect on one plane is possible. However, it is shown in this study that scanning on a 2-dimensional plane and integrating the SAFT algorithm over the object's volume of interest allow the defect visualization from different angles and at different depths, making possible a 3-dimensiona image reconstruction.

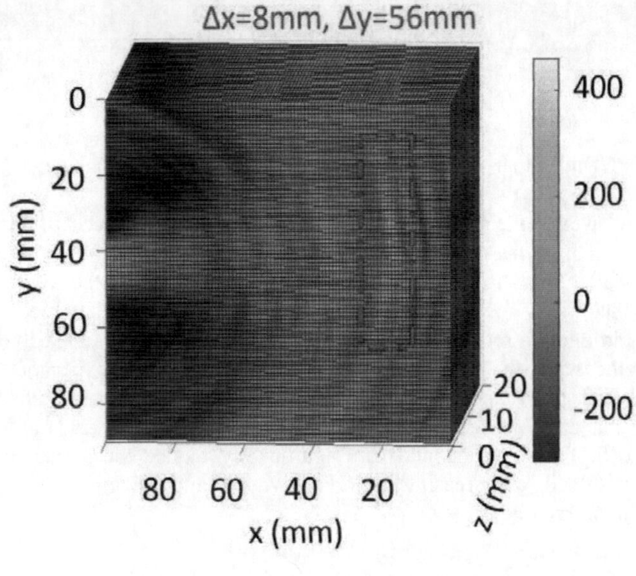

Figure 4: 3D slice for a full volume representation using a hybrid NDT approach and SAFT algorithm.

Figure 5 shows the contour map obtained as a projection on the *x-y* plane by grouping together all lines with a similar density level.

The contour map shows the reconstruction of the defect in the *x-y* plane: it is centred at $x = 20$mm, and $y = 40$mm with the corresponding size $\Delta x = 8.2$mm and $\Delta y = 56$mm. Compared with the actual size and position of the defect, the resulting reconstruction has the following errors: ($\Delta x_{error} = 2.5\%$ and $\Delta y_{error} = 6.7\%$ and a positioning error of $x_{error} = 2.5\%$ and $y_{error} = 22\%$.

IOP Conf. Series: Journal of Physics: Conf. Series **1149** (2018) 012011 doi:10.1088/1742-6596/1149/1/012011

It must be noticed that there are some shadows in the reconstruction of the defect that could be misleading. They come from the lower intensity back scattering and can be identified on the contour map.

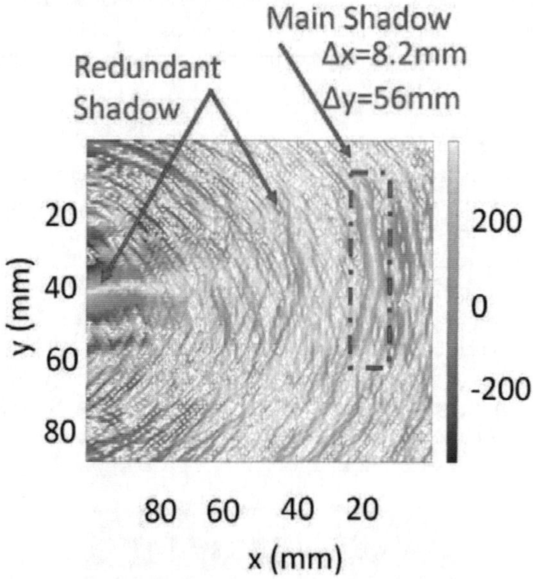

Figure 5: Contour map of the defect at *x-y* plane at z = 10mm using SAFT algorithm.

These shadows, as well as the resolution, are conditioned by some aspects of our experiment. On the one hand, the *z*-dimension of the object studied here was small compared with the *x* and *y* ones, which results in multiple reflections from the *z* boundaries, especially when the laser and the sensors are relatively far from the object. To solve this problem, the *z* dimension of the object under test should be large enough to avoid internal reflections from boundaries. This applies to all object boundaries in all directions. On the other hand, the position of the receiving sensors in the experiment was kept fixed and it only scanned the laser impact points. This configuration increased the angle between the excitation point and the position of the detected reflection by the ultrasonic sensor. Large angles reduce the performance of the SAFT algorithm and can significantly reduce the signal to noise ratio. These constraints can be resolved by increasing the object's depth and reducing the angle by scanning the receivers along with the laser beam.

5. Conclusion

The SAFT algorithm application, combined with a hybrid NDT approach composed by laser induced ultrasonic and conventional transducer receivers, has been studied. The results of the SAFT algorithm indicate the position and size of the defect, with the advantage of visualizing the defect in 3D. The projections of the SAFT planes make it easier to distinguish the location of the defect.

The defect position and size are determined with a relatively good signal to noise ratio, taking into consideration the limitations of the dimensions considered in the component under test. This means that the SAFT algorithm can give relatively good and reliable results even if the input signal and experimental conditions are not optimal.

Finally, two fixed ultrasonic receivers were used and they showed that by increasing the number of transmitters (i.e. laser scan points), the same good-quality images with a very limited number of receivers fixed at particular points can be obtained.

Future research involves more enhanced experimental conditions with better automated scan in both transmitted and received signals as well as the object's dimensions that satisfy the SAFT algorithm constraints.

Acknowledgements
Authors acknowledge the support of Ministerio de Economía y Competitividad with the Coordinated Project 'Light and sound waves in crystals, structured media and metamaterials', Ministerio de Economía e Innovación and European Union FEDER with the Project FIS2015-65998-C2-1 and FIS2015-65998-C2-2 and Conselleria d'Educación, Cultura y Deporte de la Generalitat Valenciana with the Project No. AICO/2016/060.

References
[1] Ham S, Song H, Oelze M L and Popovics J S 2017 A contactless ultrasonic surface wave approach to characterize distributed cracking damage in concrete *Ultrasonics* **75** 46–57
[2] Amjad U, Yadav S K and Kundu T 2015 Detection and quantification of pipe damage from change in time of flight and phase *Ultrasonics* **62** 223–36
[3] Kharrat M and Gaillet L 2015 Non-destructive evaluation of anchorage zones by ultrasonics techniques *Ultrasonics* **61** 52–61
[4] Masserey B, Raemy C and Fromme P 2014 High-frequency guided ultrasonic waves for hidden defect detection in multi-layered aircraft structures *Ultrasonics*
[5] Kreis T 2016 Application of Digital Holography for Nondestructive Testing and Metrology: A Review *IEEE Trans. Ind. Informatics* **12** 240–7
[6] Zhang K, Zhou Z and Zhou J 2015 Application of laser ultrasonic method for on-line monitoring of friction stir spot welding process. *Appl. Opt.* **54** 7483–9
[7] Brown M D, Allen T J, Cox B T and Treeby B E 2014 Control of optically generated ultrasound fields using binary amplitude holograms *IEEE International Ultrasonics Symposium, IUS* pp 1037–40
[8] Jen C and Wu K 2008 NDE using laser generated ultrasound and integrated ultrasonic transducer receivers *Ultrason. Symp.* 1516–9
[9] Spies M, Rieder H, Dillhöfer A, Schmitz V and Müller W 2012 Synthetic Aperture Focusing and Time-of-Flight Diffraction Ultrasonic Imaging — Past and Present 310–23
[10] Boonsang S, Zainal J and Dewhurst R J 2004 Synthetic aperture focusing techniques in time and frequency domains for photoacoustic imaging *Insight Non-Destructive Test. Cond. Monit.* **46** 196–9
[11] Stepinski T 2010 Synthetic aperture focusing techniques for ultrasonic imaging of solid objects *Synth. Aperture Radar* 438–41
[12] Jensen J A, Nikolov S I, Gammelmark K L and Pedersen M H 2006 Synthetic aperture ultrasound imaging *Ultrasonics* **44**

Diagnostics based on continuous scanning LDV methods: numerical study

S Bruinsma, D Di Maio, T Tinga

Dynamics Based Maintenance, Department of Engineering Technology, Horst 2, 7522 LW, Enschede, The Netherlands

Abstract. This paper presents a research work on diagnostics using continuous scanning laser methods. Structural Health Monitoring based on laser vibrometry can exploit the ability of a scanning laser to measure vibration responses remotely both regarding temporal and spatial resolution. Therefore, it can be a powerful approach for rapid damage detection of components presenting an unhealthy dynamic behaviour. The work is focused on numerical analysis of a cantilever beam which is subjected to damage of different severity levels. The severity will be measured regarding response phase shift for a fixed excitation frequency. The simulated nodal responses of the surface being scanned by a laser beam are selected and used for simulating the spectral responses obtained from CSLDV methods. The main objective of this work is to determine the structural integrity based on an indicator obtained by referencing the relative spectral sidebands between a pristine and damage condition.

1. Introduction

In recent years, durability and extension of life of components has become an important topic of research. Structural Health Monitoring (SHM) focuses on detection of unhealthy dynamic behaviour as damage occurs and propagates in the structure. The early detection of damage is highly interesting for for maintenance in the industry, which supported research for better diagnostic tools. These consist of destructive and non-destructive methods. Obviously, non-destructive methods are preferred, as these are most applicable in the industry for SHM. A conventional way of testing is to apply strain gauges to the component. However, these cannot be placed on rotating shafts, for example. Also, the fastening methods cannot always be applied due to high operating temperature. This can be omitted by using a non-destructive and non-contact measurement technique. Most of these techniques use either laser light, such as Holography, Speckle interferometry and Scanning Laser Doppler Vibrometry (SLDV), or a camera, like Digital Image Correlation (DIC).

DIC uses a camera and image processing to obtain dynamic information[1]. DIC requires surface preparation and a lot of post processing to obtain results on the strain. Holography is a expensive method that allows for comprehensive measurements [3]. The downside for the industry is that the part needs to be isolated and placed in this large and finely tuned test set-up, increasing downtime of the system. Interferometry needs a similar sterile environment to do the measurements [9]. SLDV is a more robust approach that allows for the possibility of in-situ measurements [2]. Continuous SLDV strengthens this application by making it a very fast measurement method.

This paper focuses on Continuous SLDV, which main advantage is fast acquisition of both spacial and temporal information. CSLDV uses a laser that scans a vibrating structure. The time signal from a sinusoidal line scan can be modulated to a frequency spectrum [4], which can be used for SHM. The level of depth in damage detection is categorised in 4 levels, as explained in [5], the definition of the different levels are displayed in Table 1. Most of the current research with this technique is on locating the damage [6], which is quantified as SHM level 2. To exploit its main advantage of speed, the level of SHM is reduced to 1 in this study, where only the detection of damage is pursued.

Table 1: Level of SHM

Level 1	Determination that damage is present
Level 2	Determination of the geometric location of the damage
Level 3	Quantification of the severity of the damage
Level 4	Prediction of the remaining service life

A numerical study is undertaken, analysing a cantilever, aluminium beam. The prismatic beam is subjected to an oscillating force at the free end. The goal of the study is to determine the sensitivity of the CSLDV method to introduced damage. To accomplish this, different approaches to defining a damage criterion are taken.

2. Method

In this chapter the methodology of the research is presented. First the construction of the model is described. The model is checked and the sensitivity to the damage is analysed. This is then used to define damage indicators that are sensitive enough to show propagation of damage.

2.1. Numerical Model

Researching the viability and sensitivity of this method is done on a simple geometry, a prismatic, cantilever beam. The dimensions of the beam are 0.4x0.04x0.01 m (l,w,h) or (x,z,y). The beam is modelled in Finite Element Program ANSYS Mechanical APDL 16.2. The beam is modelled with 200x20 (x,y) rectangular SHELL181 elements. Only membrane bending is used, and full integration. All nodes are locked in z-direction, one end is fixed in y- and x-directions as well. At the other end, an excitation force is applied in y-direction. The harmonic response is simulated from 1 to 9000 Hz with a frequency step of 1 Hz. The model shows good correspondence to analytical eigen frequency calculations. The first six eigen frequencies are compared in Table 2.

Table 2: Comparison Analytical and Numerical

	ω_1	ω_2	ω_3	ω_4	ω_5	ω_6
Analytical (Hz)	51.04	319.9	895.4	1755	2901	4334
Numerical (Hz)	51	319	889	1731	2837	4196
Error (%)	<0.1	0.3	0.7	1.4	2.2	3.2

The prismatic beam has a consistent width (z-direction) of elements of 0.04 m. Damage is introduced to the model by reducing the width of the elements, which results in a stress concentration around the damaged elements. This way of modelling allows for a wide variety of elaborate damage cases, such as delamination and interlaminar cracks. In this study, a simple damage case is defined: a transverse cut, or saw cut. The transverse cut is placed at 0.04 m from the root, starting on the lowest row of elements, made in three severities. The most extreme case is the 50% damage case, where the cut is 0.005 m, thus half the thickness (in y-direction), which corresponds to 10 elements with reduced width. Furthermore, 25% of the thickness and 10% of the thickness are evaluated, corresponding to 5 and 2 elements respectively.
The ODS for all frequencies is extracted and a curve is fitted through the shape using the $polyfit$-function in MATLAB. The x-coordinates are transformed between -1 and 1, to allow for simulation of the CSLDV time signal. This time signal is the result of equation 1, from [7].

$$v(t) = \sum_{n=0}^{p} V_{Rn} \cos^n(\Omega t) \cos(\omega t) \tag{1}$$

Where p is the order of the polynomial used to describe the ODS, V_R is the real part of the polynomial coefficients, Ω is the scan frequency of the CSLDV (in this paper 1.1 Hz.). ω is the excitation frequency.

IOP Conf. Series: Journal of Physics: Conf. Series **1149** (2018) 012012 doi:10.1088/1742-6596/1149/1/012012

2.2. Sensitivity

To establish an indicator for the damage severity levels in relation to the vibration modes three FRFs were simulates for the damage cases. The crack is placed close to the root of the beam (10% the beams length, measured from the clamped end). The 50% transverse crack is 2 mm wide and 5 mm in length, which causes a frequency shift of less than 1% from the pristine natural frequency for almost all eigen frequencies, both FRFs from pristine and damage cases are presented in Figure 1.

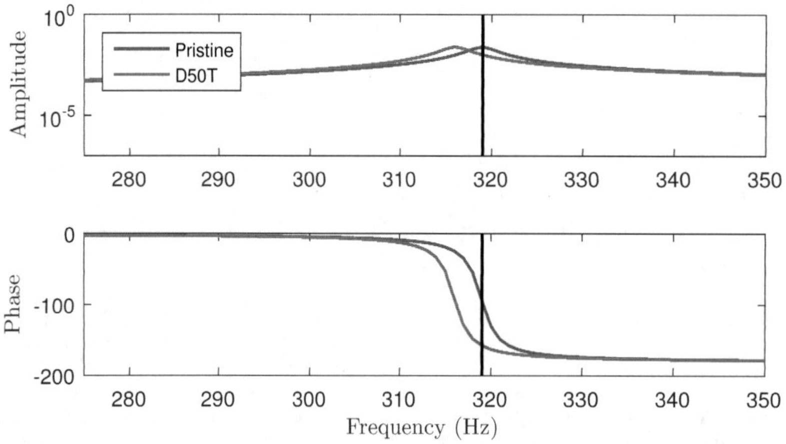

Figure 1: FRF of Pristine & D50T

This behaviour is as expected, the applied damage to the structure reduces the stiffness in the beam [8]. The eigen frequency is proportional to k/m, where k is stiffness and m is mass. The severity of the damage is commonly measured in the frequency shift of the eigen frequencies. However, for a given structural damping, the response phase of a damaged structure shows much larger deviation from a pristine condition. As seen in Figure 1 this shows a higher sensitivity due to phase behaviour at a fixed frequency, such as the eigen frequency. Table 3 shows both the frequency shift and the phase shift with respect to the resonant frequency.

Table 3: Response to damage severity

	Frequency (Hz)				Phase shift at ω^P (deg)			
Eigen	P	D10T	D25T	D50T	P	D10T	D25T	D50T
ω_1	51	51	50	50	-79	-134	-164	-170
ω_2	319	318	317	316	-94	-117	-145	-158
ω_3	889	889	888	887	-92	-97	-107	-117
ω_4	1731	1731	1731	1731	-96	-94	-92	-91
ω_5	2837	2837	2835	2833	-94	-97	-105	-115
ω_6	4196	4193	4184	4178	-96	-108	-131	-142
ω_7	5794	5785	5766	5750	-100	-120	-148	-157
ω_8	7616	7602	7571	7546	-104	-127	-154	-163

These differences are visualised in Figures 2 & 3. Where the found values are normalised to the pristine case and a relative shift is plotted.

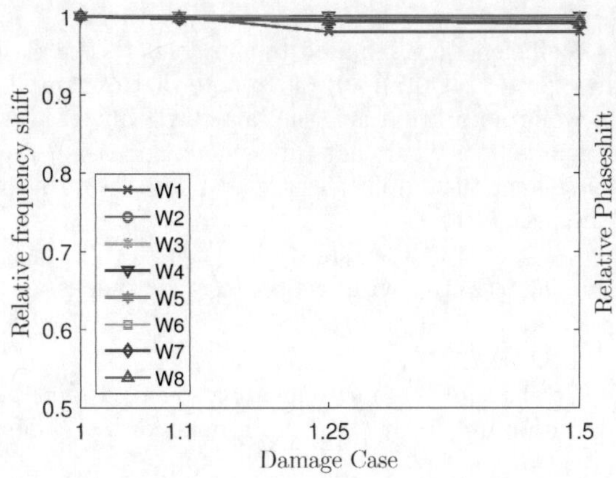

Figure 2: Frequency shift

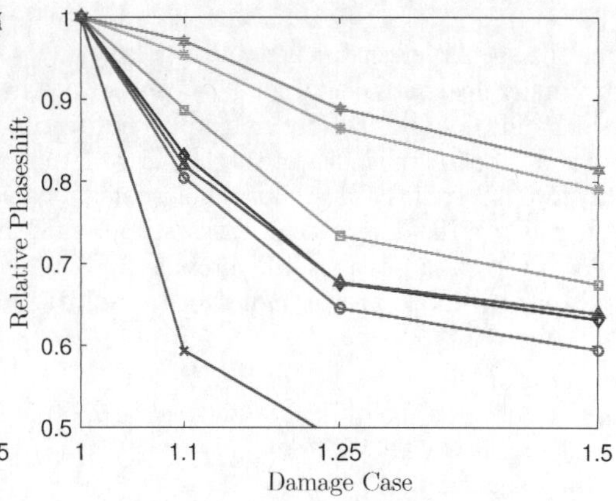

Figure 3: Phase shift

Based on such indicator, the most sensitive modes to the damage can be highlighted and closely observed during the next phase of the research.

2.3. signal processing

All the ODSs simulated by the FEM are curve-fitted with a 15th order polynomial. LDV output modulated signals are generated and processed by dFFT. the dFFT is discretised at the excitation frequency and subsequently at interval of the scan frequency, as in [10]. This results in a middle amplitude with sidebands at equal distances. The amount of sidebands is equal to the amount of polynomial coefficients. The magnitude of these sidebands carry the spatial information of the ODS. Meaning that the distribution of the amplitude of each sideband is characteristic for that system at that specific frequency.

When damage is introduced to the beam, this characteristic distribution of amplitudes over the sidebands differs, as seen in Figure 4. This is more clearly seen when a log scale on the y-axes is used.

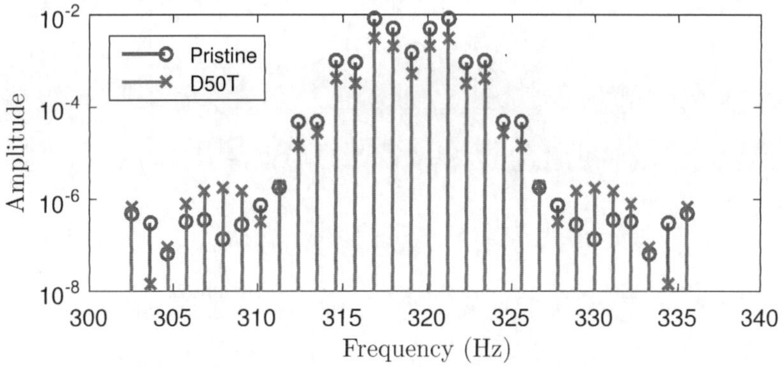

Figure 4: dFFT at 2nd eigen frequency (P and D50T)

As already introduced, 15 coefficients are used for the polynomial evaluation. However, the ODSs require a different amount of coefficients, less are needed if the shape is simple like the first bending mode. The minimal number of coefficients needed to adequately describe the ODS is investigated at the eigen frequencies of the beam. This is done by checking the MAC value of the fit for increasing number of polynomial coefficients.

Table 4: Minimal Nr. of Coefficients (MAC = 0.999)

eigen freq	1	2	3	4	5	6	7	8
Nr. of coeff.	2	4	5	8	9	10	13	14

So for the second eigen frequency (319 Hz), only 4 coefficients are required to construct the ODS. This results in many near-zero coefficients, as can be seen in Figure 4. It is important to note that the amplitude of the sidebands does not simply shift up or down, this would mean that is is the same ODS but at lower or higher (excitation) amplitude. In the higher order sidebands, it is clear that this is not the case. However, as said before, these sidebands do not hold much meaning since the amplitudes are near-zero. So it becomes critical to analyse the difference of the first four sidebands as well.
The following two subsections will provide an attempt to quantify the deviation between a pristine and damaged case by developing an indicator to rank the damage severity with respect to mode shapes.

2.4. Fingerprint

A method to observe the difference between a pristine and damaged case is to create a so-called *fingerprint*. Every sideband is scaled by the one measured at the excitation frequency (ω). This is done for each sideband (i) following equation 2.

$$FPA_i(\omega) = SBA_i(\omega + \Omega \cdot i)/SBA(\omega) \tag{2}$$

FPA_i is the fingerprint (normalised) amplitude of sideband i, obtained by dividing sideband amplitude SBA_i over the excitation or middle amplitude. Hence, both pristine and damaged case will present a typical fingerprint.

2.5. Sideband polynomial

As described before, the spectral sideband amplitudes are related to the polynomial coefficients of the wave shape of the ODS. The relation between the sideband amplitude and the polynomial coefficient is used to define a damage indicator. The difference in sideband amplitudes between the pristine case and damage case is related to the change in ODS, which is a polynomial function. The first sideband corresponds to the linear term, the second to the quadratic, and so on. This is integrated in the damage indicators defined below.

$$\text{DIP}_1 = \sum_{i=1}^{n} SBA^i_{Pi} - SBA^i_{Di} \tag{3}$$

$$\text{DIP}_2 = \sum_{i=1}^{n} SBA_{Pi} \cdot i^i - SBA_{Di} \cdot i^i \tag{4}$$

$$\text{DIP}_3 = \sum_{i=1}^{n} SBA_{Pi} \cdot i^{i-1} - SBA_{Di} \cdot i^{i-1} \tag{5}$$

$$\text{DIP}_4 = \sum_{i=1}^{n} (SBA_{Pi} \cdot i^i - SBA_{Di} \cdot i^i) \cdot \omega \tag{6}$$

Where SBA_{Pi} is the ith sideband of the Pristine case. ω is the excitation frequency. This is done for a number of sidebands up to 15.

3. Results

In this section the results of the damage indicators described in the section above are shown.

3.1. Fingerprint

The fingerprint method yields a characteristic shape for the pristine case and the damaged case as seen in Figure 5. These shapes are referred to the fingerprint of the beam.

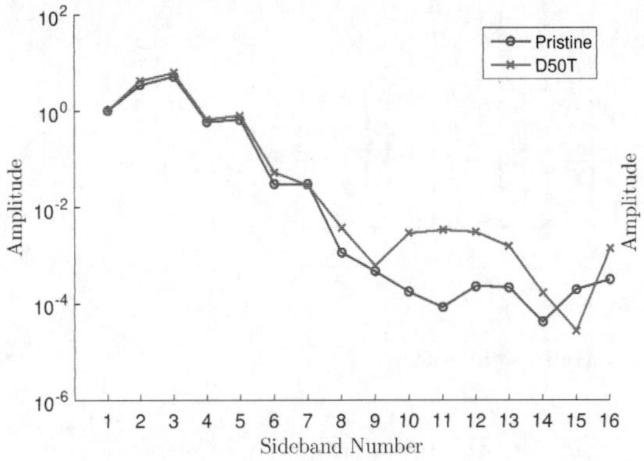

Figure 5: Fingerprint method

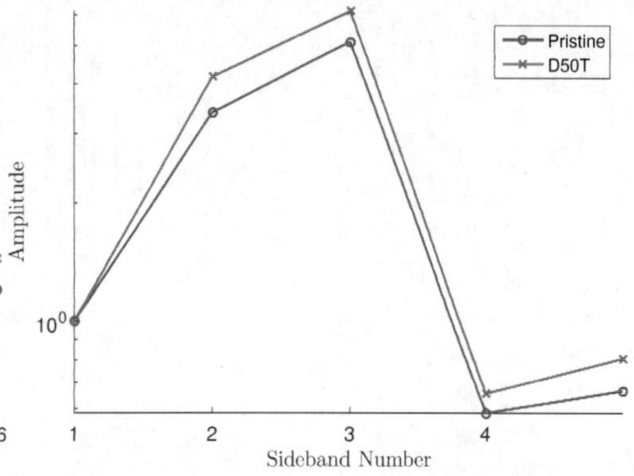

Figure 6: Fingerprint zoom

As can be seen in Figure 5, the higher order sidebands show most difference in relative amplitude. These are however most difficult to detect and in absolute numbers have very low value. The difference is visible in the first four sidebands when these are observed, see Figure 6. Detecting damage from this Figure can be done, but assigning a scalar value would be preferable. One possible damage criterion to obtain from this is the mean square error between the two fingerprints. The mean square error is 0.1 for the second eigen frequency. When this is evaluated over the entire frequency range, some spikes appear, as seen in Figure 7.

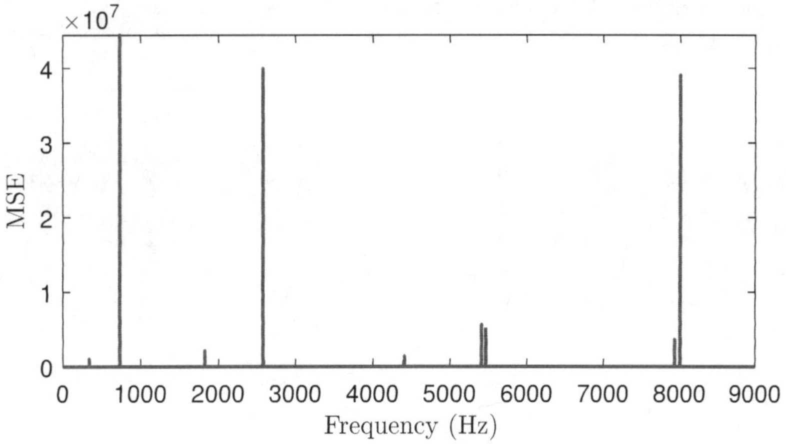

Figure 7: Mean Square Error for all frequencies

The reason for these spikes is investigated and can be explained with Figure 8.

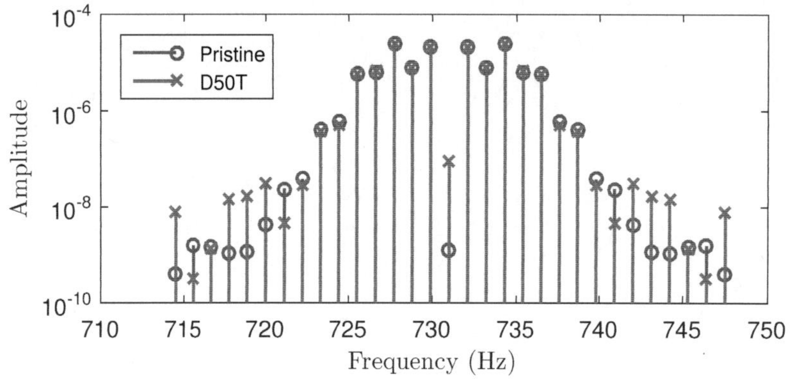

Figure 8: Problem case Mean square error

From Figure 8, it can be concluded that a large difference in the middle amplitude results in unusually large Mean Square Errors.

3.2. Sideband polynomial

The result of the sideband polynomial methods are damage indicators that are highly dependent on the frequency and the method does not normalise the sideband amplitudes. Therefore the indicators only show meaningful difference around the eigen frequency, where the vibration amplitudes are generally high. Furthermore, at the anti-resonance a spike occurs, because vibration amplitudes are very small. The frequency dependency built in to equation 6 shows more consistency regarding the value indicated by the damage severity. However, at any frequency away from the eigen frequency, the damage is not indicated. The frequency dependent indicator is view in Figure 9.

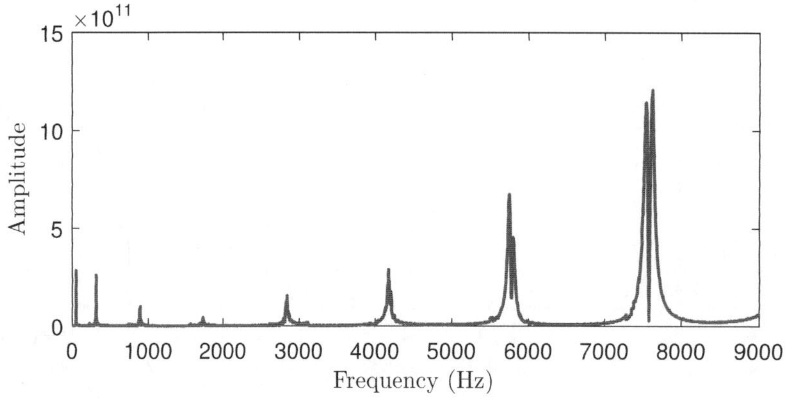

Figure 9: Sideband Polynomial Indicator

4. Conclusions

In this study, the use of the CSLDV method for SHM purposes is explored. The method is fast in assessing the ODS of a beam by using the FFT frequency spectrum. This spectrum has a characteristic shape that can be used as a reference. The fingerprint method shows potential in detecting the difference between a damaged and pristine spectrum. However, this method shows inconsistent behaviour over the frequency range, making it an unsuitable approach for this application. The other methods suffer from the same problem, a very small window of frequencies is found where a useful value can be extracted. This makes it heavily model dependent and not much more useful than the current technique of measuring the shift in eigen frequency.

The problem with the fingerprint method is that the indicator is too dependent on the amplitude of one of

the sidebands used for the calculation and therefore makes the indicator unstable. The approach of relative contribution of the sidebands does hold. When fixed frequencies are observed, the overall amplitude changes when damage is introduced, as seen in the FRF. From an application point of view, this is also the case when not the exact same force is applied. So making the sidebands a relative, or normalised factor would make it comparable again.
Instead of normalising the sideband amplitudes to one sideband, it would be better to assign partial contributions to the sidebands in relation to all other sidebands. This is further investigated, mathematically and experimentally, in an upcoming paper.

References

[1] Stéphane Avril, Marc Bonnet, Anne-Sophie Bretelle, Michel Grédiac, François Hild, Patrick Ienny, Félix Latourte, Didier Lemosse, Stéphane Pagano, Emmanuel Pagnacco, and Fabrice Pierron. Overview of Identification Methods of Mechanical Parameters Based on Full-field Measurements. *Experimental Mechanics*, 48(4):381–402, 8 2008.
[2] P. Castellini, M. Martarelli, and E.P. Tomasini. Laser Doppler Vibrometry: Development of advanced solutions answering to technology's needs. *Mechanical Systems and Signal Processing*, 20(6):1265–1285, 8 2006.
[3] I. Manuel De la Torre, María del Socorro Hernández Montes, J. Mauricio Flores-Moreno, and Fernando Mendoza Santoyo. Laser speckle based digital optical methods in structural mechanics: A review. *Optics and Lasers in Engineering*, 2015.
[4] Dario Di Maio. Use of continuous scanning LDV for diagnostics. In *Conference Proceedings of the Society for Experimental Mechanics Series*, 2016.
[5] Z A Jassim, N N Ali, F Mustapha, and N A Abdul Jalil. A review on the vibration analysis for a damage occurrence of a cantilever beam. *Engineering Failure Analysis*, 31:442–461, 2013.
[6] Jeong Tae Kim, Yeon Sun Ryu, Hyun Man Cho, and Norris Stubbs. Damage identification in beam-type structures: Frequency-based method vs mode-shape-based method. *Engineering Structures*, 2003.
[7] Milena Martarelli. Exploiting the Laser Scanning Facility for Vibration Measurements. 2001.
[8] W.M. Ostachowicz and M. Krawczuk. Analysis of the effect of cracks on the natural frequencies of a cantilever beam. *Journal of Sound and Vibration*, 150(2):191–201, 10 1991.
[9] Sridhar Reddy Guntaka, Manfred Hertwig, Torsten Flemming, and Ralf Usinger. Speckle interferometry for detection of subsurface damage in fibre-reinforced composites Speckle interferometry for detection of subsurface damage in f i bre-reinforced composites. *Meas. Sci. Technol*, 5:100–104, 1994.
[10] A. B. Stanbridge and D. J. Ewins. Modal testing using a scanning laser Doppler vibrometer. *Mechanical Systems and Signal Processing*, 1999.

IOP Conf. Series: Journal of Physics: Conf. Series **1149** (2018) 012013 doi:10.1088/1742-6596/1149/1/012013

Real-time study of high-cycle fatigue damage using the averaged speckle dynamics

A.P. Vladimirov[1,2,3], N.A. Drukarenko[1], I.S. Kamantsev[1], V.N. Trishin[1], N.A. Lukin[1]

[1] Institute of Engineering Science, Ural Department of RAS (Russian Academy of Science), 34, Komsomolskaya St., Yekaterinburg, 620049, Russian Federation.
[2] Ural Federal University, 19, Mira St., Yekaterinburg, 620002, Russian Federation.
[3] Yekaterinburg Research Institute of Viral Infections, 23, Letnyaya St, Yekaterinburg, Russian Federation.

vap52@bk.ru; targosk@yandex.ru

Summary. The theory of the method permitting a study of damage accumulation in high-cycle fatigue of reflecting and transparent objects using the time-averaged speckle images is briefly discussed. The formulas that relate the relative displacement vector of the scattering centres to the time-average intensity and the temporal autocorrelation function of this intensity at some observation point are presented. A technique identifying the component of the relative displacement vector as local plastic deformations emerge during fatigue testing is discussed. The issue of real-time recording and processing of the speckle images using the conventional personal computers and special-purpose video systems with function-oriented processors based on homogenous computing environment is discussed. Processors with homogenous architecture allow assurance of the highest possible image processing performance within the order of 10-teraflops/s. The specimens, the optical system, the equipment for fatigue testing, the USB camera and the software used in high-cycle fatigue experiments are discussed. The possibilities and prospects for application of this technique and equipment in evaluation of the remaining life of some object types are discussed.

1. Introduction

It is known that high-cycle fatigue of the material is the dominant destruction factor of mechanical parts and construction elements [1]. Nevertheless, currently there are no relevant physical models of fatigue failure, and the engineering practice lacks techniques for evaluation of material and construction durability [2, 3]. This situation is to a large extent related to absence of simple, reliable and unambiguously interpreted techniques for studying the fatigue failure without interrupting the cyclic loading process.

Previously, we established the dependences relating the time-average radiation intensity and the temporal autocorrelation function at some point of the image plane of the intermittently deformed object with an average value, variance and correlation time of the difference in the plane point displacements in the conjugated plane. It was shown on the metal specimens that if the averaging time is divisible by the cyclic deformation period, then in case of the absence of irreversible processes the speckle image sections will be invariable, and in micro-and macroscopic surface profile variation image decorrelation will take place.

The objective of this research is studying the possibility of applying the technique for quantitative evaluation of the limiting roughness values, the surface shape of steel as well as the refraction and density indexes of plexiglas. In this case, the time to the crack start should be evaluated by the variation rate of these values.

2. The conditions of the experiment

5x10x55-mm size samples made from Acruma plexiglas were the research targets. To localize the spot of probable fatigue crack initiation, a sharp V-notch 2 mm deep with a 0.25-mm tip radius was made on the specimen. The cyclic tests were performed using an up-to-date high-frequency testing machine MIKROTRON manufactured by the Rumul firm.

A monochrome VIDEOSCAN-415M-USB TV camera with a matrix containing 782´582 8.3x8.3-μm photocells was used in the experiment. The selected TV camera exposure time of 0.5 s corresponded to ≈50 loading cycles. The object was illuminated by a beam emitted by a KLM-H650-40-5 type laser module with the 0.65-μm wavelength and 40-mW capacity.

The formula for calculating the correlation coefficient of two speckle image fragments is:

$$\eta = \frac{\frac{1}{nm}\sum_{i=0}^{n-1}\sum_{j=0}^{m-1}\left(A_{ij}-\bar{A}\right)\left(B_{ij}-\bar{B}\right)}{\left(\frac{1}{nm}\sum_{i=0}^{n-1}\sum_{i=0}^{m-1}\left(A_{ij}-\bar{A}\right)^2\right)^{1/2}\left(\frac{1}{nm}\sum_{i=0}^{n-1}\sum_{i=0}^{m-1}\left(B_{ij}-\bar{B}\right)^2\right)^{1/2}}, \tag{1}$$

where A_{ij} is the digital value of the illumination intensity in a *mxn*-pixel size region in real time, B_{ij} is the digital value of the illumination intensity in the same region at another time point, and *i* and *j* are the element numbers in the plane of the *x*- and *y*-axes, respectively.

3. Theory

It is supposed that the speckle pattern varies due to the following three processes:

- slow translational motion of the specimen along the ox-axis,
- intermittent oscillation of the specimen also along the ox-axis with a hard-set amplitude and T period,
- random variation of optical wave paths sounding the research target.

According to the selected model, the waves with a random initial phase first pass through the $2a_s$-diametre region of the object with the centre in point *A*, where $2a_s$ is the linear resolution of the lens. Then the waves pass through a small diaphragm near the lens and reach point A' in the object image plane conjugated to point *A*. We will determine the optical path length for the *j*th wave using formula:

$$u_j = \int_{l_j}\left[n_j(l) - n_0\right]dl, \tag{2}$$

where $n_j(l)$ –is the distribution of the refraction index in the phase object along the path of the *j*th wave, l_j is the path length of the *j*th wave in the object, and n_0 is the refraction index in the absence of the phase object; the integration is performed along the wave path.

Admitting that all the random values are independent, and the random values of the differences in the wave pair phases are time-correlated, we obtained an expression for time-average radiation intensity $\tilde{I}$ and temporal autocorrelation function $R_{1,2}(t_1, t_2)$ of this intensity at an arbitrary observation point *q* of the observation plane [4,5]:

$$R_{1,2}(t_1,t_2) = I_0^2 N(N-1)C_0^2 \cos[(\langle x_2\rangle + \langle\tilde{x}_2\rangle - \langle x_1\rangle - \langle\tilde{x}_1\rangle)] \times e^{-\frac{1}{2}k_{11}-\frac{1}{2}k_{22}+k_{12}} e^{-\frac{1}{2}\tilde{k}_{11}-\frac{1}{2}\tilde{k}_{22}+\tilde{k}_{12}}, \tag{3}$$

where I_0, N and C_0 are constants, $x = k\mu$, $k=2\pi\gamma/\lambda$, μ is the difference of the optical wave pair paths average by the $2a_s$-diametre region and time-average, $\langle x_1\rangle$ and $\langle x_2\rangle$ are the values of *x* average by the object ensemble (representation ensemble) at time points t_1 and t_2, respectively, k_{11} and k_{22} are the variance of value x at time points t_1 and t_2, respectively, and k_{12} is the temporal autocorrelation function of value x.

Values $\langle\widetilde{x}_1\rangle, \langle\widetilde{x}_2\rangle, \widetilde{k}_{11}, \widetilde{k}_{22}, \widetilde{k}_{12}$ in formula (3) have the same meaning as $\langle x_1\rangle, \langle x_2\rangle, k_{11}, k_{22}, k_{12}$. But they characterize variation of the optical wave pair paths emerging only in case of a rough surface motion. If the surface roughness is homogeneous, i.e. $\langle\widetilde{x}_1\rangle = \langle\widetilde{x}_2\rangle, \widetilde{k}_{11} = \widetilde{k}_{22}$ for normed autocorrelation function (3) we obtain that:

$$\eta(t_1, t_2) = \eta(u_x)\cos[(\langle x_2\rangle - \langle x_1\rangle)] \times e^{-\frac{1}{2}k_{11} - \frac{1}{2}k_{22} + k_{12}}, \quad (4)$$

$$\eta = \eta(u_x) = e^{-\widetilde{k}_{11} + \widetilde{k}_{12}(u_x)} \quad (5)$$

where $\eta(u_x)$ is a temporal autocorrelation function corresponding to displacement of a transparent plate in plane *XY*.

For the reflecting object formula (4) has the same form, but $\eta(u_x)$ in it is the correlation coefficient in translational motion u_x of the object:

$$\eta(u_x) = \frac{2a_s - u_x}{2a_s} \quad (6)$$

For the reflecting specimen x can be found using formula:

$$x = k\Delta u, \; \Delta u = \Delta\vec{u}\,(\vec{l}_s + \vec{l}\,) \quad (7)$$

where $\Delta\vec{u}$ is the mean difference in the point displacement vectors located at the edge of the $2a_s$-diametre region, and $\vec{l}_s$ and $\vec{l}$ are unit vectors that determine the illumination and the observation directions respectively.

If a non-stationary process $x=x(t)$ is characterized by vanishing variances k_{11} and k_2, then:

$$\eta(t_1, t_2) = \cos\langle x_2\rangle. \quad (8)$$

Next, let the refraction index in the monitored section remain invariable by the specimen thickness L and its variations occur only in the direction of the specimen length (*ox* axis). If the refraction index grade is constant in the section, then, taking (2) into account, instead of (8) we have:

$$\eta = \cos\left[\left(\frac{2\pi}{\lambda}\right)L \times \Delta n\right], \quad (9)$$

where Δn is the difference of the refraction indexes at average characteristic distance Δx along the *ox* axis. Let us note that $\Delta x \leq 2a$. Thus, using formula (9), we can determine value Δn.

4. Program interface

The data received form the testing machine and the speckle images were processed in real time in the LabVIEW environment using original software.

IOP Conf. Series: Journal of Physics: Conf. Series **1149** (2018) 012013 doi:10.1088/1742-6596/1149/1/012013

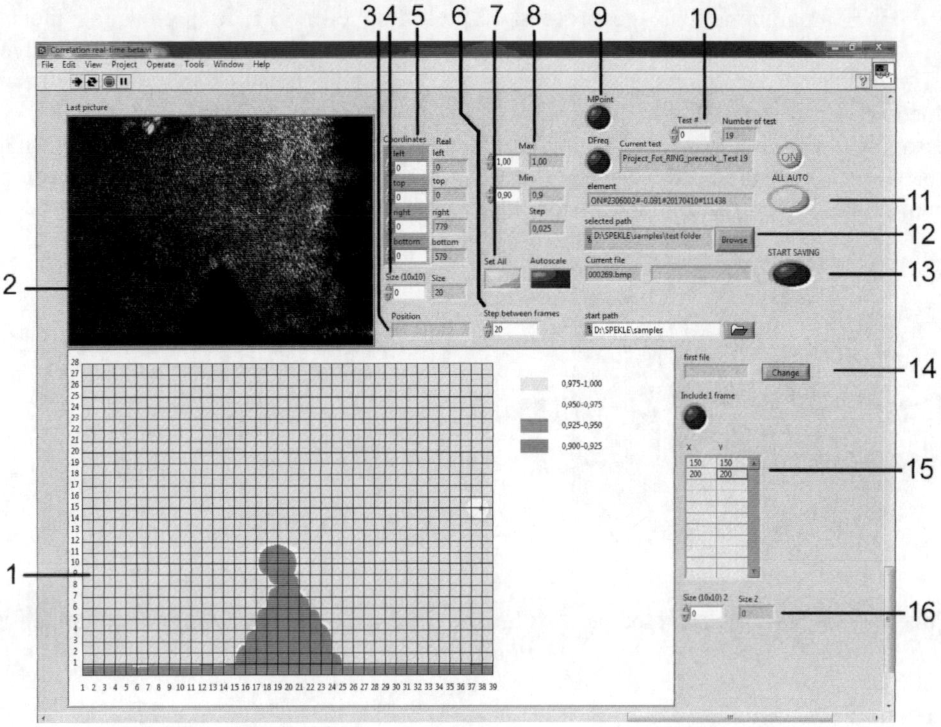

Figure 1. 1-value η distribution display window, 2-window displaying the current speckle image of the object, 3-coordinate of the cursor position on the screen, 4-window for entering the fragment size to divide the image when displaying the distribution of η or processing, 5-window for control region coordinate selection, 6-window for entering the frame count between the initial and the current frame in mode 1, 7-button selecting the entire speckle image region as the control and object observation zone, 8-window for presetting the minimum and maximum correlation coefficient values in η distribution, 9 call-button of dependence graphs η(N), 10-selection of the active experiment number with the window displaying the experiment description, 11-button for data saving every 1,000 cycles, 12-selecting a directory for saving speckle images, 13-button switching over to forceful saving of all the data,14-selection of the first frame for plotting dependence $\eta(N)$ in the preset sections, 15-window for presetting the coordinates of the section centres, 16-window for section size selection.

5. Special-purpose video system

In our tests we used single-flow software on a conventional personal computer for real-time data processing. For two frames of 782x582 pixels per second with a 10x10 data correlation distribution window it was sufficient. More complicated tasks will call for paralleling and probably special-purpose processors. One of such computing instruments named homogeneous computation environment (HCE) is being created by a team of Russian engineering designers [6].

HCE is special-purpose computation systems consisting of identical processor elements (PE). Each element is configured for performing an arithmetic or a logical function as well as for exchange protocol implementation with the adjacent units. The homogeneity of HCE architecture permits increasing the computational power by a simple increase in the number of the processing elements (scalability). It also permits reaching the highest possible performance with the preset restrictions on the equipment cost, energy consumption and weight, which makes the HCE applicable in real-time systems.

The most complicated task in HCE programming is reflecting of the algorithm to the array of the processors. HCE programming is the process of presetting every PE to performing the relevant processing and transmission operations necessary to reflect the dataflow graph of the objective to the PE grid. The geometric peculiarities of the programming are the necessity to plot the dataflow spread paths in the two-dimensional PE environment.

Processors with the homogeneous architecture allow reaching the highest possible performance of image processing within the order of 10 teraflops/s. Currently, a team of engineers is developing an algorithm of processing correlation images for studying the fatigue process.

Figure 2. Minitera-2 HCE prototype.

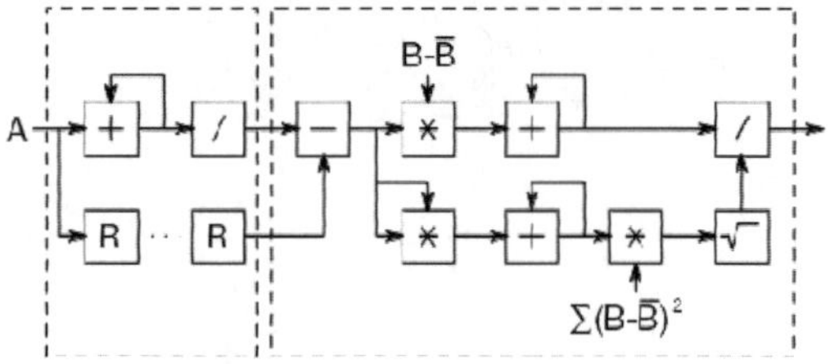

Figure 3. Streaming correlation computation algorithm.

6. Results and discussion

The procedure of detection and monitoring on a transparent specimen made from plexiglas was performed using two different optical schemes:

- flatwise;
- an angle q to the *x-axis*

First, the object was observed flatwise; value η variations near the notch were monitored in real time. As soon as the first variations in the correlation coefficient distribution appeared, testing in this geometry was stopped. Visual examination of the specimen did not detect any presence of a macrocrack. Analysis of the surface profiles recorded by an optical profilometre before and after the testing with the height resolution ≥ 3 nm showed absence of any surface relief alterations

The next stage of the experiment consisted in observation of the zone at an angle during the high-cycle test that was conducted to the specimen destruction. As the crack emerged, the irreversible deformation zone size increased by an order and had an area of several square millimetres. Visual observation showed a crack that covered half of the specimen thickness. Analysis of three-dimensional surface profiles near the crack did not detect any variations of roughness or surface profile.

Let us select a system of rectangular coordinates where the point of reference coincides with the notch tip, the *oy*-axis is directed along the notch axis, and the *ox*-axis is parallel to the specimen axis. We obtained the dependence of value η on coordinate x for the frames corresponding to the crack formation stage. The dependence was obtained by scanning the specimen image by the *4x4*-pixel region via the centre of optical inhomogeneity formed prior to the crack.

It was found that when coordinate x is vanishing, dependence η *(x)* decreases from the value close to 1 to the minimum, and then in the inhomogeneity image centre the value of η increases to a value close to 1. This peculiarity of dependence η *(x)* can be explained by formula (4). By symmetry one can suppose that the difference of the optical paths of the space-averaged wave pairs (averaged by $2a_s$-diametre spaces) and the time in the inhomogeneity centre equals zero. Then the cosine in formula (3) will equal 1, and η will depend only on the values of vanishing k_{11} and k_{22}. So, variation of the environment refraction index at the pre-destruction stage can be regarded as a determined process altering value η by the law of cosines.

IOP Conf. Series: Journal of Physics: Conf. Series **1149** (2018) 012013 doi:10.1088/1742-6596/1149/1/012013

The limiting value of the refraction index was found using the speckle images recorded before and after the crack start. It was supposed that value Δx was equal to the linear dimension of the lens equalling 10 μm. First, the edges of the irreversible process region were determined by the distribution of value η in the object image plane. Then a rectangular region from the periphery of the specified region to the crack was selected. The region was split into uniform subregions. It was supposed that grade dn / dx is constant within a small region. Using formula (9), refraction index n_1 was found on the border between the first and second parts. The refraction indexes on the edges of the other sections were found in a similar way.

For transition from the refraction index to the density index we used the Lorenz-Lorentz formula. We determined value n beyond the inhomogeneity limits using the polarization optical method, and we defined the density by the specimen volume and mass. Using these data, we determined the specific refractivity of plexiglas. The application of these methods allowed to establish that the maximum relative irreversible density changes occur during the fatigue of plexiglas, not less than 3.9×10^{-3}. Let us note that the limiting refraction index values measured using the speckle technique and the polarization optical method coincided well.

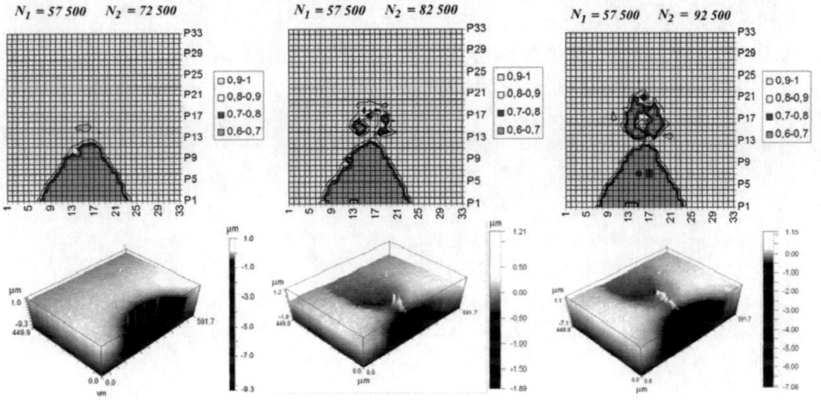

Figure 4. Distribution of value η at different stages of the experiment and three-dimensional profiles near the notch.

Strong profile variation as well as the surface form variation caused by the local plastic deformation were recorded on the steel specimens. The limiting roughness value determined by value η, R_a=28nм (R_a=25nm measured using a WYKO NT-1100 profilometre), the limiting value of surface slope ratio γ obtained by value η, equalled 0.01 (0.088 which was predicted theoretically) .

Thus, knowing the limiting values of density variations $\Delta\rho/\rho$, roughness, plastic deformations (by the surface slope ratio) and the variation rates of these parametres make it possible to evaluate the remaining object life to the initiation of the crack growth.

7. Conclusion

Irreversible processes occurring in high-cycle fatigue loading conditions were studied in real time on samples with sharp notches made from Acruma plexiglas and those made from steel using the developed speckle technique.

Using the developed technique and the ellipsometry method, we studied the refraction and density index variations in plexiglas and roughness and surface profile variations in steel. It was shown that the limiting value of the relative density variation in plexiglas was 3.9×10^{-3}, the limit roughness value R_a=28 nм, and the limiting variation of the surface slope ratio γ in the plastic deformation zone was 0.01.

The variation rate of these parametres and their limiting values make it possible to evaluate the remaining life of the target objects to the crack initiation

8. References

[1] Lasar J., Hola M., Cip O. Differential interferometry for real-time measurement in high cycle fatigue metal testing// Book of abstracts. Conference PhotoMechanics 2015, 2015, Delft, Netherlands, p. 64

[2] Novikov I.I., Iermishin V.A. Fizicheskaiia mehanika realnyh materialov (Physical mechanics of real materials). M: Nauka, 2004. 328p. (Rus)

[3] Terentiiev V.F. Ustalost metallicheskih materialov (Fatigue of metal materials). M.: Nauka. 2002. 248 p. (Rus)

[4] A.P. Vladimirov. Dynamic Speckle Interferometry of High-Cycle Material Fatigue: Theory and Some Experiments // AIP Conference Proceedings. – 2016. – Vol. 1740. – 040004. – DOI: 10 1063/1 4962663.

[5] A. P. Vladimirov. Speckle metrology of dynamic macro- and microprocesses in deformable media // Optical Engineering. – 2016. – Vol. 55. – No. 12. – 121727. – p. 1-10. – DOI: 10.1117/1.OE.55.12.121727.

[6] Trishin, V., Lookin, N., & Filimonov, A. (2017). Software development technology for homogeneous computing environments. CEUR Workshop Proceedings, 1990, 60-71.

Analysis of blade tip timing data from fan blades with synchronous and non-synchronous vibration

C Fan[1], P Russhard[2], A Wang[1], Y Chen[1], W Dong[1]

[1] School of mechanical engineering, Shanghai Jiao Tong University, 800 Dongchuan Road, Shanghai, China
[2] EMTD Ltd, NG8 2RD, Nottingham, United Kingdom

sjtu_fcw@sjtu.edu.cn

Abstract. Blade tip timing (BTT) as an alternative method is capable of directly measuring the blade tip displacement and inferring the vibration characteristics of rotating blades. Its use is growing in popularity due to its ease of application and the ability to deliver information concerning all blades in a rotor. Although the interpretation of the collected data remains an issue, this paper puts forward a process of using such data to analyse the vibrational characteristics of fan blades under different operating conditions and find the blade endurance margin combined with finite element analysis (FEA).
A test system with eight casing-mounted optical probes and one OPR (once per revolution) sensor is built to collect the blades' arrival time of an axial fan. Results indicate that flutter occurs at engine order (*EO*) 1.5 with a combination of three nodal diameters (2, 3, 4), however, nodal diameter 2 is dominant. A noticeable response ($EO \approx 0.5$) is considered as stall. Because every blade vibrates with a frequency lower than its natural frequency (eigenfrequency), and all of them undergo a similar vibration form, continuously rising and falling, which caused by the rotating stall cells. The 1st flex mode – *EO*2 crossing is found at 48% rotor speed and it's consistent with the results of FEA. Finally, the blade endurance in Goodman diagram demonstrates that flutter poses a higher risk in comparison with the synchronous vibration.

1. Introduction

As regards the turbomachinery, there are two kinds of vibrations, which are referred to as the synchronous and non-synchronous vibration. The former one is defined by the blade vibrational frequency being an integer multiple of the rotor speed, and it's normally due to the excitation of upstream stator blade-rows. Similarly, the non-synchronous vibration means the blade vibrational frequency is non-integral with respect to rotor speed, and it's a result of aerodynamic instabilities, i.e., rotation stall and flutter. Both synchronous and non-synchronous vibrations can be the causes of blade failure. Current engine development involves a variety of compliance tests to ensure the safety of products. Over the years, strain gauges have been the first option for the measurement of blade stress for turbomachinery. Recently, BTT, a non-contact measurement method, has become a more promising alternative technique for inferring blade tip displacement using the arrival time of blades measured by a number of casing-mounted probes. Compared with the traditional measurement technique, one of its greatest strengths is the ability to measure the blade tip displacements of all rotor blades in one stage.

A typical BTT system has several probes mounted in the casing to obtain blades' arrival time, and the OPR sensor located above the shaft, which provides the reference time and blade indexing information. Compared with capacitance and eddy current probes, the optical probe has the best spatial resolution [1]. Therefore, it is selected in this case and it's readily available. The optical probe has two paths, one for laser light emission, and the other one to receive reflections from the blade tip. A timing system will record the time of arrival (TOA) of the blade. If there is no vibration, the blade arrival time can be calculated just by the rotor frequency (Ω) and probe installation angle (θ_i), and it is defined as the theoretical time (equation (1)). However, the blade vibration always exists, so the vibrating blades will reach the probe slightly earlier or later as shown in figure 1. With the knowledge of the rotor frequency and blade tip radius (r), the blade tip displacement (d_i) at each probe can be calculated by the time difference between measurement and theory as in equation (2).

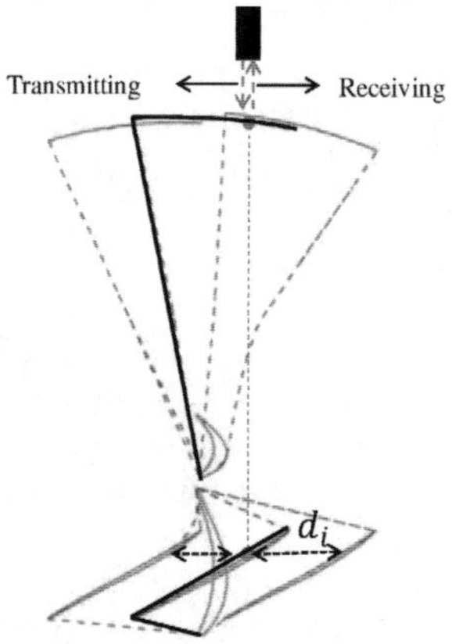

Figure 1. Blade tip displacement.

$$t_{\text{theory}} = \frac{\theta_i}{(\Omega \times 2\pi)} \quad (1)$$

$$d_i = \Omega \times (t_{\text{measure}} - t_{\text{theory}}) \times 2\pi r \quad (2)$$

There are many published methods with different configurations of probes and algorithms in order to identify the parameters of blade vibration. They can be divided into two groups according to the blade vibrational characteristics. For synchronous vibration, Zablotskiy and Korostelev [2] developed a "One parameter" method in the 1970s, which can obtain the vibrational amplitude based on its phase-frequency characteristics. Zielinski, Ziller [3] and Heath [4] proposed a two-parameter plot method used two probes to evaluate the blade amplitude and frequency, but it's sensitive to noise and has low anti-interference [5]. By using several probes circumferentially equispaced in the casing, Gallego-Garrido put forward a new class of methods for two simultaneous resonances based on an autoregressive framework, and it's validated by the experiment [6][7]. For non-synchronous vibration, the known techniques are single-blade and all-blade spectrum analysis presented by Heath. He suggested the resonance frequencies cannot be identified reliably using the single-blade spectra technique, on account of a restricted number of samples; and all-blade spectrum analysis is under the assumption that the bladed-disk assembly vibrates in nodal diameter modes [8].

In order to deal with the synchronous and non-synchronous vibration together, Russhard [9] developed a process to manipulate the raw data of blades' arrival time and he successfully extracted the vibrational amplitude, frequency, and phase angle of every blade, which can shorten the development cycles of turbomachinery [10]. The complete equation to describe blade vibration at any probe is expressed as:

$$d_i = P_i + (a_0 + a_1 \sin(EO \times \theta_i) + a_2 \cos(EO \times \theta_i)) + (b_1 \sin(feo \times \theta_i) + b_2 \cos(feo \times \theta_i)) + \tilde{n} \quad (3)$$

Where P_i represents individual probe steady positional offset, and $\tilde{n}$ is the noise term. The integer (*EO*) and non-integer (*feo*) are defined by the ratio of blade vibrational frequency to the rotor frequency. a_0, a_1, a_2, b_1, b_2 are unknown parameters that need to be calculated.

However, knowing blade vibrational parameters is not enough. The main goal of blade test is to find the stress distribution of every blade and ensure that each of them is qualified for application. Therefore, based on Russhard's work, this paper presents a complete process to explain how to interpret BTT data from fan blades under different vibrational types as shown in figure 2. The first part includes all the preparatory work: FEA and raw data processing, etc. In phenomena identification section, both synchronous and non-synchronous response can be observed directly by the traveling wave plot. Then, the traditional sine fitting (Least Square Model) method is used to calculate the amplitude and phase through the estimated blade frequency, which can be obtained in Campbell diagram. The second part mainly focuses on how to use the measured displacement of one point to deduce the distribution of blade stress for a specific mode.

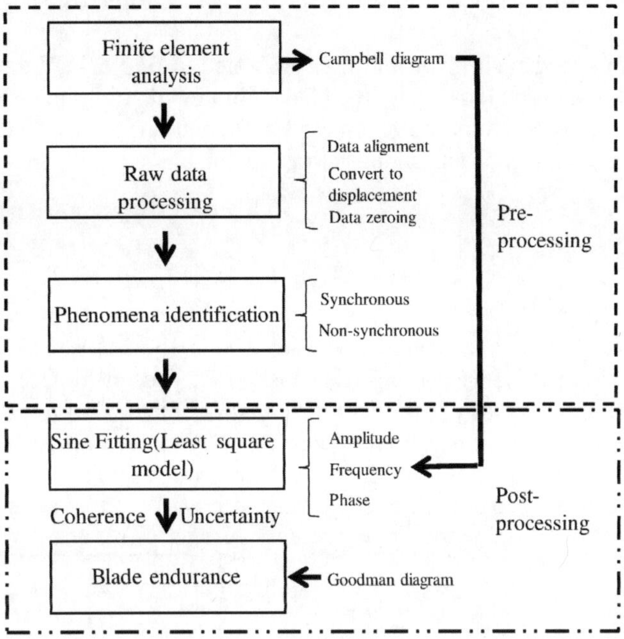

Figure 2. Flow chart of BTT data processing.

2. Measurement setup

Figure 3 shows the BTT testing system in the laboratory. Eight probes are mounted in the casing and one OPR sensor is employed for measuring the rotor speed. The first blade passing from probe 1 is labeled as blade 1, and the remaining blades are stamped successively against the rotating direction. All the probes are located in one plane perpendicular to the shaft. The laser wavelength is 625nm, and a minimum number of fibers are used allowing higher measurement resolutions to be obtained; in addition, the input bandwidth and timing resolution of timing card are 3MHz and 15ns respectively. When the laser signal triggers the timing unit, a time value will be created and stored for further processing. The operating temperature keeps below 250℃ to ensure that all is well.

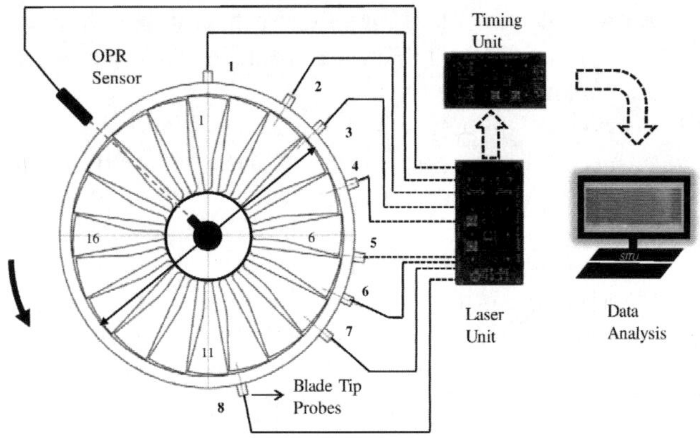

Figure 3. BTT testing set-up.

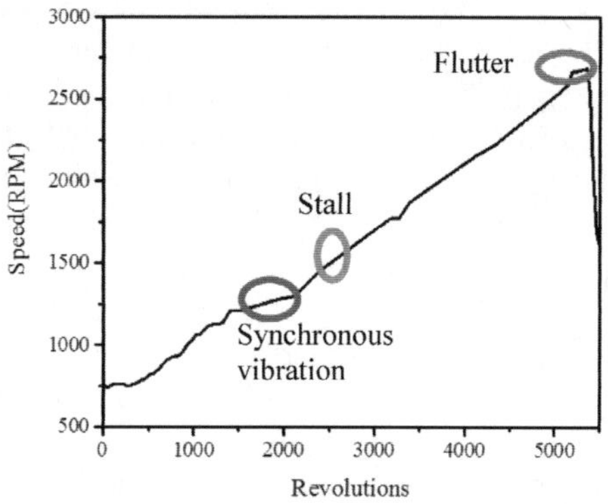

Figure 4. The change of rotor speed.

The change of rotor speed against revolution is presented in figure 4. With the slow increase in the speed, there are three marked regions where the blade vibrational amplitude is very large, in particular the third one. The reason for speed dropping after 5300 revolutions approximately is to protect blades. Data analysis only throws the emphasis on those three regions.

3. Methodology and results

3.1. Finite element analysis

The finite element analysis is a convenient way to obtain the blade natural frequencies and the corresponding mode shapes, and the probe axial position is chosen in accordance with those mode shapes. It should avoid the node at which there is no displacement. Given the blade axial movement, the probe axial position is selected near trailing edge. Furthermore, in order to precisely pinpoint the location of laser spot for the realization of the conversion from displacement to stress, 120 nodes are arranged in the chord direction so that the averaged distance between two nodes is about 2mm.

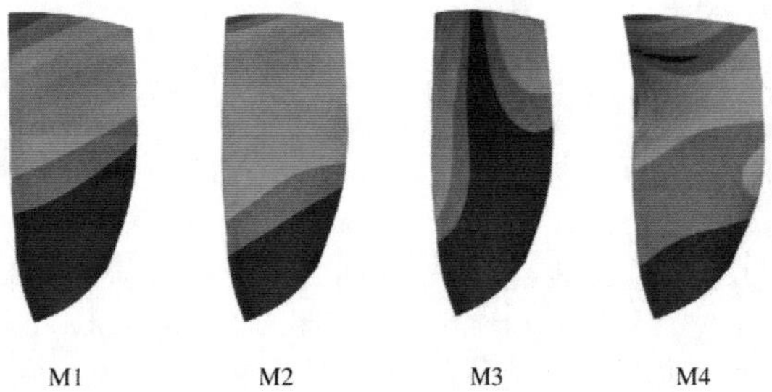

Figure 5. Mode shapes.

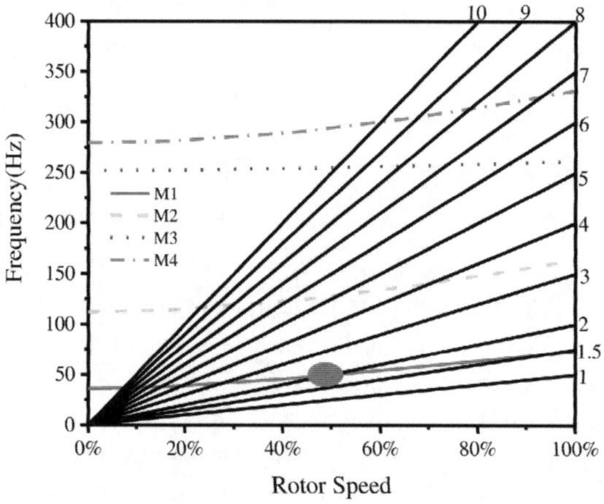

Figure 6. Campbell diagram.

Figure 5 shows the first four modes of a blade obtained by ANSYS® modal analysis under the pre-stressing force as a result of rotor speed. Campbell diagram comprises the eigenfrequencies and *EO* lines (black lines) as in figure 6, and the intersections (e.g. red point) show the possible resonant

IOP Conf. Series: Journal of Physics: Conf. Series **1149** (2018) 012014 doi:10.1088/1742-6596/1149/1/012014

position. Therefore, it is clear to observe where the synchronous vibration might happen with the help of Campbell diagram before conducting any experiments.

3.2. *Raw data processing*

The circumferential spacing of the probes results in misaligned data arrangement by default. Therefore, the first collected arrival time by each probe corresponds to a different blade number as shown in figure 7. The first subscript of TOA represents the revolution, and the second one indicates which blade is passing from the probe.

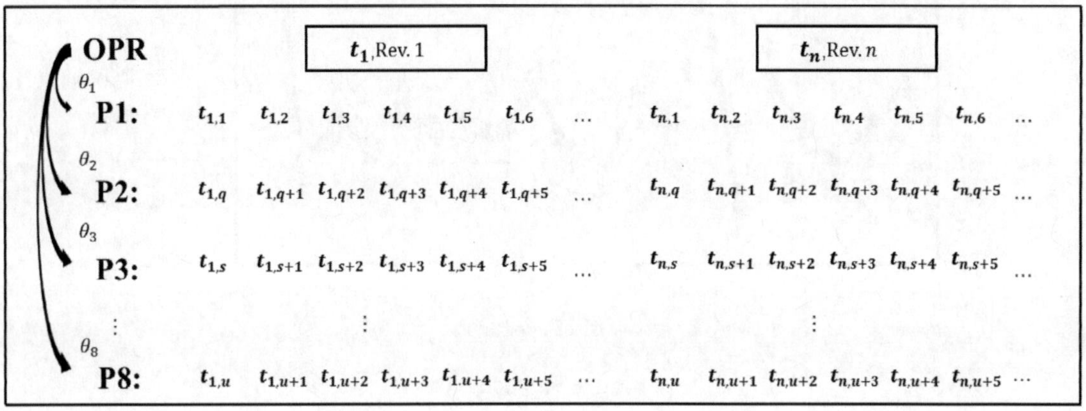

Figure 7. Data storage.

Taking probe 1 as a reference, the number of blades (N) between two probes can be calculated by:

$$\alpha = \frac{360°}{N_b} \tag{4}$$

$$N = \left\lfloor \frac{\theta_i - \theta_1}{\alpha} \right\rfloor \tag{5}$$

Where N_b is the number of blades, and α implies the pitch of cascade.

Therefore, the correct timing order for each probe is to move N elements to the front respectively in each revolution as shown in figure 8.

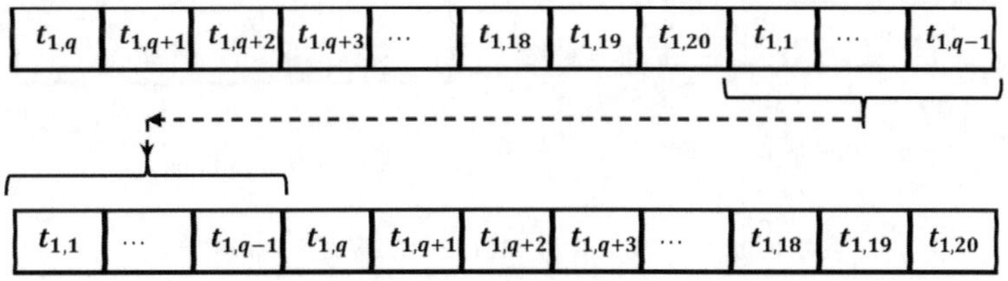

Figure 8. Data alignment (P2).

3.2.1. Stack plot. In order to validate the procedure of data alignment, the stack plot representing the blade steady state displacement is introduced. By applying equation (2), TOA data can be converted into the displacement. In figure 9, because each blade in a rotor has a different position to others due to mechanical tolerances. This leads to the steady-state displacement measured in the non-vibrating blade to be a value other than zero. Although the displacements of each probe have some

IOP Conf. Series: Journal of Physics: Conf. Series **1149** (2018) 012014 doi:10.1088/1742-6596/1149/1/012014

discrepancies, the displacement trend line of each probe is similar. This result is in accord with the expectation, and the data has been aligned.

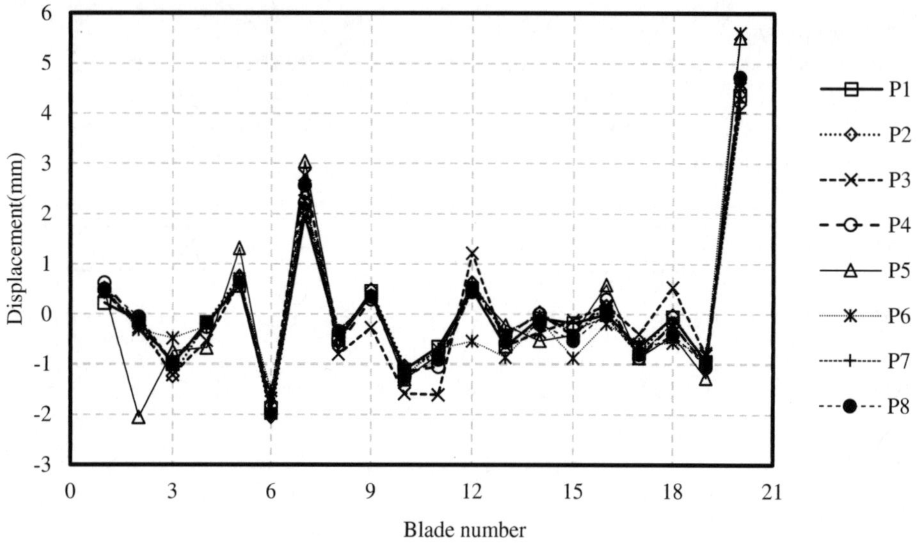

Figure 9. Stack plot.

3.2.2. Data zeroing. With the help of OPR sensor, it's convenient to identify the start of each revolution. Hence, the blade displacements are calculated by this reference time instead of t_{theory}. However, this manipulation can create a large DC offset (figure 10) by contrast with the blade vibrational displacement itself. Before the next step, the data must be zeroed.

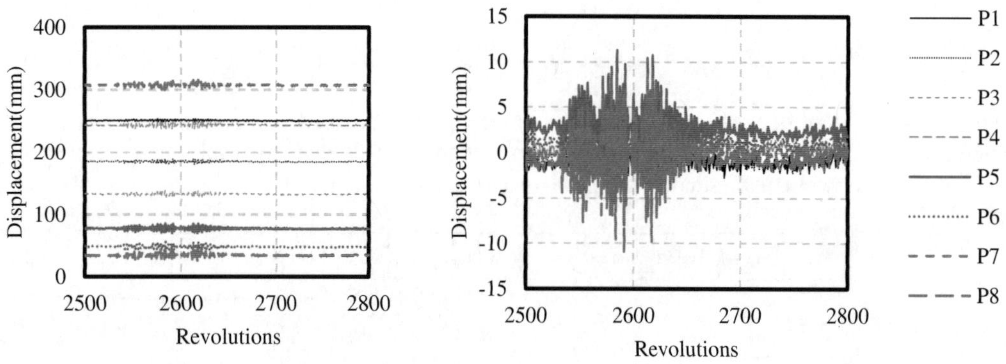

Figure 10. Data before and after zeroing.

There are many different methods to conduct the procedure of data zeroing. In this case, the piecewise linear fitting is used for each probe's data as their respective DC offset. The major advantage of this method is that it only removes the DC offset and has no effects on dynamic content. Accordingly, only blade vibration information and noise term are left after data zeroing. The error caused by this zeroing procedure will be discussed in "*Sine fitting*" section.

3.3. Phenomena identification

A method to detect a response in BTT data is termed as the traveling wave plot, which produces similar data to that of inter-stage pressure transducers, only with better fidelity. Because those transducers are located in the flow field, they carry various unwanted signals with different frequencies. The traveling wave plot in figure 11 is developed by creating a series of FFTs (fast Fourier transforms) using all blades' data from a BTT probe which has been processed to remove the differences in the inter-blade spacing observed in the stack plot.

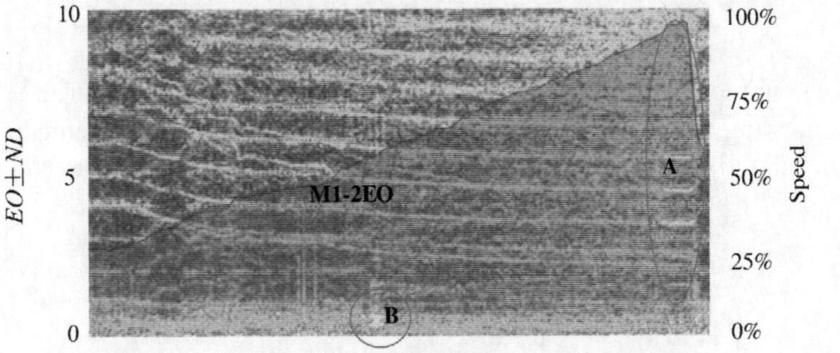

Figure 11. Traveling wave plot.

The abscissa axis divisions of traveling wave plot represent the number of FFT revolutions. Each FFT revolution contains several revolutions, which are aligned to the probe and number of blades to preserve the phase relationship for nodal diameter calculation. The ordinate axis is made up of engine order and nodal diameter, which can be deduced from:

$$f_P = f_b \pm ND \times \Omega \tag{6}$$

Where f_P (observed frequency) and f_b (blade frequency) represent the vibration frequency in stationary and rotating coordinate systems respectively. The sign associated with the nodal diameter indicates the direction of propagation of traveling wave. (+: forward-traveling wave; −: backward-traveling wave). Multiplying $1/\Omega$ on both sides of equation (6):

$$F_P = EO \pm ND \tag{7}$$

The nodal diameter (*ND*) can be expressed by:

$$ND = \frac{\varphi_2 - \varphi_1}{\theta_2 - \theta_1} = \frac{\Delta\varphi}{\Delta\theta} \tag{8}$$

Where $\Delta\varphi$ is the difference between the two-phase angles at the peak frequency, which can be obtained by a cross-power analysis using the data from two probes, and $\Delta\theta$ is the circumferentially included angle of those probes [11]. *ND* is an integer value. If the observed frequency by a BTT probe is also an integer, which means this should be a synchronous response ($EO \in Z$^+), otherwise it is a non-synchronous one. Therefore, both synchronous and non-synchronous vibration can be observed in traveling wave plot directly. In order to find the true blade frequency, it still requires data from an additional probe to determine the nodal diameter and remove it from the observed frequency. With the knowledge of *EO* and rotor speed, the resonance point and blade modal shape can be identified in Campbell diagram.

Using these methods, it is possible to identify a synchronous vibration associated with M1-*EO*2 and two non-synchronous events marked in the red circle, A and B. According to the blade frequency, it's known that the first one is a response with $EO<1$, possibly a stall; whereas, the second one has $EO \approx 1.5$ with a number of associated nodal diameters, possibly a stall flutter.

3.4. Sine fitting (Least square model)

In the raw data processing section, the DC offset has been removed, therefore, only the vibrational and noise terms are left. It's worth mentioning that the P_i contains lots of information, not only the blade

IOP Conf. Series: Journal of Physics: Conf. Series **1149** (2018) 012014 doi:10.1088/1742-6596/1149/1/012014

lean, untwist and axial movement, but also the axial position of the blade. To extract those parameters requires other techniques and functions that are topics of research at other universities. It's not the target of this paper. With the known engine order associated with a vibration event and the installation angles of probes, the blade amplitude and phase can be calculated by sine fitting method. The Pearson correlation coefficient (R) is used to quantify the degree of correlation between the calculated and zeroed blade displacements [12].

$$R = \left(\frac{1}{N_P - 1}\sum\left(\frac{D_i - \bar{D}}{S_D}\right)\left(\frac{d_i - \bar{d}}{S_d}\right)\right)^2 \tag{9}$$

In which N_P is the number of probes and D_i represents the calculated displacement; S_D (S_d) and $\bar{D}$ ($\bar{d}$) are the corresponding standard deviation and average value of displacements. The difference between the calculated and zeroed displacements will be classified as an error, and evaluated by the uncertainty (u):

$$u = \left(\frac{\max\{|d_i - D_i|\} - \min\{|d_i - D_i|\}}{2}\right) \div \max(D_i) \times 100\% \tag{10}$$

3.4.1. Non-synchronous vibration. As suggested by the traveling wave plot, a cross-power analysis using two probes data and the FFT are performed in area A (figure 12). The actual nodal diameters corresponding to the three relatively high peaks are calculated by equation (8).

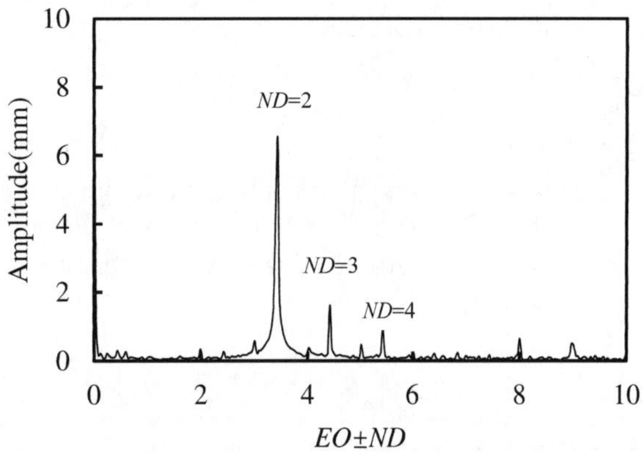

Figure 12. Fourier spectrum (FFT Revs=5300).

Table 1. The observed and true blade frequency.

Observed		True	
Order	Frequency (Hz)	Order	*ND*
3.5	153.1	1.5	2
4.5	197.4	1.5	3
5.5	241.8	1.5	4

Table 1 shows the observed and true blade frequency including the corresponding nodal diameters. All nodal diameters are positive, which indicates they should belong to the forward traveling wave. Since there are more than one nodal diameter, the observed frequencies have several different values while blade frequency is a constant. Although the blade vibrational frequency at the current situation is not an integer multiple of the rotor speed, it's equal to the blade natural frequency of M1 according to the results of FEA. Therefore, it's concluded that flutter is encountered, and some synchronous responses in Fourier spectrum imply this is not pure flutter.

The FFT result of area B is presented in figure 13. There are two main spikes having nodal diameter 0, therefore the corresponding blade vibrational frequencies are far less than its natural frequency. It's a very distinctive phenomenon, and the Fourier spectrum containing several frequency components also indicates the internal flow fields are quite complex at this time.

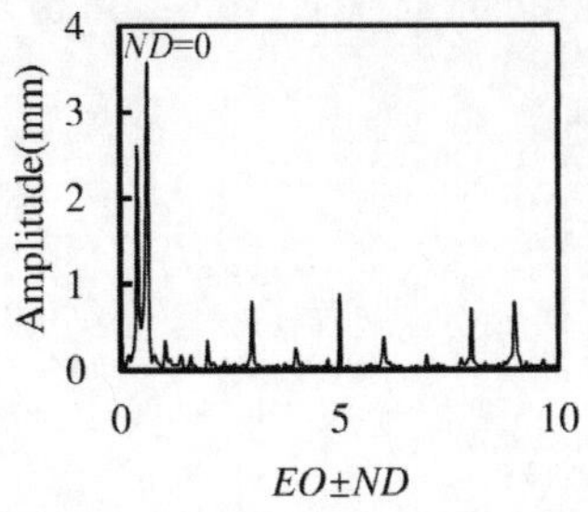

Figure 13. Fourier spectrum (FFT Revs=2610).

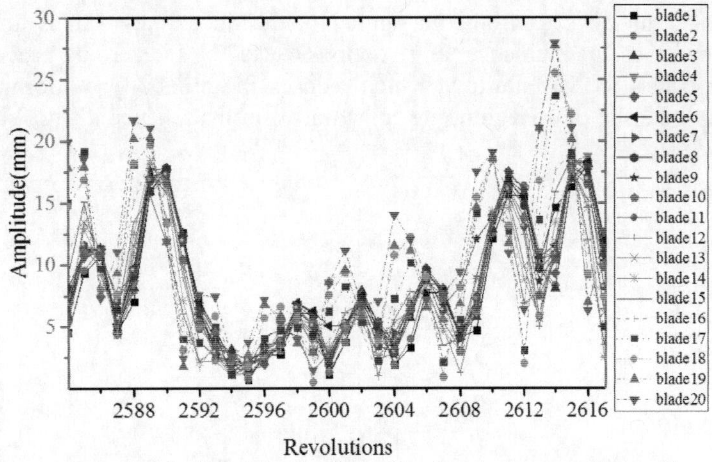

Figure 14. The change of amplitudes during rotating stall.

Figure 14 shows the blade maximum amplitudes (peak-to-peak) are quite large, which exceeds our design objective. All blades experience a similar vibration form, continuously rising and falling, but there is a delay, not all blades reaching their own peak value at the same time. This vibrational behaviour can be explained by the movement of rotating stall cells. The propagation direction of stall cells is in the reverse direction to the rotor, hence every blade suffers from loading and unloading alternately. In addition, there are several blades that have relatively larger amplitudes and need to be checked further.

3.4.2. Synchronous vibration. Regarding the synchronous vibration, figure 15 shows the amplitude of each blade, together with their corresponding fitting coherence and *EO*. Compared with the non-synchronous vibration, the maximum vibration amplitude is smaller. The coherence values are in excess of 0.9 in the region from 1600 to 1950 revolutions. It can be found that *EO*2 crosses M1 at 48% rotor speed from the Campbell diagram, which demonstrates the results of experiment and FEA are consistent.

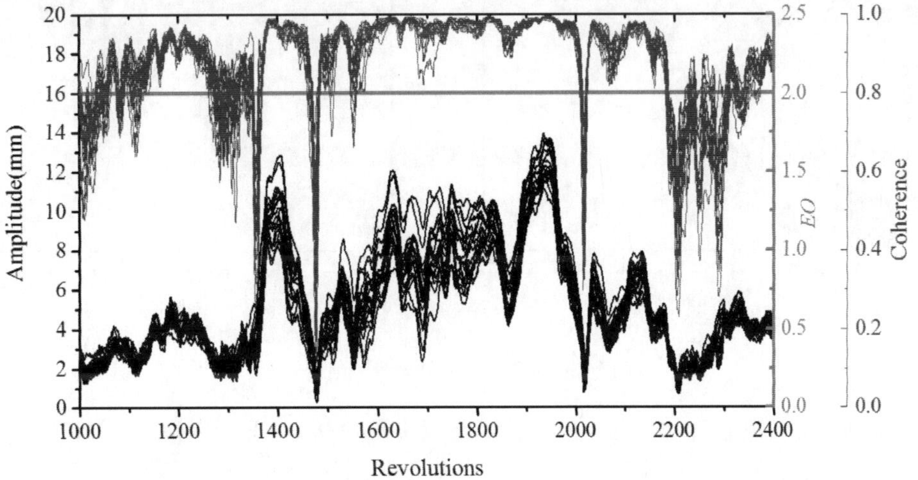

Figure 15. Synchronous vibration from 1000 to 2400 revolutions.

As can be seen from the figure 16, the uncertainty values are below 5% for the majority of revolutions. At high uncertainty regions caused by the zeroing process as well as measurement error, the coherence is quite low, which reduces the reliability of fitting results to some extent. But the amplitudes at those regions are comparatively small, hence it's not significant for the evaluation of the blade stress.

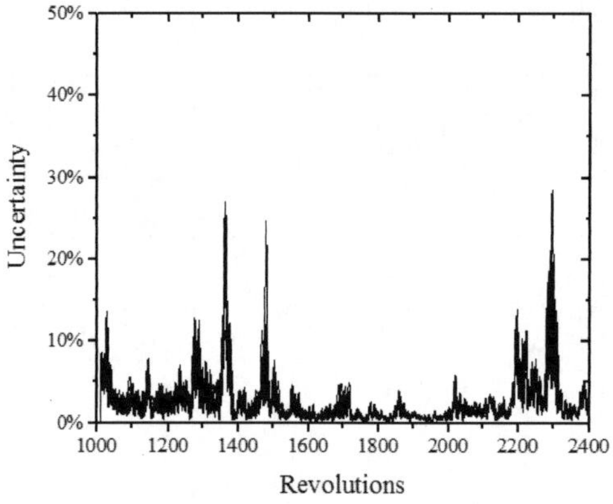

Figure 16. Uncertainty.

3.5. Blade endurance

The FEA is capable of providing the information of the non-dimensional displacement and stress of each node. It's assumed that the relationship between nodal displacement and its averaged stress is linear. Accordingly, a scaling factor (k) of nodal displacement can also be used for the nodal stress.

$$k = \frac{D_{\text{FEM}}}{D_{\text{BTT}}} \tag{11}$$

In Eqn. (11), D_{FEM} and D_{BTT} refer to the displacement calculated by FEA and BTT system respectively at the same node. The blade real stress of each node can be obtained by its corresponding modal stress multiplied by the inverse of k.

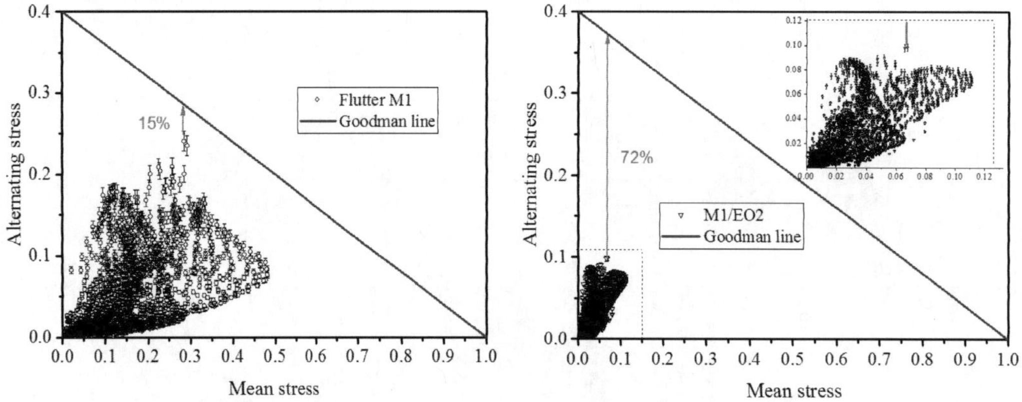

Figure 17. Goodman diagram.

Then the stress clouds of flutter and synchronous vibration for blade mode 1 can be plotted in figure 17. The blue line is known as the Goodman line that represents the cycle number of the blade failure. The intercept of X and Y axis indicates the material's ultimate yield strength and alternating stress (fully reversed) respectively. All nodal stresses have been normalized by the ultimate yield strength, and there is an error bar for each of them. Because the amplitude obtained from the curve fitting for each individual response is described with an uncertainty value. It's obvious that the flutter margin (15%) is much less than that of synchronous vibration.

4. Conclusions

With the help of FEA, blade tip timing technique is able to obtain the blade stress without using strain gauges, which enhances the application of BTT. This paper presents a process that explains how to analyze the vibration characteristics of the fan blades by the raw TOA data in detail. Key conclusions are:

- The use of a standard process is important to reduce the processing errors introduced by data zeroing and improve the accuracy of calculated blade stresses.
- In this case, flutter occurs with a dominant nodal diameter 2 at 92% rotor speed approximately, and 1st flex mode – 2 engine order crossing is found at 48% rotor speed. In the high response region, the coherence between calculated and measured displacement is beyond 0.9, which ensures the reliability of the analysis method.
- The blade endurance in Goodman diagram shows that the flutter has less margin compared with the synchronous vibration. That's the reason why we have to avoid a flutter scenario in reality.

References

[1] García I, Beloki J, Zubia J, et al. An optical fiber bundle sensor for tip clearance and tip timing measurements in a turbine rig[J]. Sensors, 2013, 13(6): 7385-7398.

[2] Zablotskiy I Y, Korostelev Y A. Measurement of Resonance Vibrations of Turbine Blades with the Elura Device[R]. FOREIGN TECHNOLOGY DIV WRIGHT-PATTERSON AFB OH, 1978.

[3] Zielinski M, Ziller G. Noncontact vibration measurements on compressor rotor blades[J]. Measurement Science and Technology, 2000, 11(7): 847.

[4] Heath S. A new technique for identifying synchronous resonances using tip-timing[J]. ROLLS ROYCE PLC-REPORT-PNR, 1999.

[5] Dimitriadis G, Carrington I B, Wright J R, et al. Blade-tip timing measurement of synchronous vibrations of rotating bladed assemblies[J]. Mechanical Systems and Signal Processing, 2002, 16(4): 599-622.

[6] Gallego-Garrido J, Dimitriadis G, Carrington I B, et al. A Class of Methods for the Analysis of Blade Tip Timing Data from Bladed Assemblies Undergoing Simultaneous Resonances—Part II: Experimental Validation[J]. International Journal of Rotating Machinery, 2007, 2007.

[7] Gallego-Garrido J, Dimitriadis G, Wright J R. A class of methods for the analysis of blade tip timing data from bladed assemblies undergoing simultaneous resonances—Part I: theoretical development[J]. International Journal of Rotating Machinery, 2007, 2007.

[8] Heath S, Imregun M. A survey of blade tip-timing measurement techniques for turbomachinery vibration[J]. Journal of engineering for gas turbines and power, 1998, 120(4): 784-791.

[9] Russhard P. Development of a blade tip timing based engine health monitoring system[D]. Thesis, Faculty of Engineering and Physical Sciences, School of Mechanical Aerospace and Civil Engineering, Manchester University, 2010.

[10] Russhard P. The Rise and Fall of the Rotor Blade Strain Gauge[M]//Vibration Engineering and Technology of Machinery. Springer, Cham, 2015: 27-37.

[11] WATKINS W, CHI R. Noninterference blade-vibration measurement system for gas turbine engines[J]. Journal of Propulsion and Power, 1989, 5(6): 727-730

[12] Russhard P. Timing analysis: U.S. Patent 8457909 B2[P]. 2009-10-22.

Experimental investigation by Laser Ultrasonics for train wheelset flaw detection

A. Cavuto[a], M. Martarelli[b], G. Pandarese[a], G.M. Revel[a], E.P. Tomasini[a]

[a] Università Politecnica delle Marche – Dipartimento di Ingegneria Industriale e Scienze Matematiche, Via Brecce Bianche, 60131, Ancona, IT
[b] Università degli Studi e-Campus, Via Isimbardi, Novedrate (CO), IT

E-mail: a.cavuto@univpm.it

Abstract. The transport safety is a key factor in the rail field and to ensure it the monitoring of train components structural integrity must be performed regularly and following strict regulations. Consequently, the development of Non-Destructive Testing (NDT) procedures is a topic of great interest and impact, it aiming at the early identification of cracks and inhomogeneities on train axle and wheels which propagation can causes faults and accidents undermining the passenger and crew safety. Ultrasound based nondestructive testing is currently applied in ordinary maintenance procedures, although they are applied to axle and wheels dismounted from the train wheelset. In fact, the technique requires ultrasonic probes to be in contact with the object under test and therefore the fretting surface between wheel and axle, where fatigue phenomenon induces cracks enucleation and propagation, cannot be monitored. The possibility of applying NDT systems to the whole wheelset, without dismounting it in its components, will allow drastically reducing the inspection time and increasing the inspection frequency. The present paper proposes a feasibility study for the development of a Laser-Ultrasonic Testing (LUT) procedure to improve performances of train wheelset ultrasonic inspection. The method exploits one or an array of air-coupled ultrasonic probes, which detects ultrasonic waves generated by a high-energy pulsed laser. Thanks to the non-contact nature of both the response measurement sensor and the excitation system, the experimental set-up is extremely flexible and it allows extremely speeding up the inspection time.

Keywords: Train wheelset diagnostics, NDT, air-coupled ultrasound, Laser Ultrasonics.

PACS: 46.40.-f Vibrations and mechanical waves; 46.80.+j Measurement methods and techniques in continuum mechanics of solids; 07.10.-h Mechanical instruments and equipment.

1. Introduction

In the last few years, the demand for new non-contact systems for remote control of the structural integrity of the main train components related to the safety of the exercise, such as axles, wheels and tracks, is growing. The procedures based on NDT methods are constantly evolving to meet the European standards

EN 15313 [[1]] and the directives of the National Agency for Safety of Railways (ANSF) dictating the regulation for the safety of railway traffic.

The axles and wheels are often the main responsible for breakdowns and accidents, being the most stressed in both static and dynamic conditions. In particular, the wheels are subjected to high loads caused by wheel-rail contact [[2]], leading to the generation and propagation of different types of defects, Figure 1 [[1]].

At the state of the art, manual ultrasound techniques are used for the detection of radial defects, through the use of eight double angled probes (settled in the so-called "Tandem" arrangement) and distributed along a circumference, [[3]]. Being those techniques manual, the occurrence of human error is extremely likely. Automated systems equipped with phased arrays [[4]] are currently unattractive, given the high costs associated with the acquisition of instrumentation and the adaptation of test benches to these systems. The main disadvantages of phased arrays are:

- difficult adaptation to different wheel profiles (Velaro, TGV, Shinkansen, Bombardier etc.),
- encumbrance that limits their integration into the system for the monitoring of the wheelset while the train is moving,
- complexity of the measurement set-up and in particular of the system for the accumulation, injection and recovery of the coupling means, necessary in the application of contact probes.

Figure 1. Typical morphology of railway wheel defects.

To overcome these problems, LUT systems have been developed which, by applying high power laser sources and air receiving probes, represent completely non-contact techniques, flexible in installation and configuration. Air-coupled ultrasound inspection has already been successfully used in many industrial applications e.g. NDT on curved composites elements [5], NDT on light thin historical vaults [6], density measurement of ceramic tiles [7], NDT on thin metallic laminated [8, 9], etc.. In [10] and [11] a LUT procedure has been defined for the inspection of railway components, axles and wheels respectively. A FE model has been also developed to compute the ultrasound wave propagation in thermoelastic regime and allowing to guide the design of the experimental set-up and the analysis of the results. It has been shown that it is possible to work with low energy waves in the thermoelastic-ablative limit maintaining a satisfactory contrast level for the detection of the defect. However, at present LUT methods have been only applied to dismounted wheelset, i.e. for the monitoring of axles and wheels separately. This is due to the fact that the energy of the ultrasound waves travelling into the axle is strongly attenuated when passing through discontinuities such us the fretting region between axle and wheel. Since that region is the one in which there is the greatest probability of presence of defects, in order to ensure an accurate inspection the wheel dismounting from the axle is required.

In this paper, following the indication of a previous work [12], demonstrating by a numerical model that a high power laser is able to generate ultrasound wave sufficiently energetic to propagate across the fretting region, LUT has been applied to a whole wheelset and its suitability has been confirmed by the experimental results.

2. The wheelset and experimental set-up

2.1 The tested wheelset

Trenitalia provided a train wheelset (Figure 2) in whose axle two typical fatigue cracks were simulated, see Figure 3:

- Defect A located on the axle surface at XX mm from the fretting region,
- Defect B located on the fitting region between the axle and the wheel.

Figure 2. Trenitalia train wheelset.

a) Defect A

b) Defect B

Figure 3. Defects simulated in the wheelse - Defect A (a) and Defect B (b).

2.2 *Experimental set-up*

The train wheelset was mounted on a support that made it possible to control its rotation, in order to be able to scan the axle surface along a circumference. The probes were installed on a frame where they could move along the axial direction. Therefore, the complete lateral surface of the axle could be inspected by the laser ultrasonic system, see Figure 4.

The laser ultrasonic system was made up of a pulsed laser source, a Nd-Yag IR laser (1064 nm), pulses of 12 ns duration and 82 mJ energy, from Continuum, and a 500 kHz air-coupled ultrasound piezoelectric probe from Ultran Group (model NCT210, with 20 mm diameter of active area). The probe was inclined with respect the normal to the axle surface of 33 deg, as shown in Figure 5 and Figure 6. This angle has been determined from a numerical analysis of the ultrasonic wave propagation from the generation point, were the laser impinges the axle, through the material and the air gap between the axle surface and the receiving probe [10]. The ultrasound probe conditioning system was a DPR 300 Pulser/Receiver from JSR Ultrasonics. The ultrasound signals were amplified with a gain level of 69 dB and acquired with a high speed Digitizer board NI PXI-5122 (100 MHz bandwidth). The laser beam was guided towards the axle under test by means of an arm connected to the pulsed laser cavity as shown in Figure 4.

Data were collected by sampling at 100 MSamples/s and performing 4 averages on each measurement point in order to improve the Signal to Noise Ratio (SNR).

The laser ultrasonic system was installed with the laser source and receiving probe far apart, on opposite sides with respect to the wheel, as shown in Figure 4. A collimated laser beam with a diameter of about 8.5 mm was used to keep the ultrasonic waves generation within a thermo-elastic regime [13].

The axle was made to rotate by an electric motor and the rotation angle was measured by an integrated encoder. A circumferential scan was performed along an angle of 63 deg with an angular resolution of 1.5 deg.

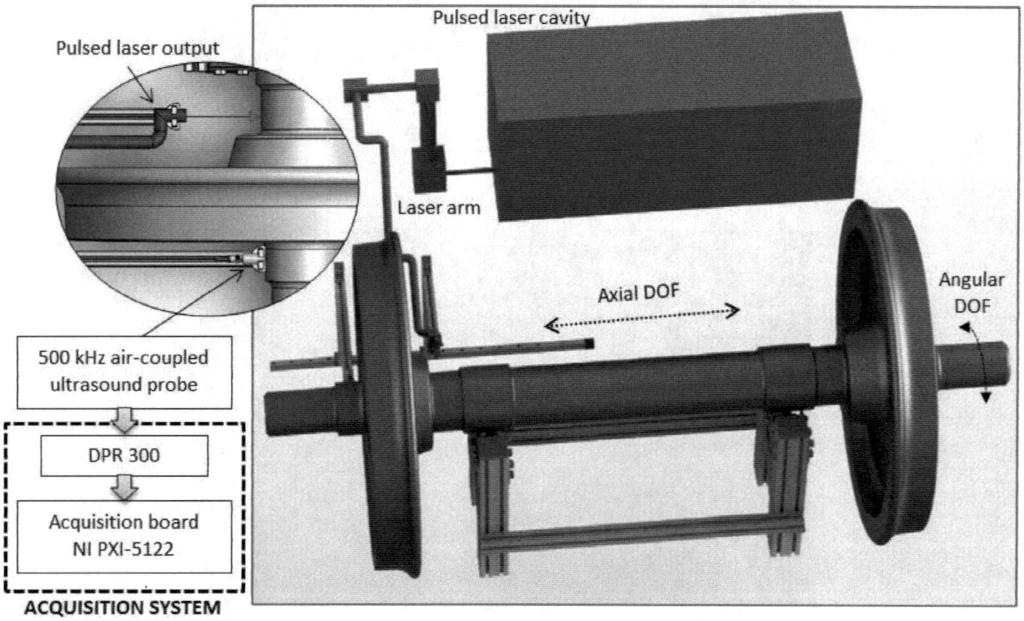

Figure 4. LUT experimental set-up.

3. Results

For each configuration, a series of ultrasound time histories with a duration of 180 μs was acquired at every scanning position along the arc considered. The experimental set-up (i.e. laser and probe positions), the axle section tested and the defect location are shown in the subplot (a) of Figure 5 and Figure 6 for defect A and B respectively. The subplot (b) of each figure reports a close-up around the excitation position and the damaged area. In the draw it is also sketched the path of the ultrasound waves travelling into the material once generated by the laser pulse impinging onto the axle surface. The time history acquired in the first angular scanning position is plotted in the subplot (c). Subplot (e) illustrates the time histories at two angular positions located within and outside the damaged region. As far as it concerns the defect A, it is evident that the presence of the defect acts as a filter for the bulk wave Figure 5 (b). This is confirmed also by the B-scan reported in Figure 5 (d) where the extension of the cracks is clearly visible. Observing the time history close-up around the bulk waves and the B-scan recorded for defect B, Figure 6 (e) and (d) respectively, it can be noticed that part of the bulk waves are transmitted even in presence of the defect and specifically the fastest one. This can be explained by the fact that the defect located in the fitting region between wheel and axle does not hinder the passage to all the ultrasound waves. As it is sketched in the scheme of Figure 6 (b), the direct bulk wave (solid blue arrow in the sketch), the fastest

IOP Conf. Series: Journal of Physics: Conf. Series **1149** (2018) 012015 doi:10.1088/1742-6596/1149/1/012015

one, passes undisturbed by the defect, while the scattered waves (dashed blue arrow in the sketch) are reflected backwards.

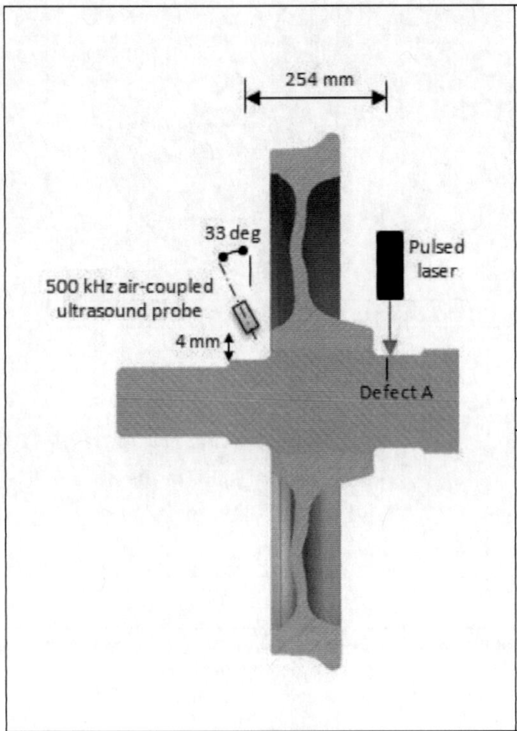

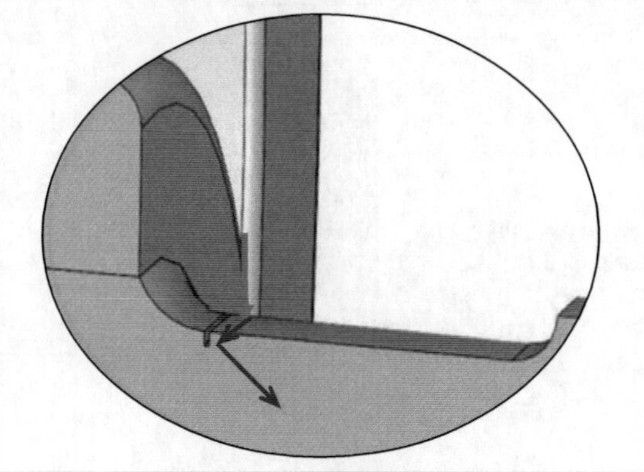

(b) close-up around the damage location

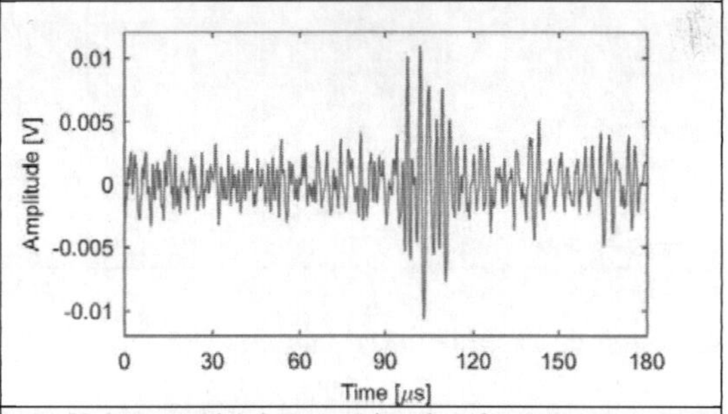

(a) Experimental set-up

(c) Voltage, which is proportional to the pressure wave, as a function of time recorded by the 500 kHz air-coupled ultrasound transducer located in front of the axle fitting

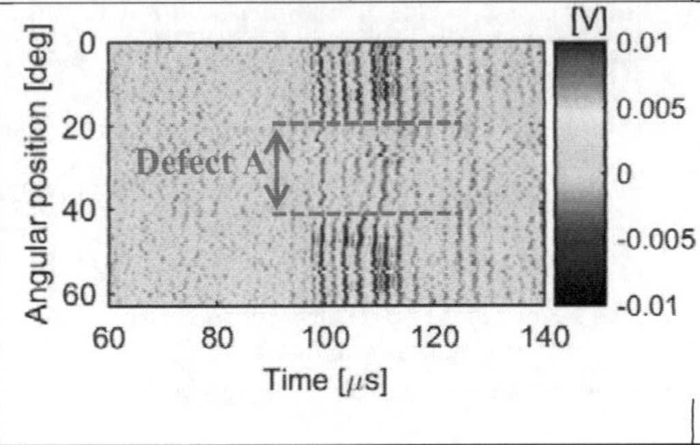

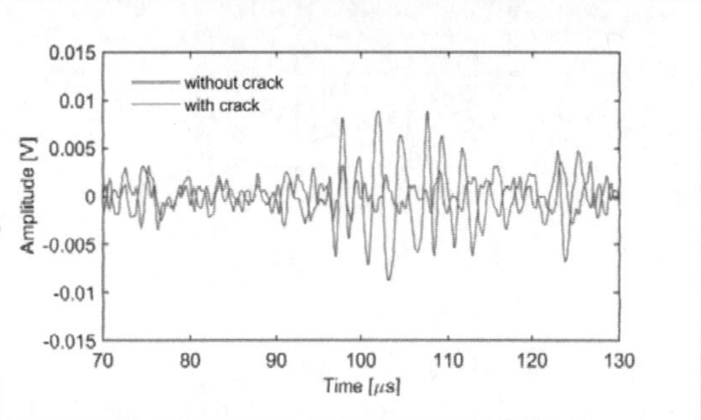

(d) B-scan

(e) Time history close-up on the bulk waves

Figure 5. Experimental set-up scheme (a), time history (b), B-scan (c) and time history close-up on the bulk waves (d) - Defect A.

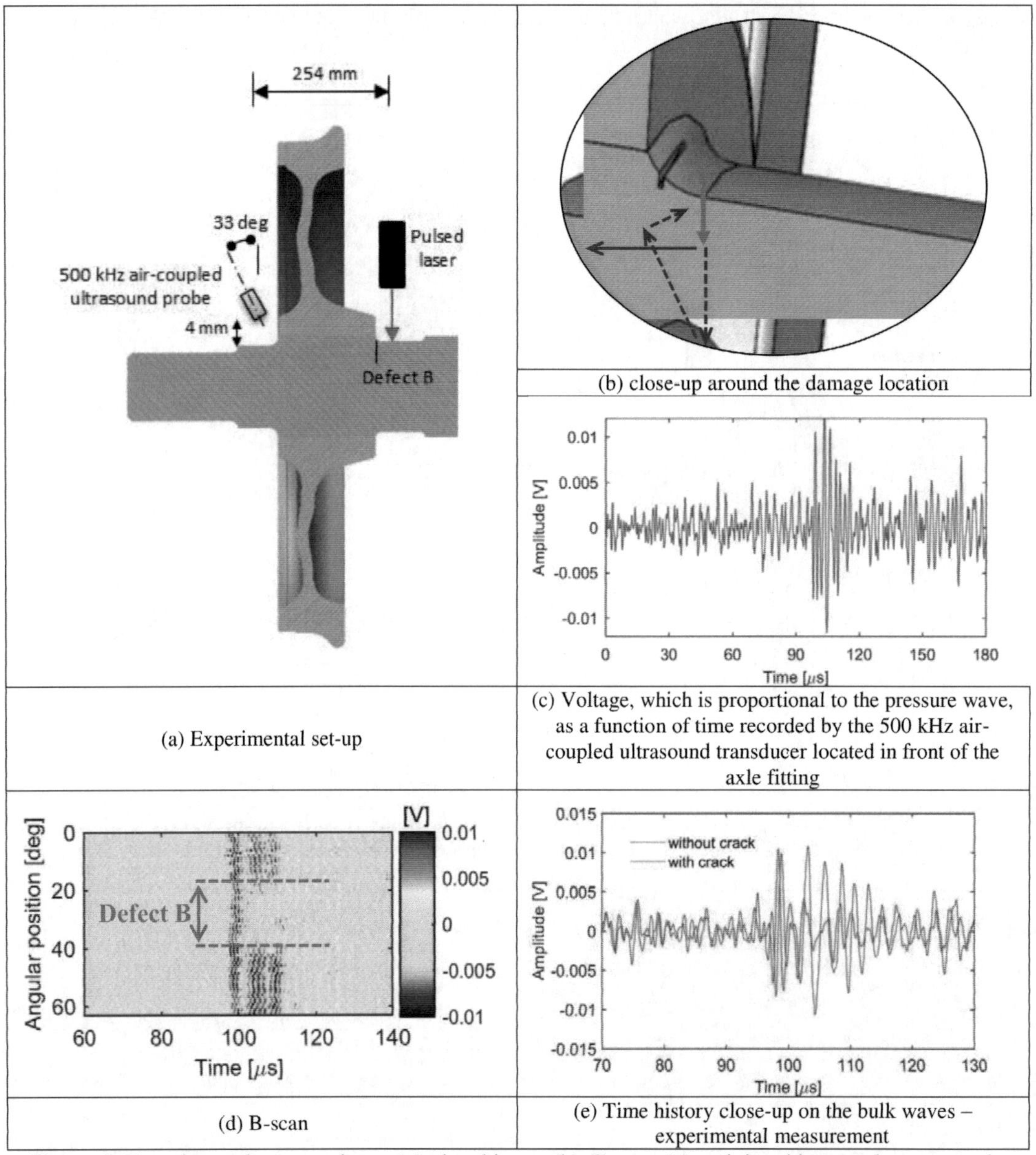

(a) Experimental set-up

(b) close-up around the damage location

(c) Voltage, which is proportional to the pressure wave, as a function of time recorded by the 500 kHz air-coupled ultrasound transducer located in front of the axle fitting

(d) B-scan

(e) Time history close-up on the bulk waves – experimental measurement

Figure 6. Experimental set-up scheme (a), time history (b), B-scan (c) and time history close-up on the bulk waves (d) - Defect B.

The RMS plots over the time axis (abscissa) of the B-scans has been calculated for defect A on a time window between 90 and 120 μs and for defect B on a time window between 102 and 115 μs in order to cut the signal related to the direct wave which is insensitive to the presence of the defect. Table 1 shows the RMS plots and the contrast between the RMS in the damaged and in the undamaged areas, which can be seen as a damage index. That index was calculated considering the minimum values of the RMS (in the damaged area) and the floor values RMS (in the undamaged area). Both the defects exhibit a high contrast.

Table 1. Normalized RMS and Damaged/Undamaged RMS Contrast.

<table>
<tr><td>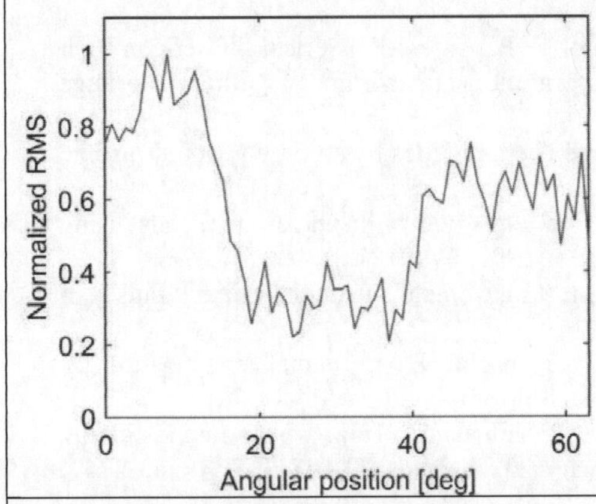
</td><td>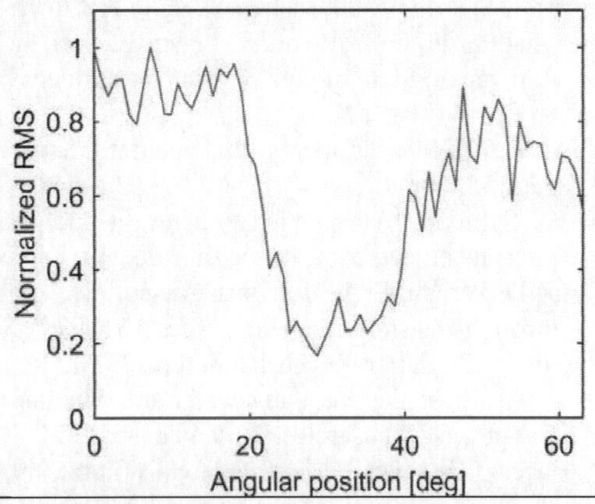
</td></tr>
<tr><td>Defect A – RMS Contrast – 11.3 dB</td><td>Defect B – RMS Contrast – 12.1 dB</td></tr>
</table>

4. Conclusions

The paper demonstrates the applicability of the LUT based on ultrasonic waves generation by pulsed laser and air coupled ultrasound probes for damage identification on train wheelset. The technique has been applied to a wheelset provided by Trenitalia where typical fatigue defects have been created in region where those defects frequently occur. Several works have demonstrated the power of such technique for the diagnostic of train axle and wheel. The standard practice for the defect identification in rail industry is based on ultrasound methods. However both the ultrasound based NDT technique and the LUT requires dismounting the wheel from the axle when performing the test. This is due to the fact that the ultrasonic wave is strongly attenuated when passing through the fretting area between axle and wheel. By optimizing the laser energy of the impinging pulse generating the ultrasonic wave and the position of the detecting probe collecting the wave transmitted through the material and the air, it has been demonstrated that the LUT can be applied also to the whole wheelset and that the resulting signal exhibits a very good contrast (a RMS contrast of 11.3 dB for defect A and 12.1 dB for defect B) allowing to clearly identify the defect position.

The good results achieved with the experimental procedure proposed in this paper incentivize the authors to develop a dedicated FE numerical model to optimize the measurement procedure with the aim to prove its suitability for the remote inspection of railway axles in service.

References

[1] Railway applications – In-service wheelset operation requirements – In-service and off-vehicle wheelset maintenance, in EN 15313:2010.
[2] D. Milkovic, G. Simic, Z. Jakovljevic, J. Tanaskovic, V. Lucanin, Wayside system for wheel–rail contact forces measurements, Measurement 46 (2013) 3308–3318.
[3] L. Labbadia, L. Aruta, M. Sarti, Evoluzione dei CND in ambito Trenitalia, in: Proceedings of AIPnD, Florence, Italy, 2011.

[4] J. Peng, X. Gao, Y. Zhang, Z. Wang, C. Peng, Y. Tan, K. Yang, B. Zhao, An ultrasonic technology study for subsurface defect in railway wheel tread, in: Proceedings of 11th European Conference on Non-Destructive Testing (ECNDT), Prague, Czech Republic, 2014.
[5] G.M. Revel, G. Pandarese, A. Cavuto, Advanced ultrasonic non destructive testing for damage detection on thick and curved composite elements for constructions, J. Sandwich Struct. Mater. 15 (1) (2013) 1–20, http://dx.doi.org/10.1177/1099636212456861.
[6] E. Quagliarini, G.M. Revel, S. Lenci, E. Seri, A. Cavuto, G. Pandarese, Historical plasters on light thin vaults: State of conservation assessment by a Hybrid ultrasonic method, J. Cultural Heritage 15 (2014) 104–111.
[7] G.M. Revel, Measurement of the apparent density of green ceramic tiles by a non-contact ultrasonic method, Exp. Mech. 47 (5) (2007) 637–648.
[8] D.W. Schindel, Air-coupled generation and detection of ultrasonic bulk waves in metals using micromachined capacitance transducers, Ultrasonics 35 (1997) 1799181.
[9] W.M.D. Wright, D.A. Hutchins, Air-coupled ultrasonic testing of metals using broadband pulses in through-transmission, Ultrasonics 37 (1999) 19–22.
[10] Cavuto, M. Martarelli, G. Pandarese, G.M. Revel, E.P. Tomasini, Experimental investigation by laser ultrasonics for high speed train axle diagnostics, Ultrasonics 55 (2015) 48–57.
[11] A. Cavuto, M. Martarelli, G. Pandarese, G.M. Revel, E.P. Tomasini, "Train wheel diagnostics by laser ultrasonics", Measurement, Volume 80, February 2016, Pages 99-107, ISSN 0263-2241, http://dx.doi.org/10.1016/j.measurement.2015.11.014.
[12] C. Mineo, D. Cerniglia, A. Pantano, Numerical study for a new methodology of flaws detection in train axles, Ultrasonics 54 (3) (2013) 841–849.
[13] P. Castellini, G.M. Revel, L. Scalise, R.M. De Andrade, Experimental and numerical investigation on structural effects of laser pulses for modal parameter measurement, Opt. Lasers Eng. 32 (6) (2000) 565–581.

IOP Conf. Series: Journal of Physics: Conf. Series **1149** (2018) 012016 doi:10.1088/1742-6596/1149/1/012016

Laser-Doppler vibrometer microscope with variable heterodyne carrier

Robert Kowarsch[1], **Rigele Te**[1], **and Christian Rembe**[1]

[1] Institute of Electrical Information Technology, Department of Applied Metrology, Clausthal University of Technology, Leibnizstraße 28, 38678 Clausthal-Zellerfeld, Germany

E-mail: `kowarsch@iei.tu-clausthal.de`

Abstract. The generation of a heterodyne carrier frequency for coherent detection via offset-locking in an optical phase-locked loop was demonstrated in communication, spectroscopy and other fields. In state-of-the-art laser-Doppler vibrometers (LDV), common acoustooptic devices (Bragg cells) allow efficient heterodyning with a static frequency. Moving mechanical parts vibrate in many technical applications, e.g. moving cars, trains, robots, ect.. In such applications, a large laser-Doppler-shift generated by the movement impairs the measurement of the relevant vibrations. A variation of the frequency offset of the interfering laser beams can suppress the Doppler effect of high-speed movements with slow speed variations and adjusts the signal spectrum to the optimal value. Thus, the electronic bandwidth can be set to an optimal value within the modulation bandwidth of the relevant vibrations. However, Bragg cells are not well applicable to shift laser frequencies without affecting the interferometer adjustment. In this paper, we propose offset locking of two independent lasers with an optical phase-locked loop to exploit the possibility of flexible heterodyne-carrier generation for laser-Doppler vibrometry. We demonstrate a LDV setup with offset locked visible DBR semiconductor lasers for a flexible heterodyning up to 200MHz and present first measurements.

1. Introduction

Laser-Doppler vibrometers (LDV) are a group of indispensable instruments to measure vibrational information without contact in many applications in research, development and production [1, 2]. LDVs exploit the laser-Doppler effect which linearly transduces a motion into a phase modulation of the radiation, which is scattered back by the moving or vibrating object [1]. Therefore, a fairly-stable laser frequency (or wavelength) as the standard is required which is usually provided by most laser sources. The phase modulation at the laser frequency (as the fundamental carrier at hundreds of THz) cannot be measured directly. Thus, it is down-mixed by interference with a second laser beam and coherently detected with a fast photodetector. In heterodyne interferometers, the laser radiation for down-mixing comprises a frequency-shifted copy of the original laser radiation. Therefore, the vibration-induced phase modulation (Bessel spectrum) emerges around a heterodyne carrier frequency in the spectrum of the photodetector signal. Especially in field applications (air-born, vehicle-born or hand-held applications), a superposed high velocity with slow variations can generate large laser-Doppler frequency shifts which shifts the complete Bessel spectrum [2]. This effect can also impair tracking-LDV measurements on large rotating structures (e.g. wind turbines). Typical heterodyne LDVs incorporate a fixed frequency shift up to several hundreds of MHz between the interfering beams by acoustooptic (AO) devices [1, 2]. Carrier-tracking demodulation by PLL [3] or post-processing can eliminate the disturbing motions [4]. Therefore, the complete vibration spectrum must remain within the electronic bandwidth to enable distortionless reconstruction of the underlying vibration. However, the motion-induced frequency shift may shift the Bessel spectrum out of measurement bandwidth or to regimes where a higher noise level or spurious lines impair the sensitivity of the LDV. For the full exploitation of the measurement bandwidth for vibration detection, a variable and stabilized difference frequency in coherent detection can compensate the motion-induced frequency shift. The goal is to enable a variable frequency difference between the interfering laser beams, to maintain the central frequency of the Bessel spectrum at the center of the measurement bandwidth. In conventional AO devices, the diffraction angle is directly dependent on the drive frequency [5]. Thus, with these devices a flexible change of frequency shift is not possible without undesirable misalignment.

In this work, we pursue the generation of a variable heterodyne carrier frequency for LDV by frequency-offset locking of two semiconductor lasers in an optical phase-locked loop. In general, heterodyning by offset-locking in an optical phase-locked loop (OPLL) was shown in the 1960s for remote sensing and communication [6, 7] and the principle of coherent receiver (with CO_2 laser) was patented in the mid-1970s for optical communication systems [8]. Hitherto, the generation of a well-defined frequency and phase relation between two independent laser sources was demonstrated in many applications [9–18]. With the emergence of semiconductor lasers with intra-cavity wavelength selection, reliable laser sources become available with linewidths of kHz to MHz. Especially at visible wavelengths, only recent developments provide distributed Bragg reflector (DBR) semiconductor lasers with fast tuning by injection current and temperature over a wide wavelength range [19].

In this paper, we discuss the advantages and challenges of a flexible generation of heterodyne carrier frequency for laser-Doppler vibrometry by OPLL. This technique provides the capability of eliminating dominant motions of the vibrating sample and adapting the heterodyne frequency carrier to the measurement problem. Decisive parameters of the OPLL are presented in conjunction with the application to laser-Doppler vibrometry. As an application, we demonstrate our LDV microscope setup with two visible DBR semiconductor lasers at 632 nm which is capable to provide a variable heterodyne carrier frequency up to 200 MHz by offset-locking via OPLL. Further, we present first measurements of resonances of a piezoelectric transducer to show the capability and limitations of the novel application to laser-Doppler vibrometry. Finally, we discuss the vibration sensitivity.

2. Signal in heterodyne vibrometry

2.1. Coherent detection of laser-Doppler shift

LDVs contain a typical laser interferometer with a broadband acquisition of the light phase difference [1]. For the interference, the radiation of usually the same coherent light source is split into two copies generating a reference and measurement beam. The measurement beam contains phase changes or fluctuations associated with the measured quantity. In LDVs, these phase changes $\varphi_{\mathrm{vib}}(t)$ over time t due to laser-Doppler effect are principally caused by changes of the geometric length of the measurement beam (reference path length remains constant). For a displacement $s(t)$ of the sample (parallel to the incident measurement beam) relative to the LDV (containing sender and receiver), the observed phase modulation is [2]

$$\varphi_{\mathrm{vib}}(t) \approx \frac{4\pi}{\lambda} s(t) \tag{1}$$

with the wavelength λ of the light. The approximation in Eq. 1 is sufficient for estimating the laser-Doppler effect even at high modulation frequencies [20].

IOP Conf. Series: Journal of Physics: Conf. Series **1149** (2018) 012016 doi:10.1088/1742-6596/1149/1/012016

The fundamental carrier of this phase modulation (by laser-Doppler effect) is the light frequency $f_{\rm m}$ (430 to 770 THz at visible light) which cannot be measured directly. This phase modulation of the measurement beam is linearly shifted to the measurement bandwidth of the LDV via interference (coherent detection) with the reference beam at the frequency $f_{\rm r}$. The resulting current at the photodetector (PD) is

$$i_{\rm PD}\left(t\right) = 2\chi K_{\rm PD}\sqrt{P_{\rm m}P_{\rm r}}\cos\left[2\pi\left(f_{\rm m} - f_{\rm r}\right)t + \Delta\varphi\left(t\right)\right] \tag{2}$$

with the interference efficiency χ, the PD sensitivity $K_{\rm PD}$, the differential phase fluctuation $\Delta\varphi$, and the power $P_{\rm m}$ of the measurement beam at the photodetector ($P_{\rm r}$ respectively for reference beam). Thus, any differential phase modulation is linearly shifted to the difference frequency $f_{\rm m} - f_{\rm r}$. In heterodyne LDV, a non-zero, static difference frequency (intermediate frequency) of the interfering laser beams is applied. In analogy to communication technology, this intermediate frequency is denominated as heterodyne carrier frequency $f_{\rm c} = |f_{\rm m} - f_{\rm r}|$ [2]. The advantages of heterodyning for LDV are the choice of a low noise spectral range and the unambiguous discrimination of the direction of the vibration or motion [2].

2.2. *Bandwidth requirement and tracking of low-frequency motions*

A decisive specification of heterodyne LDVs is the necessary measurement bandwidth which is dependent on the vibration-signal bandwidth induced by laser-Doppler effect. For the estimation, a sample is moving (in direction of the LDV) at high-speed with slow variations with the motion $s_{\rm lf}\left(t\right)$ superposed by an harmonic vibration with frequency $f_{\rm vib}$ giving

$$s\left(t\right) = s_{\rm lf}\left(t\right) + \hat{s}\cos\left(2\pi f_{\rm vib}t + \varphi_{\rm vib}\right) \tag{3}$$

with the vibration amplitude $\hat{s}$ and the vibration phase $\varphi_{\rm vib}$. The interferometer signal according to Eqs. 1, 2, and 3 can be expanded with the Jacobi-Anger identity [21]. Thus, the typical Bessel spectrum emerges in the spectrum of the vibrometer signal [2] of the form

$$i_{\rm PD}\left(t\right) = 2\chi K_{\rm PD}\sqrt{P_{\rm m}P_{\rm r}}\sum_{n=-\infty}^{\infty} J_n\left(\frac{4\pi\hat{s}}{\lambda}\right)\cos\left[2\pi\left(f_{\rm c} + \frac{2}{\lambda}\frac{{\rm d}s_{\rm lf}\left(t\right)}{{\rm d}t} + nf_{\rm vib}\right)t + n\varphi_{\rm vib} + \varphi_0\right] \tag{4}$$

with the Bessel function of first kind J_n and the (integer) order n and constant phase φ_0. The amplitudes of the equidistant spectral lines are determined by the modulation index $M_{\rm vib} = 4\pi\hat{s}/\lambda$ and the frequency spacing corresponds to the vibration frequency $f_{\rm vib}$. According to Carson's rule [22] the required measurement bandwidth, which contains > 99% of the spectral power (of the harmonic vibration), is $B_{\rm vib} > 2\left(M_{\rm vib} + 1\right)f_{\rm vib}$.

In Eq. 4, the central frequency f_{J0} of the Bessel spectrum ($n = 0$) observes an instantaneous frequency shift $\Delta f\left(t\right) = 2{\rm d}s_{\rm lf}\left(t\right)/\left(\lambda\,{\rm d}t\right) = 2v_{\rm lf}\left(t\right)/\lambda$ in respect to the heterodyne carrier frequency $f_{\rm c}$ with the (low-frequency) speed $v_{\rm lf}\left(t\right)$. For an undisturbed reconstruction of the motion, the heterodyne carrier frequency must be chosen, that (for the maximum values of $v_{\rm lf}$, $\hat{s}$ and $f_{\rm vib}$) no overlap of the spectra occurs ($f_{\rm c} > B_{\rm vib}/2$). Therefore, it follows

$$f_{\rm c} > \Delta f + \frac{B_{\rm vib}}{2} = \frac{2v_{\rm lf}}{\lambda} + \left(\frac{4\pi\hat{s}}{\lambda} + 1\right)f_{\rm vib} \tag{5}$$

with the assumption of bidirectional (maximum) values $v_{\rm lf}$. Additionally, the measurement bandwidth $B_{\rm el}$ has to be chosen that the transfer function and the phase shift for all frequency components is flat. The heterodyne carrier frequency f_c is usually placed in the center of the measurement bandwidth, which has to be larger than twice the heterodyne carrier frequency ($B_{\rm el} > 2f_{\rm c} > B_{\rm vib}$) consequently.

Thus, any instantaneous frequency shift $\Delta f(t)$ induced by (unwanted) low-frequency motion costs measurement bandwidth which cannot be utilized for detection of vibrations. Further, the vibration spectrum may be shifted to frequencies with high noise level or spurious lines. Therefore, a variable heterodyne carrier frequency $f_{\mathrm{c}}(t)$ enables the flexible choice for the spectral region of the measurement bandwidth. Further, it allows to compensate frequency shift due to low-frequency motions of the sample (directly at the PD) to maintain a constant center frequency of the vibration spectrum ($f_{J0} = f_{\mathrm{c}}(t) + 2v_{\mathrm{lf}}(t)/\lambda \approx \mathrm{const.}$).

2.3. Reconstruction of vibration

Standard FM decoders from communication technology (e.g. IQ demodulator) can be utilized to reconstruct the motion [2]. For a complete motion recovery, the local oscillator is driven with the heterodyne carrier frequency f_{c}. If compensation of low-frequency motions via a variable heterodyne carrier is applied, the stabilized center frequency f_{J0} of the vibration spectrum has to be fed to the demodulator. Therefore, it is beneficial to synchronize both frequencies and to allow only discrete frequency steps for compensation [3].

In ISO 16063-41 the calibration with the raw vibration spectrum is recommended to avoid any influence of demodulation [23]. If the narrow-band approximation is applicable, the carrier-to-sideband ratio CSR directly delivers the modulation index

$$M_{\mathrm{vib}} \approx 2\frac{J_1(M_{\mathrm{vib}})}{J_0(M_{\mathrm{vib}})} = \frac{2}{\sqrt{\mathrm{CSR}}} \tag{6}$$

with the Bessel function of first kind J_n and (integer) order n. The carrier-to-sideband ratio is $\mathrm{CSR} = P_{\mathrm{c}}/P_{\mathrm{SSB}}$ with the carrier power P_{c} and the power of the first sideband P_{SSB}.

Further, the following relation from the magnitudes of adjacent Bessel lines can be deduced [2] with the derivation property of the Bessel function for broad-band phase modulation

$$M_{\mathrm{vib}} = \frac{2nJ_n(M_{\mathrm{vib}})}{J_{n-1}(M_{\mathrm{vib}}) + J_{n+1}(M_{\mathrm{vib}})} \tag{7}$$

2.4. Influence of phase noise on coherent detection

In LDVs, phase fluctuations of the interfering beams within the electronic bandwidth are transduced to the phase-modulation of the signal current (Eq. 2). If these phase fluctuations are correlated to a certain degree (of coherence), only the non-correlated difference of the phase fluctuation contributes to the phase noise in the detected vibrometer spectrum. Beyond the photodetector bandwidth, this differential phase fluctuations are integrated and degrade interference contrast (χ in Eq. 2). Regarding free-running laser sources, which are not derived from stable atomic transition (like HeNe), the laser emission frequency may fluctuate. However, with common lasers and good drive electronics, frequency fluctuations below several tens of MHz can be achieved, which changes the wavelength of the measurement beam (as reference length for the displacement) only insignificantly.

If the phase fluctuation (Eq. 4) can be split into laser-Doppler-induced phase modulation $\varphi_{\mathrm{vib}}(t)$ and differential phase noise $\Delta\varphi_{\mathrm{N}}(t)$, the signal current for a harmonic vibration $\hat{s}\cos(2\pi f_{\mathrm{vib}}t + \varphi_{\mathrm{vib}})$ and a small differential phase modulation $\Delta\varphi_{\mathrm{N}}(t) = \hat{\varphi}_{\mathrm{N}}\cos(2\pi f_{\mathrm{N}}t + \varphi_{\mathrm{N}})$ becomes

$$i_{\mathrm{PD}}(t) \approx 2\chi K_{\mathrm{PD}}\sqrt{P_{\mathrm{m}}P_{\mathrm{r}}}\times \sum_{n=-\infty}^{\infty} J_n(M_{\mathrm{vib}})\left\{\cos\left[2\pi(f_c + nf_{\mathrm{vib}})t + n\varphi_{\mathrm{vib}}\right] \pm \frac{\hat{\varphi}_{\mathrm{N}}}{2}\cos\left[2\pi(f_{\mathrm{c}} + nf_{\mathrm{vib}} \pm f_{\mathrm{N}})t + n\varphi_{\mathrm{vib}} \pm \varphi_{\mathrm{N}}\right]\right\} \tag{8}$$

The relation shows that the each spectral line of the intended vibration spectrum gets two sidebands with the differential phase modulation of both laser sources. Therefore, it is obvious that small differential phase noise is desirable for an undisturbed vibration spectrum.

3. Generation of variable heterodyne frequency carrier

3.1. State of the art

There are multiple technical realizations for the introduction of a frequency shift between the reference and the measurement beam (derived a common laser source) with paratellurite Bragg cells, Lithium-Niobate frequency shifters or serrodyning [5, 24]. Further, a moving or vibrating reference mirror or direct frequency modulation of the laser (in an unbalanced interferometer) can achieve heterodyning [25].

The most popular realization is the frequency shift by paratellurite acoustooptic (AO) frequency shifters [2]. The AO design is usually optimized for efficiently diffracting into first order by operating in the Bragg regime (Bragg cell). Conservation of energy and momentum requires that the first-order beam is frequency shifted by acoustic drive frequency f_{c} and diffracted with an angle θ_{diff}. Therefore, the acoustic drive frequency corresponds to the heterodyne carrier frequency f_{c} in most instruments with one Bragg cell. However, drive frequency and diffraction angle are directly coupled

$$\sin\left(\theta_{\mathrm{diff}}\right) = \frac{\lambda}{2 v_{\mathrm{ac}}} f_{\mathrm{c}} \tag{9}$$

with the specific sound velocity v_{ac} of the AO material [5]. Therefore, any change in drive frequency results in an angular deviation which destroys the sensitive alignment of the LDV. Even for collinear designs, the drive frequency has to be tuned carefully to achieve energy and momentum conservation. Therefore, common acoustooptics systematically does not provide a reasonable capability of a variable frequency shift.

To overcome the problems of conventional AO frequency shifters, we incorporated two phase-locked semiconductor lasers which are stabilized to an offset frequency by an optical phase-locked loop (OPLL) for the generation of a variable frequency heterodyne carrier $f_{\mathrm{c}}(t)$.

3.2. Heterodyning by optical phase-locked loop

A good interference requires the same transverse mode (spatial coherence) and polarization. Further, a well-defined phase relation (temporal coherence) between the interfering beams results in a high interference contrast [26]. These requirements can be met by two independent laser sources, where the mutual temporal coherence of both beams is established by an optical phase-locked loop (OPLL) (scheme in Fig. 1). Therefore, a photodetector (PD) detects the frequency difference $f_{\mathrm{m}} - f_{\mathrm{r}}$ between the (free-running) master laser (ML) and the tunable slave laser (SL). A phase-sensitive detector (PSD) compares this frequency difference to the frequency $f_{\mathrm{LO,el}}$ of a local oscillator $\mathrm{LO}_{\mathrm{el}}$. The resulting error is fed back to the tunable SL by a loop filter with the transfer function $F(s)$.

In lock, the SL inherits the coherence properties of the ML within the loop bandwidth f_{ol} (coherence cloning). The loop bandwidth therefore has to be wider than the summed linewidth of the lasers [27], which is typically lower than several MHz for typical semiconductor lasers. In lock, the SL frequency matches the frequency of the ML (optical local oscillator $\mathrm{LO}_{\mathrm{opt}}$) shifted by the frequency of the electric local oscillator $\mathrm{LO}_{\mathrm{el}}$. Hence, the $\mathrm{LO}_{\mathrm{el}}$ directly defines the frequency difference and, thus, the heterodyne carrier frequency (Eq. 2) which is

$$f_{\mathrm{c}} = |f_{\mathrm{m}} - f_{\mathrm{r}}| = |f_{\mathrm{ML}} - f_{\mathrm{SL}}| = |f_{\mathrm{ML}} - (f_{\mathrm{ML}} \pm f_{\mathrm{LO,el}})| = f_{\mathrm{LO,el}} \tag{10}$$

with the frequency f_{ML} of the ML, f_{SL} of the SL, and $f_{\mathrm{LO,el}}$ of the $\mathrm{LO}_{\mathrm{el}}$ (absolute due to symmetry of cosine).

IOP Conf. Series: Journal of Physics: Conf. Series **1149** (2018) 012016 doi:10.1088/1742-6596/1149/1/012016

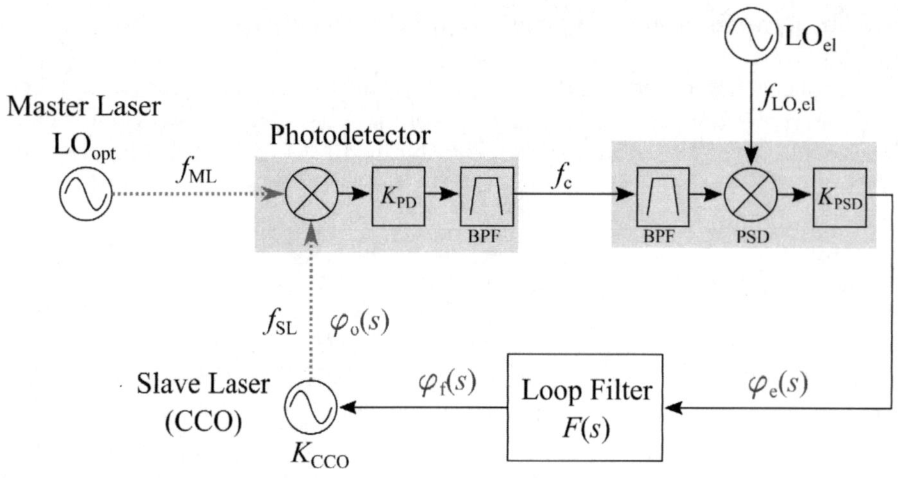

Figure 1. Scheme of an optical phase-locked loop (OPLL) with a frequency-offset lock of the tunable SL (current-controlled oscillator, CCO) to the frequency of the ML (optical local oscillator LO$_{opt}$). The phase-sensitive detector (PSD) detects the phase error between the beat frequency and the LO$_{el}$ frequency. The error signal $\varphi_e(s)$ is fed back via the loop filter with transfer function $F(s)$ to the tunable SL. In lock, the frequency difference of the SL in respect to the ML is locked to the frequency of the LO$_{el}$ (BPF - bandpass filter). The optical part is shown dotted-red, the rest is electronics. Small-signal phase approximation in lock is blue.

3.3. Decisive parameters of optical phase-locked loop

The OPLL is a nonlinear control system, which can be linearized around the operating point in lock state for small phases [$\sin(\varphi) \approx \varphi$] [28]. The decisive figures of merit are the open-loop gain $|G_{ol}(0)|$ and loop bandwidth f_{ol}. Within the bandwidth, the OPLL reduces differential phase noise and therefore generates phase relation (mutual coherence) between the two lasers in lock. For an open-loop transfer function $G_{ol}(s) = \varphi_o(s)/\varphi_e(s)$, the differential-phase-noise power spectral density (in lock state) can be estimated (for small phase noise) to [29]

$$S_{\Delta\varphi}^{2-\text{sided}}(f) \approx \left|\frac{1}{1+G_{ol}(j2\pi f)}\right|^2 \left[S_\varphi^{\text{ML,fr}}(f) + S_\varphi^{\text{SL,fr}}(f)\right] \tag{11}$$

with the phase-noise power spectral density of the free-running ML $S_\varphi^{\text{ML,fr}}(f)$, and the free-running SL $S_\varphi^{\text{SL,fr}}(f)$ (in steady-state $s = j2\pi f$). The contribution of the LO$_{el}$ is neglected since a signal generator with low phase noise is utilized.

Regarding variations of the heterodyne frequency carrier, the hold-in range f_H defines the frequency range where the loop can track modulations of the local oscillators (ML or LO$_{el}$)

$$\Delta f_H = \pm\frac{1}{2\pi}\lim_{s\to 0} sG_{OL}(s) \approx \pm\frac{1}{2\pi}K_{CCO}K_{PD}2\chi\sqrt{P_m P_r}K_{PSD}F(0) \tag{12}$$

with the DC gains K_{CCO}, K_{PD}, K_{PSD} and the assumption of first-order-integrating CCO [28].

From Eqs. 11 and 12, it becomes clear that a high open-loop gain $|G_{ol}(0)|$ and wide bandwidth f_{ol} are desirable, which results in a minimum differential-phase noise and large hold-in range in lock. These goals are, however, restricted by the stability criterion of a sufficient phase margin to 180° at loop gains > 1. Regarding intra-loop delays a wide loop bandwidth is demanding. Thus, any propagation length within the loop are to be reduced to a minimum to achieve a

IOP Conf. Series: Journal of Physics: Conf. Series **1149** (2018) 012016 doi:10.1088/1742-6596/1149/1/012016

broadband suppression of phase noise. Eq. 12 shows that a proper operation of the OPLL is dependent on the gains along the whole plant (SL $\rightarrow$ PD $\rightarrow$ PSD) and even the powers of the interfering beams. Additional to the requirement of modulation frequency within the hold-in range, the stable OPLL operation relies on modulation indices $M_{\mathrm{FM}} < 2.4$ (first zero of J_0 Bessel line) [28]. At modulation frequencies near the loop bandwidth, error amplification (see servo bumps) occurs which impairs the stability of the OPLL. Thus, a wide loop bandwidth is beneficial (even for low-noise lasers) to be capable of a fast variations in the heterodyne carrier frequency for compensating dominant Doppler shifts.

For achieving the desired open-loop transfer behavior, the loop filter can be designed to correct the phase response by lead/lag compensation [28] (compare Fig. 2(a)). At MHz frequencies, the thermal cut-off frequency [30] in current-induced frequency variations of semiconductor lasers and further nonlinearities [27] impede wider loop bandwidths. Facing free-running lasers, a systematic approach of measuring the transfer function of the plant can be accomplished only with high efforts, since the instantaneous state is fairly near the set-point where linearization is possible. Thus, estimations of the open-loop transfer function $G_{\mathrm{OL}}(s)$ in lock state have to be conducted.

3.4. Mutual coherence

The relation in Eq. 12 shows that the powers of the interfering beams are directly related to the open-loop transfer function $G_{\mathrm{ol}}(s)$. Since the measurement power P_{m} strongly varies with the reflectivity of the sample (typically over orders of magnitude), the laser-Doppler vibration signal on the beat PD would be filtered (within the OPLL bandwidth) or would induce loop instabilities. Thus, we propose a separate (vibrometer) PD to detect the phase fluctuations from the laser-Doppler effect at the vibrating sample. The measurement beam for the LDV is provided by the offset-locked slave laser beam. The master laser beam is utilized as reference beam.

The additional interference of the beams on the vibrometer PD has to consider the (generated) degree of mutual coherence between the (complex) EM waves of the locked lasers which is defined [26] over vibrometer delay $\tau_{\mathrm{d}} = \Delta z/c$ as

$$g_{12}(\tau_{\mathrm{d}}) = \frac{\langle E_{\mathrm{m}}^{*}(t)\, E_{\mathrm{r}}(t+\tau_{\mathrm{d}})\rangle}{\langle E_{\mathrm{m}}(t)\, E_{\mathrm{m}}^{*}(t)\rangle \langle E_{\mathrm{r}}(t)\, E_{\mathrm{r}}^{*}(t)\rangle} \propto \frac{\langle E_{\mathrm{m}}^{*}(t)\, E_{\mathrm{r}}(t+\tau_{\mathrm{d}})\rangle}{\sqrt{P_{\mathrm{m}} P_{\mathrm{r}}}} \tag{13}$$

with the path difference Δz between the position of the beat and the vibrometer detector. The ensemble average (symbolized by $\langle\cdot\rangle$) is equal to the temporal average for ergodic processes. The asterisk denotes the complex conjugate. The (complex) degree of mutual coherence is related to the heterodyne-frequency-carrier amplitude, since the interference efficiency (see Eq. 2) is

$$\chi \propto |g_{12}(\tau_{\mathrm{d}})| \tag{14}$$

if all other sources degradation of interference efficiency can be neglected (angular misalignment, spatial mutual coherence, state of mutual polarization etc.). The position of maximum temporal coherence is generated by the OPLL at the position of the beat PD. At this position, the maximum heterodyne-frequency-carrier amplitude is achieved. Any path difference Δz to this position impairs the interference efficiency.

For many applications, a significant path difference between the PD positions might be inevitable (e.g. microscope). If the path difference exceeds certain limits, the degradation in temporal coherence leads to a collapse of the heterodyne carrier [31] to the summed linewidth of the (free-running) lasers. This effect of coherence collapse for LDV is subject to further research.

4. Setup of laser-Doppler vibrometer with heterodyning by OPLL

4.1. Realization of optical phase-locked loop

For our optical setup, we chose two visible distributed-Bragg-grating (DBR) GaAs semiconductor lasers (EYP-DBR-0633) which enable intrinsically modehop-free emission at of 632 nm with a wide tuning range of > 300 GHz. The tuning coefficient for injection current is approximately 700 MHz/mA and the injection-current modulation bandwidth is > 100 MHz. The free-running lasers show a summed linewidth of < 500 kHz. To achieve highest of optical output power, the set-point of both lasers was optimized with the knowledge of wavelength tuning behavior (over temperature and injection current) and optical output power. The wavelength mapping of both semiconductor lasers, was conducted with a wavelength meter (HighFinesse WS6-600). A commercial laser driver (Toptica SYS DC 100) stabilizes the current and temperature set-point of both semiconductor lasers which achieves reproducibility (after heat-up) of the frequency difference below ±100 MHz. For improved emission-wavelength stability, Faraday isolators protect each semiconductor laser from back-reflection from the experiment and fiber (Fig. 3).

For the PSD and loop filter, we chose commercial analog electronics (Toptica mFALC) allowing a maximum difference frequencies and, thus, heterodyne carrier frequencies of 200 MHz. The loop filter enables lead/lag compensation to a 4th-order OPLL which was optimized in experiment (Bode plot in Fig. 2(a)). The derivative behavior of the loop filter was set to a phase lead of +22° at 1.7 MHz to enhance the loop bandwidth to higher frequencies. We observed stable operation with a loop bandwidth of maximum 3 MHz. Parallel to the loop filter, a slow (unlimited) integrator controls the SL set-point to compensate for drift between the semiconductor lasers.

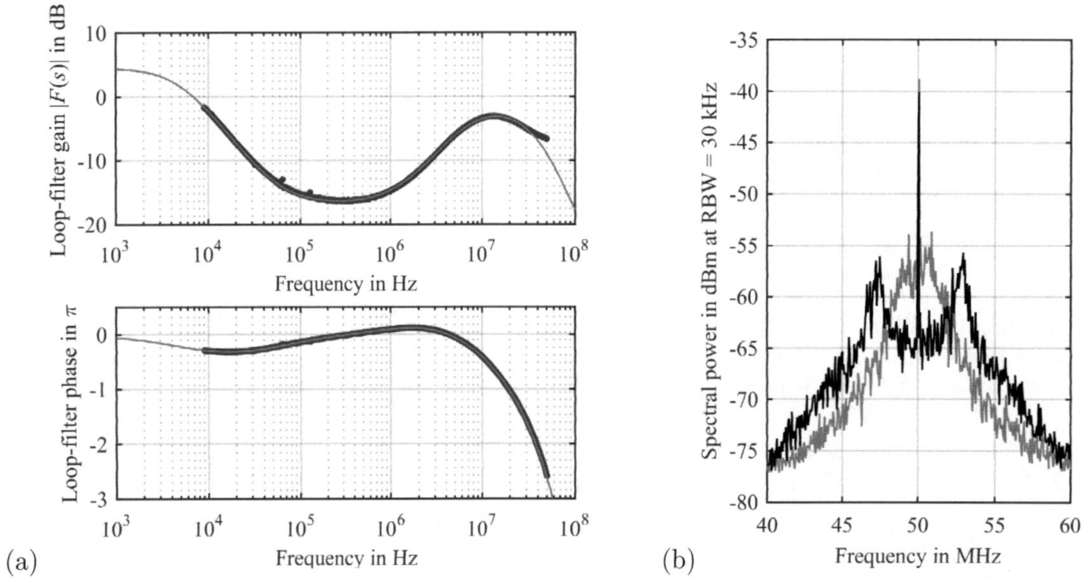

Figure 2. (a) Bode plot of the loop filter $F(s)$ optimized for the stabilization of the two DFB semiconductor lasers. The maximum phase lag is −56.7°. The derivative behavior creates a weak gain maximum at 10 MHz and a phase lead of +22° (dots are measurements, line is fitted model). (b) Beat spectrum at the beat photodetector of the 50 MHz heterodyne frequency carrier at low (grey) and high gain (black). The phase noise is reduced by the OPLL to a pedestal around the heterodyne carrier frequency. At critically-stable gain, noise peaks (servo bumps) at a frequency offset corresponding to loop bandwidth (here at 3 MHz) are developed.

In accordance with Eq. 11, the differential phase noise is suppressed within the loop bandwidth (Fig. 2(b)). This reduces the Lorentzian line-shape of the free-running semiconductor lasers to a pedestal around the heterodyne carrier frequency. For high gains, excessive noise peaks (servo bumps) appear at frequencies where the phase shift is nearly 180° (low phase margin) which occurs around at the loop bandwidth for stable OPLLs [31].

The stabilized heterodyne carrier frequency can be varied or modulated by the local oscillator up to the loop bandwidth in the MHz range. We observed tracking (without losing lock) of the LO frequency-modulated with sinusoidal deviations of $\Delta f \geq 100\,\mathrm{kHz}$ at $f_{\mathrm{mod}} = 1\,\mathrm{MHz}$ (modulation index $M_{\mathrm{FM}} = \Delta f / f_{\mathrm{mod}} = 0.1$). Which is consistent with the concept of the hold-in frequency Δf_{H} (Sec. 3.3).

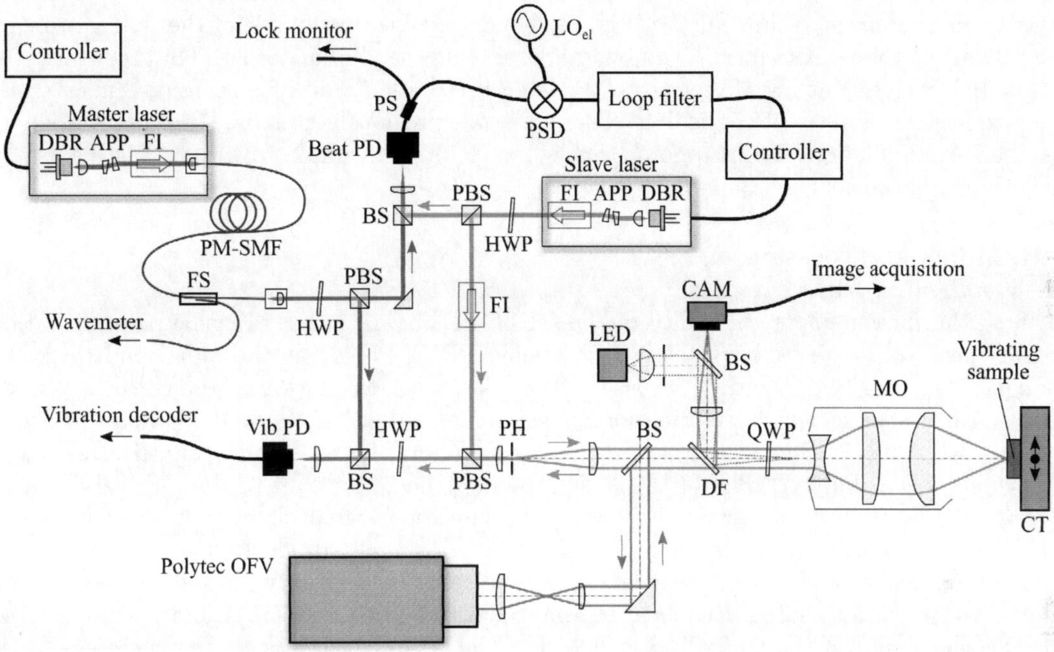

Figure 3. Setup of the LDV microscope with generation of heterodyne carrier frequency via offset-lock in an optical phase-locked loop (OPLL). As reference vibration measurement system a Polytec OFV-353S is coupled into the measurement path. DBR is semiconductor laser, APP is anamorphic prism pair, FI is Faraday isolator, PM-SMF is polarization-maintaining single-mode fiber, FS is fiber splitter, HWP is half-wave plate, QWP is quarter-wave plate, PBS is polarizing beam splitter, BS is beam splitter, Beat PD is beat photodetector, PS is power splitter, LO is local oscillator, PSD is phase-sensitive detector, Vib PD is vibrometer photodetector, PH is pinhole, DF is dichroic filter, MO is microscope objective, CAM is overview camera, LED is light-emitting diode, and CT is cross table

4.2. Heterodyne laser-Doppler vibrometer microscope

With the realized OPLL-LDV microscope (see Fig. 3) the position of the vibrometer detector is inevitably different from the beat detector. The path difference accumulates to $\Delta z \approx 1.5\,\mathrm{m}$ or a vibrometer delay $\tau_{\mathrm{d}} = \Delta z / c \approx 5\,\mathrm{ns}$ (Sec. 3.4). Despite the expected carrier collapse, the carrier-to-noise ratio at both detectors is similar which is probably due to the large OPLL bandwidth. For good OPLL stability, an additional Faraday isolator protects the beat PD from strong specular reflections from the vibrating sample.

A scanning-microscope setup is deployed to acquire the two-dimensional (out-of-plane) operating deflection shape (ODS) of microscopic samples with sufficient lateral resolution. The capability of the LDV microscope (with the commercial vibrometer) was already demonstrated with energy-harvesting MEMS [32]. For good resolution, the measurement spot size on the vibrating sample has to be smaller than the nodal distance of an ODS (Nyquist criterion). Since the spot size is directly proportional to wavelength, we chose visible lasers at 632 nm for our LDV microscope (Fig. 3). For spatial filtering of the elliptical beam shape of the semiconductor lasers, a polarization-maintaining single-mode fiber is applied for the reference beam. To efficiently exploit the emitted power, we circularized the beams with anamorphic prism pairs. Spatial filtering of the measurement beam by fiber was relinquished to avoid negative influence on the stabilization. However, a confocal pinhole in the beam expander optics allows to spatially filter the measurement beam. Further, the fundamental transverse mode of the reference beam spatially filters the measurement beam by interference on the vibrometer PD. For the future goal of RF-MEMS testing at high frequencies, the lateral resolution limit of conventional microscopy needs to be overcome systematically. Therefore, we previously proposed the application of absorbance modulation in thin photochromic layers on the sample to achieve the necessary resolution enhancement [33, 34].

5. Results and discussion

5.1. Vibration measurements

To show the measurement capability of the OPLL-LDV, we conducted single-point vibration measurements on a piezoelectric disc (PI Ceramic PRYY+1119) at the radial and thickness resonance ($f_{\mathrm{rd}} = 224\,\mathrm{kHz}$ and $f_{\mathrm{th}} = 11.4\,\mathrm{MHz}$). The theoretic thickness-resonance amplitude is $\hat{s}_{\mathrm{th}} = 1.6\,\mathrm{nm}$ (piezoelectric deformation coefficient $d_{33} = 0.4\,\mathrm{nm/V}$ and excitation amplitude $\hat{u}_{\mathrm{exc}} = 4\,\mathrm{V}$ at $50\,\Omega$). Since there are no vibration calibration objects in the MHz regime, a commercial LDV (Polytec OFV-353S) is coupled into the beam path of the OPLL-LDV as a reference vibrometric measurement system. The vibration signal directly at the vibrometer PD is analyzed with a spectrum analyzer (HP 8591E) assuming an ideal phase response in the observed frequency band. The raw-signal evaluation is further necessary, since the demodulation bandwidth of the associated decoder electronics (Polytec OFV-2500-2) is limited to 3.2 MHz. The frequency shift of the OFV-353S is 40 MHz. The OPLL-LDV was set to generate a static heterodyne frequency carrier at 50 MHz.

Since the wavelengths of both LDVs are similar (HeNe laser at 632.9 nm of OFV-353S) and the measurement points are aligned, the raw laser-Doppler signals after photodetection are comparable between the OPLL-LDV and the Polytec OFV. For the thickness resonance f_{th} we measured a carrier-to-sideband ratio $\mathrm{CSR}_{\mathrm{OFV}} = (38 \pm 1)\,\mathrm{dBc}$ and $\mathrm{CSR}_{\mathrm{OPLL-LDV}} = (40 \pm 2)\,\mathrm{dBc}$ which corresponds to a vibration amplitude $\hat{s}_{\mathrm{th}} = (1.2 \pm 0.2)\,\mathrm{nm}$ (applying Eq. 6). The signal-to-noise ratio at this resonant frequency was approximately 48 dB (at 1 Hz resolution bandwidth) for the OPLL-LDV. The measured thickness-resonance amplitude at both LDVs showed a significant difference compared to theoretic value. The reason might be a lower piezo impedance, extra load of the solder drops, and/or slight frequency deviations from resonance.

For the vibration measurement of the radial resonance, a Bessel spectrum emerges around the heterodyne carrier frequency (Fig. 4(a)) in the photodetector spectrum of both LDVs. The calculated modulation index for the radial vibration at 224 kHz is $M_{\mathrm{vib}}^{\mathrm{OFV}} = 1.34 \pm 0.04$ and $M_{\mathrm{vib}}^{\mathrm{OPLL-Vib}} = 1.40 \pm 0.06$ (applying Eq. 7, averaged from eight Bessel lines). This modulation index correspond to a vibration amplitude of approximately $\hat{s}_{\mathrm{rd}} \approx 70\,\mathrm{nm}$ at the measurement point. The cause for the deviation between the LDV measurements is probably due to misalignment and different spatial resolution (measurement spot sizes).

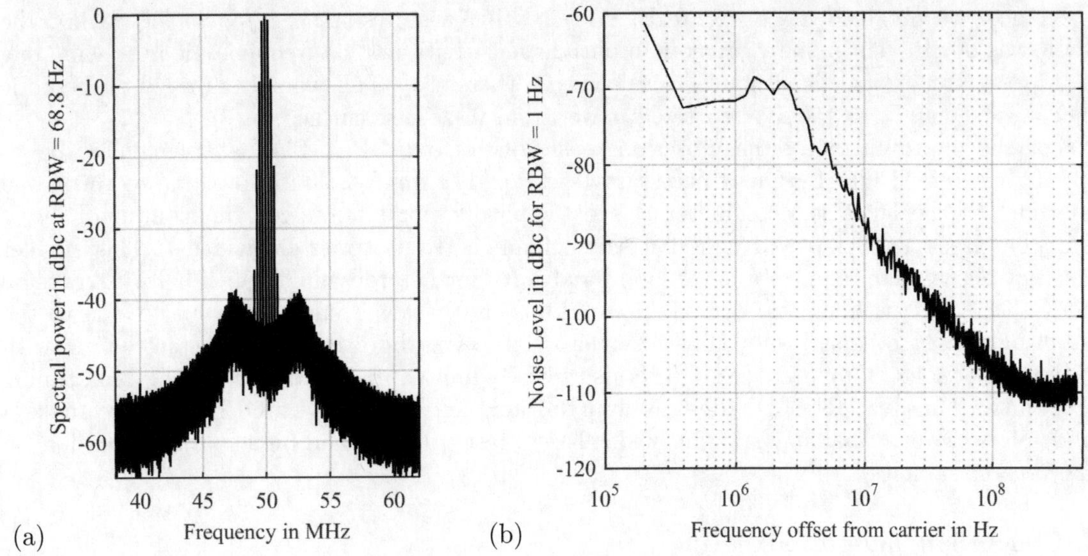

Figure 4. (a) Vibration spectrum measured with OPLL-LDV of a piezoelectric disc vibrating at the radial resonance at $f_{\mathrm{rd}} = 224\,\mathrm{kHz}$. The heterodyne carrier frequency of the OPLL-LDV was set to 50 MHz. (b) Spectral distribution of the (upper-sideband) relative noise level of the OPLL-LDV over frequency-offset from heterodyne carrier (inverse values correspond to SNR).

5.2. *Vibration sensitivity*

The differential phase noise of the heterodyne frequency carrier directly limits the vibration sensitivity of the OPLL-LDV (Eq. 11). Therefore, the vibration-amplitude sensitivity is usually expressed as noise-equivalent vibration amplitude $\hat{s}'_{\mathrm{ne}}$ at 1 Hz resolution bandwidth (RBW). Applying narrow-band approximation, the noise-equivalent vibration amplitude (before demodulation) is deduced in analogy to Eq. 6 and the frequency offset from heterodyne carrier directly corresponds to the vibration frequency ($f_{\mathrm{vib}} = f - f_{\mathrm{c}}$), yielding

$$\hat{s}'_{\mathrm{ne}}\left(f - f_{\mathrm{c}}\right) = \frac{\lambda}{2\pi\sqrt{\mathrm{SNR}'\left(f - f_{\mathrm{c}}\right)}} \tag{15}$$

with the signal-to-(single-sideband) noise ratio SNR′ at 1 Hz RBW. Note that this definition solely considers the unmodulated carrier which contains the same total power as any FM-signal with arbitrary modulation index M_{vib}.

With the assumption that phase noise is dominant over amplitude noise, the achieved (phase) noise level is estimated from the beat signal with a spectrum analyzer (shown in Fig. 4) [35]. For our realization, the pedestal around the heterodyne carrier frequency defines the sensitivity of the instrument for vibration frequencies $f_{\mathrm{vib}} \leq 50\,\mathrm{MHz}$ (see Fig. 4(b)). With the assumption of a Lorentzian line-shape (of the summed linewidth of the lasers) the pedestal slope is approximately $-30\,\mathrm{dB}$ per decade (at frequency offsets larger than the summed laser linewidth). At vibration frequencies larger than $f_{\mathrm{vib}} > 50\,\mathrm{MHz}$ the noise level due to relative intensity noise (RIN) of both semiconductor lasers (at $-110\,\mathrm{dB/Hz}$) limits the vibration sensitivity. Thus, for our OPLL-LDV the heterodyne carrier frequency must be set to $f_c \geq 50\,\mathrm{MHz}$ to achieve comparable phase-noise level at both sidebands.

The pedestal noise level is $\mathrm{SNR}' \geq 60\,\mathrm{dB}$ at offset frequencies lower than the loop bandwidth, which corresponds to a noise-equivalent vibration amplitude of $\hat{s}'_{\mathrm{ne}}\left(f_{\mathrm{vib}} < 3\,\mathrm{MHz}\right) \leq$

$100\,\mathrm{pm}/\sqrt{\mathrm{Hz}}$ (applying Eq. 15). For the first four Bessel lines of the observed radial resonance, we obtained $\mathrm{SNR}' > 20\,\mathrm{dB}$. At the thickness resonance ($f_{\mathrm{th}} = 11.4\,\mathrm{MHz}$), the SNR' was 90 dB. Thus, the vibration-amplitude sensitivity at this frequency is approximately $\hat{s}'_{\mathrm{ne}}\,(f_{\mathrm{th}}) \approx 5\,\mathrm{pm}/\sqrt{\mathrm{Hz}}$. The achieved sensitivity is, therefore, 60 dB worse compared to theoretic shot noise level ($P_{\mathrm{m}} \approx 150\,\mu\mathrm{W}$ scattered power from piezo-disc surface).

For the improvement of the vibration sensitivity of our OPLL-LDV at frequencies below 50 MHz, increased OPLL gain is necessary (see Eq. 11) which is ambitious due to the broad linewidth of the DBR semiconductor lasers. Typically single-frequency semiconductor lasers and fiber lasers with infrared emission show significantly narrower linewidths. This smaller inherent phase-noise requires a lower loop bandwidth for heterodyning, which helps to increase the OPLL gain (without loss of stability). However, the lower spatial resolution due to the wavelength may be disadvantageous for microscopy. On the other hand, heterodyning with visible gas or solid-state lasers (e.g. HeNe) is typically limited in difference frequency and tuning capability. Thus, capable LDV designs with variable heterodyning by OPLL have to make a trade-off between available narrow-linewidth lasers, fast tracking, and high spatial resolution (in microscopic samples).

6. Conclusion and outlook

We showed the benefits of a variable heterodyne frequency carrier for laser-Doppler vibrometry generated by an offset-lock of two lasers by an optical phase-lock loop. The frequency of the local oscillator directly defines the offset-lock frequency and, thus, the heterodyne carrier frequency. Any variation in the heterodyne carrier frequency does not impair the alignment of the instrument which is the problem for conventional acoustooptic frequency shifters. The heterodyne carrier frequency can be modulated to compensate dominant Doppler shifts due to relative motions between the sample and the instrument, e.g. for air-born, vehicle-born or handheld applications. In principle, the OPLL can follow modulation frequencies lower than its loop bandwidth with modulation indices < 2.4.

As a possible realization, we demonstrated a laser-Doppler-vibrometer microscope with a variable heterodyne frequency carrier. Therefore, we employed commercial electronics and visible DBR semiconductor lasers emitting at 632 nm with a summed linewidth of $< 500\,\mathrm{kHz}$. A stable offset-lock with a loop bandwidth up to 3 MHz was achieved for heterodyne frequency carrier up to 200 MHz. For a heterodyne carrier at 50 MHz, the capability of the laser-Doppler vibrometer setup was evaluated by measurements on a piezoelectric disc at its resonances. This vibration measurements were verified by a commercial laser-Doppler vibrometer. For vibration frequencies below 50 MHz the pedestal of the residual (differential) phase-noise limits the vibration sensitivity of our vibrometer. The achieved vibration sensitivity at vibration frequencies below the loop bandwidth of 3 MHz is $100\,\mathrm{pm}/\sqrt{\mathrm{Hz}}$. Higher loop gain could increase the vibration sensitivity (up to the loop bandwidth) to the cost of loop stability.

Further improvements in sensitivity require the choice of laser sources with inherently narrower linewidth. The achieved vibration sensitivity in the low MHz regime is far away from shot-noise level which is reached by heterodyne LDVs with an acoustooptic frequency shifter and a single narrow-linewidth laser source. However, in applications, where compensation of large motions is desirable, the concept of heterodyning by OPLL enables flexible wideband-tracking capability to the disadvantage of lower vibration sensitivity.

Beyond the phase-noise pedestal ($> 50\,\mathrm{MHz}$), the relative intensity noise of the semiconductor lasers currently limits the vibration sensitivity. Thus, in the next steps we plan to increase the capability of the vibrometer microscope for generation of heterodyne frequency carriers in the GHz regime, where common acoustooptic frequency shifters are inefficient. Such laser-Doppler vibrometers can enable vibration detection at high frequencies for RF-MEMS testing.

IOP Conf. Series: Journal of Physics: Conf. Series **1149** (2018) 012016 doi:10.1088/1742-6596/1149/1/012016

References

[1] Drain L E 1980 *The Laser Doppler technique* (A Wiley-Interscience publication) (Chichester: Wiley)
[2] Rembe C, Siegmund G, Steger H and Wörtge M 2007 Measuring MEMS in Motion by Laser Doppler Vibrometry *Optical Inspection of Microsystems* Optical science and engineering ed Osten W (Boca Raton, FL: CRC/Taylor & Francis) pp 245–292
[3] Buchhave P 1975 *DISA Information* 15–20
[4] Halkon B J and Rothberg S J 2017 *Optics and Lasers in Engineering* **91** 16–23
[5] Xu J and Stroud R 1992 *Acousto-optic devices: Principles, design, and applications* A Wiley Interscience publication (New York: Wiley)
[6] Jacobs S 1962 Technical note on heterodyne detection in optical communication: TRG-168-TDR-1
[7] Teich M C 1968 *Proceedings of the IEEE* **56** 37–46
[8] Graves R E and Stokes L S 1975 Optical heterodyne receiver with phase or frequency lock (US 3975628A)
[9] McManamon P F 2012 *Optical Engineering* **51** 060901
[10] Satyan N, Vasilyev A, Rakuljic G, Leyva V and Yariv A 2009 *Optics Express* **17** 15991–15999
[11] Kikuchi K 2016 *Journal of Lightwave Technology* **34** 157–179
[12] Deninger A J, Göbel T, Schönherr D, Kinder T, Roggenbuck A, Köberle M, Lison F, Müller-Wirts T and Meissner P 2008 *Review of Scientific Instruments* **79** 044702
[13] Kwee P, Bogan C, Danzmann K, Frede M, Kim H, King P, Pöld J, Puncken O, Savage R L, Seifert F, Wessels P, Winkelmann L and Willke B 2012 *Optics Express* **20** 10617–10634
[14] Santarelli G, Clairon A, Lea S N and Tino G M 1994 *Optics Communications* **104** 339–344
[15] Holzwarth R, Udem T, Hänsch T W, Knight J C, Wadsworth W J and Russell P S 2000 *Physical review letters* **85** 2264–2267
[16] Wynands R, Coste O, Rembe C and Meschede D 1995 *Optics Letters* **20** 1095
[17] Tanaka H and Sonehara T 2002 *Review of Scientific Instruments* **73** 1998–2010
[18] Sternkopf C, Diethold C, Gerhardt U, Wurmus J and Manske E 2012 *Measurement Science and Technology* **23** 074006
[19] Blume G, Schiemangk M, Pohl J, Feise D, Ressel P, Sumpf B, Wicht A and Paschke K 2013 *IEEE Photonics Technology Letters* **25** 550–552
[20] Rembe C 2008 Retardation effects in laser-Doppler measurements *8th Int. Conference on Vibration Measurements by Laser and Noncontact Techniques - AIVELA* (*SPIE Proceedings* vol 7098) ed Tomasini E P (SPIE) p 709813
[21] Mäusl R 1992 *Analoge Modulationsverfahren* 2nd ed (*Telekommunikation* vol 1) (Heidelberg: Hüthig)
[22] Carson J R 1922 *Proceedings of the I.R.E* **10** 57–64
[23] 2011 *Methods for the calibration of vibration and shock transducers - Part 41: Calibration of laser vibrometers* ISO 16063-41:2011(E)
[24] Lauermann M, Weimann C, Knopf A, Elder D L, Heni W, Palmer R, Korn D, Schindler P, Koeber S, Alloatti L, Yu H, Bogaerts W, Dalton L R, Rembe C, Leuthold J, Freude W and Koos C 2014 Integrated Silicon-Organic Hybrid (SOH) frequency shifter *Optical Fiber Communication Conference* (Washington, D.C.: OSA) p Tu2A.1
[25] Wolf E 2004 *Progress in Optics* 1st ed (*Progress in Optics* vol 46) (Elsevier textbooks)
[26] Saleh B E A and Teich M C 2013 *Fundamentals of Photonics* 2nd ed Wiley series in pure and applied optics (Chicester: Wiley)
[27] Liang W 20 May 2008 *Study of optical phase lock loops and the applications in coherent beam combining and coherence cloning* Dissertation (ph.d.) California Institute of Technology
[28] Gardner F M 1979 *Phaselock techniques* 2nd ed (New York: Wiley)
[29] Yariv A 2005 *Optics Letters* **30** 2191
[30] Petermann K 1991 *Laser diode modulation and noise* (*Advances in optoelectronics* vol 3) (Dordrecht: Kluwer)
[31] Satyan N, Liang W and Yariv A 2009 *IEEE Journal of Quantum Electronics* **45** 755–761
[32] Kowarsch R, Janzen J, Rembe C, Cho H and Choo H 2017 *tm - Technisches Messen* **84** 131–137
[33] Kowarsch R and Rembe C 2016 Modellierung der Auflösungssteigerung mittels photochromer Schichten für die nanoskopische Laser-Doppler-Vibrometrie *XXX. Messtechnisches Symposium* ed Zimmermann S (De Gruyter Oldenbourg) pp 85–95
[34] Kowarsch R, Geisler C, Egner A and Rembe C 2018 *Optics Express* **26** 5327–5341
[35] Riehle F 2004 *Frequency standards: Basics and applications* (Weinheim: Wiley-VCH)

IOP Conf. Series: Journal of Physics: Conf. Series **1149** (2018) 012017 doi:10.1088/1742-6596/1149/1/012017

Both the phase modulation and frequency shift evaluation on laser vibrometer

Haolin Sun, Zhiguo Liang, Jun Yang, Dazhi Zhang and Huafeng Li

National Key Laboratory of Science and Technology on Metrology & Calibration, Changcheng Institute of Metrology and Measurement, Huanshan Village, Wenquan Town, Haidian District, Beijing 100095, China

Haolin Sun's e-mai address : sunhaolin24@163.com
Zhiguo Liang's e-mail address : Lzg304@sina.com
Jun Yang's e-mail address : yangjun@cimm.com.cn
Dazhi Zhang' e-mail address : zhangdazhi@cimm.com.cn
Huafeng Li's e-mail address : lihuafeng@cimm.com.cn

Abstract. Sinusoidal phase-modulated signal light through the interferometer can produce a beat signal. Moreover, its amplitude monotonically changes with the signal light frequency. So the beat signal amplitude can be used to measure laser-Doppler frequency shift. In addition to the beat signal, the phase-modulated signal also contains a direct current signal, and it still contains a large amount of frequency shift information, but the information is not utilized, resulting in the waste of information. In this paper, this kind of phase-modulated laser-Doppler frequency shift measurement method is improved to simultaneously utilize the useful information in the DC and beat signal for the frequency shift measurement. The specific method is to use the ratio of beat signal amplitude to DC signal amplitude to define a new parameter used in frequency shift measurement. The signal light intensity terms in DC and beat signal can be eliminated, so the improved measurement method does not need to measurement the signal light intensity, which makes its structure further simplified. By comparing the frequency change curves between the newly defined parameter and the beat signal amplitude theoretically, we find that they have the same distribution rule. In theory, by comparing the measurement sensitivity curves, it is proved that the improved phase-modulated frequency shift measurement method has higher measurement sensitivity and dynamic range than the un-improved one.

1. Introduction

Laser-Doppler frequency shift measurement method can be used to acquire information (e.g. vibration and speed) of the target. The phase-modulated frequency-shift measurement method has low requirements on light source and optical system and thereby it is easy to be realized. Extracting amplitude (or phase) of phase-modulated beat signal based on coherent detection method can reduce the working bandwidth of the system effectively, thus decreasing the noise power and bringing extremely high measurement accuracy. Nevertheless, The DC component of phase-modulated signal also contains abundant useful information, but they are not used effectively, resulting in the information waste.

IOP Conf. Series: Journal of Physics: Conf. Series **1149** (2018) 012017 doi:10.1088/1742-6596/1149/1/012017

In this paper, this kind of phase-modulated Doppler-frequency-shift measurement method is improved to simultaneously utilize the useful information in the DC and beat signal for the frequency shift measurement, by using the FM signal stimulation, the measurements of phase shift are gotten. The method in this paper can be used to evaluating the frequency shift on laser vibrometer.

2. The phase-modulated frequency shift measurement method

2.1. Principle of phase modulation

The simple-frequency signal after sine phase modulation can be expressed as:

$$E_{in} = E_0 \exp\left[i(wt + \beta \sin \quad t\right] \tag{1}$$

Where E_0 and w are the field intensity and angular frequency of the signal light. Ω and β are the sine-phase-modulated angular frequency and modulation index.

Let the modulation signal light run through the interferometer and the photoelectric detector was used for measurement. The output electric signal can be expressed as:

$$i = i_d + i_\Omega + i_{2\Omega} \tag{2}$$

Where i_d is the DC signal. i_Ω is the beat signal of the once modulation frequency (generated by the carrier wave and two first-order sidebands). $i_{2\Omega}$ is the beat signal of the double modulation frequency 2Ω (generated by the positive and negative first-order sidebands. It was overlooked in this paper because it is weak).

The period of sine phase modulation was defined T_Ω=1/Ω, it can be seen from the Equation (2) that at the integral time T=2nT_Ω(n=1,2,3,⋯) of the out signal $i(t)$ of the detector, the integral results of AC components $i_\Omega(t)$ and $i_{2\Omega}(t)$ are zero. In this way, the DC component i_d can be extracted:

$$i_d(w) = \frac{1}{T}\int_0^T i(t)dt \propto E_0^2 h'(w) \tag{3}$$

Since the sine-phase modulation frequency Ω is known, it is easy to generate two orthogonal reference signals with the same frequency of beat signal $i_\Omega(t)$:

$$\begin{cases} i_{rs}(t) = \sin(\Omega t) \\ i_{rc}(t) = \cos(\Omega t) \end{cases} \tag{4}$$

Through the correlation operation between two reference signals in the Equation (4) and $i(t)$, it can be get:

$$R_{srs(c)}(\tau) = \frac{1}{T}\int_0^T i(t)\cdot i_{rs(c)}(t-\tau)dt \tag{5}$$

The correlation operation of three components of i(t), namely, i_d, i_Ω, and $i_{2\Omega}$, was implemented. When the integral time is T=nT_Ω(n=1,2,3,⋯), the integral results of i_d and $i_{2\Omega}$ are zero. Then, the equation (5) can be rewritten as:

$$\begin{cases} R_{srs}(\tau) \propto \dfrac{E_0^2}{2}|A_0|\cos(\Omega\tau - \phi_0) \\ R_{src}(\tau) \propto \dfrac{E_0^2}{2}|A_0|\sin(\Omega\tau - \phi_0) \end{cases} \tag{6}$$

it can be seen from the Equation (6) that:

$$E_0^2|A_0| \propto 2\sqrt{\left[R_{srs}(\tau)\right]^2 + \left[R_{src}(\tau)\right]^2} \tag{7}$$

$$\phi_0 = \text{-arctan}\left[R_{src}(0) / R_{srs}(0)\right] \tag{8}$$

The DC signal $E_0^2 h'$ and the amplitude $E_0^2|A_0|$ and phase ϕ_0 of the beat signal of once modulation frequency can be gained from the above signal processing method. Meanwhile, the light intensity E_0^2 of signal light can be measured, thus enabling to realize Doppler-frequency-shift.

2.2. *Improvement method*

To use useful information in i_d and i_Ω simultaneously, a new frequency demodulation parameter was defined. Its absolute value $|A_n(w)|$ can use $R_{srs}(\tau)$, $R_{src}(\tau)$ and i_d to get :

$$\begin{aligned}|A_n(w)| &= \frac{2\sqrt{[R_{srs}(\tau)]^2+[R_{src}(\tau)]^2}}{i_d} \\ &\propto \frac{E_0^2|A_0(w)|}{E_0^2 h'(w)} \\ &= \frac{|A_0(w)|}{h'(w)}\end{aligned} \tag{9}$$

It can be seen from the Equation (8) that in the definition process of $|A_n(w)|$, the signal light intensity E_0^2 in i_d and i_Ω is cancelled simultaneously. This implies that the frequency-shift measurement based on $A_n(w)$ can solve above mentioned problems by indirect use of E_0^2 of i_d only, but doesn't need to measure E_0^2 directly. Moreover, since $A_n(w)$ is the function of $A_0(w)$ and $h'(w)$, it is used in the measurement, this means that the frequency-shift information in i_d and i_Ω of the phase-modulated signal are used simultaneously.

3. Simulations

For the comparison, different shapes of shift variation curves of A_n and A_0 are shown in Fig.1(a). It can be seen from Fig.2(a) that there's one upper peak and one low peak on the curve of $A_n(w)$ and the monotonic changes of the curve between these two peaks run through the origin. The overall shape of this section is same with that of the curve of $A_0(w)$. This reveals that frequency-shift measurement based on the curve of $A_n(w)$ is feasible just like the original phase modulation method based on the curve of $A_0(w)$, including using the origin as the working point.

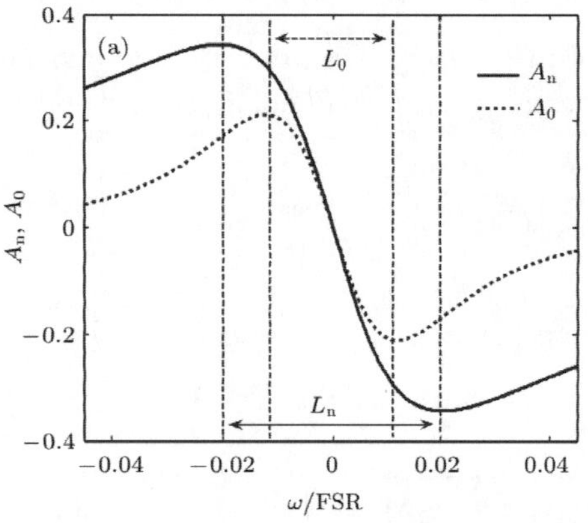

(a) $A_n(w)$ and $A_0(w)$

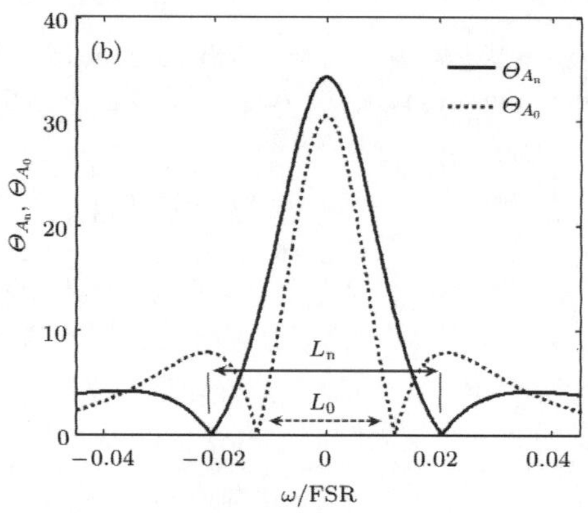

(b) $\theta_{A_n(w)}$ and $\theta_{A_0(w)}$

Fig.1 Curves of A_n and A_0 and their absolute sensitivities θ_{A_n} and θ_{A_0} changing with frequency

There are some important differences between the curves of $A_n(w)$ and $A_0(w)$. The peak-peak width L_n of the curve of $A_n(w)$ is larger than the peak-peak width L_0 of the curves of $A_0(w)$ and the curve of $A_n(w)$ slope is larger than that of the curve of $A_0(w)$. These determine the larger dynamic range and higher sensitivity of frequency-shift measurement based on the curve of $A_n(w)$ compared with those based on the curve of $A_0(w)$.

For the quantitative comparison on dynamic ranges and sensitivity of frequency-shift measurement based on the curves of $A_n(w)$ and $A_0(w)$, the variation curves of their absolute sensitivities $\theta_{A_{n(0)}(w)}$ are shown in Fig.1 (b), where $\theta_{A_{n(0)}(w)}=\left|dA_{n(0)}(w)/dw\right|$. Both the $\theta_{A_n(w)}$ and $\theta_{A_0(w)}$ curves reach the peak at the working point (origin) and then decrease to zero gradually with the further increase of shift. The frequency-shift position of $\theta_{A_n}=\theta_{A_0}=0$ in Fig 1(b) is corresponding to the shift position of peaks of the curves of $A_n(w)$ and $A_0(w)$ in Fig.1 (a), which determines the dynamic range of frequency-shift measurement. The shift range $(-\frac{L_n}{2}\sim\frac{L_n}{2})$ of the curve of $A_n(w)$ is larger than the range $(-\frac{L_0}{2}\sim\frac{L_0}{2})$ of curve of $A_0(w)$, and in this shift range, $\theta_{A_n}>\theta_{A_0}$.

The phase-modulated frequency-shift measurement method was improved by defining a new frequency demodulation parameter. It not only can retain the original working mode of phase modulation, but also doesn't need to measure the signal light intensity, thus simplifying the system structure and reduce the noise channel. Furthermore it can increase the dynamic range and sensitivity of measurement.

4. Experiment and result

The experimental apparatus of improved phase-modulated frequency measurement method is shown in Fig.2.

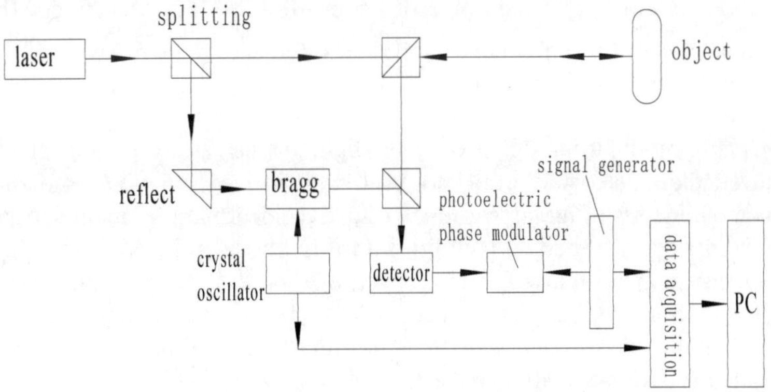

Fig.2 Experimental research device schematic

By making the output light frequency change at the step length of 3MHz in the range of w_0-150MHz~ w_0+150MHz. After every change of emergent light frequency, the data acquisition card collect signals output by the photoelectric prober at the sampling frequency of 500MHz, subsequently, sampling data of the photoelectric prober were calculated, getting the values of DC signal E_0^2h' and the amplitude $E_0^2\left|A_0\right|$, phase ϕ_0 of the beat signal. Finally, the measured values of A_0 and A_n under different frequency can be gained by the data above and the Equation (6)~(9).

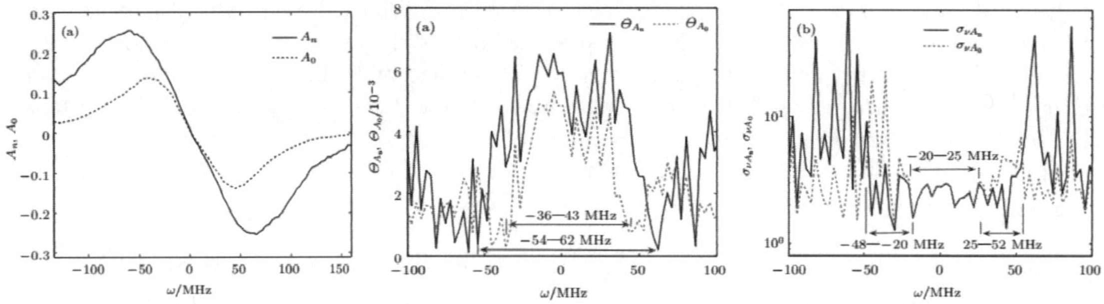

(a) A_n and A_0 (b) $\theta_{A_n(w)}$ and $\theta_{A_0(w)}$ (c) $\sigma_{vA_n}(w)$ and $\sigma_{vA_0}(w)$

Fig.3 Measurement average values curves of A_n and A_0, and curves of frequency-shift measuring sensitivities $\theta_{A_{n(0)}(w)}$ and standard deviations $\sigma_{vA(0)}(w)$

A_n and A_0 at each frequency-shift position were measured by many times and the corresponding average values were calculated. It can be seen by comparing Fig.1 (a) and Fig.3 (a) that the measured curves of $A_n(w)$and $A_0(w)$ are very similar with the theoretically calculated curves witch prove the accuracy of theoretical analysis.

Partial differential operation of the measured curves of $A_n(w)$ and $A_0(w)$ in Fig.3 (a) was carried out by numerical calculation, through which the absolute sensitivity ($\theta_{A_n(w)}$and $\theta_{A_0(w)}$) curves of shift measurement were gained (Fig.3(b)). The dynamic range (about -54~62MHz) for shift measurement of A_n is larger than that of A_0 (about -36~43MHz). In the whole dynamic range, there's $\theta_{A_n(w)} > \theta_{A_0(w)}$, this agrees with the distribution law of theoretical curves in Fig.1 (b). It can be seen from distributions of curves($\sigma_{vA_n}(w)$and $\sigma_{vA_0}(w)$) that σ_{vA_n} and σ_{vA_0} are basically same in a small frequency shift range of -25~25MHz, but σ_{vA_n} is smaller in the range of about -48~-20MHz and 25~52MHz. This proves that A_n contributes higher accuracy in frequency shift measurement compared with A_0.

5. Conclusions

DC component of phase-modulation signal includes light intensity information and frequency shift information. However, these information are not used well in previous phase-modulation frequency shift measurement method, resulting in waste of effective information. In this paper, the original phase-modulation frequency shift method is improved to use the useful information in DC component of phase-modulation signals reasonably. The improved method requires no light intensity detector. It not only has simplified structure, but also reduces nose entrance into channels. Experimental results proved the accuracy of the theory, and by using the improved method, the dynamic measurement range and measurement accuracy are all increased.

References

[1] Köning R, Wimmer G, Witkovský V. Ellipse fitting by nonlinear constraints todemodulate quadrature homodyne interferometer signals and to determine thestatistical uncertainty of the interferometric phase. *Measurement Science& Technology*, 2014, 25(11):115001.

[2] Pozar T, Možina J. Enhanced ellipse fitting in a two-detector homodyne quadrature laser interferometer. *Measurement Science & Technology*, 2011,22(8):085301.

[3] Jeng Hu P C, Zhu J H, Guo X B. Compensation for the Variable Cyclic Error in Homodyne Laser Interferometers. Sensors, 2015, 15(2):3090-3106.

[4] KIM C, LI G, LI J. Numerical analysis on effective electric field penetration depth for interdigital impedance sensor. *Journal of Physics*, 2013, 418(1):2020.

[5] Methods for the calibration of vibration and shock transducers Part 41:Calibration of laser vibrometers:ISO16043-41[S]. 2011.

[6] C. Vuye, S. Vanlanduit, F. Presezniak, G. Steenackers, P. Guillaume. Optical measurement of the dynamic strain field of a fan blade using a 3D scanning vibrometer. *Optics and Lasers in Engineering*, 2011, 49(7): 988-997.

[7] Matthew S. Allen, Michael W. Sracic. A new method for processing impact excited continuous-scan laser Doppler vibrometer measurements. *Mechanical Systems and Signal Processing*, 2010, 24(3): 721-735.

[8] Shibata N, Na kazo no A, Inoue, Y. Interference between two orthogonally polarized modes traversing a highly brief ringent air-silica microstructure fiber. *Light wave Technology*, 2005, 23(3): 1244-1252.

[9] Shaoli Zhu, Wei Zhou, Yunfeng Song. Detecting oscillation amplitude and defects of hard disk rotating in high speed by laser Doppler technique. *Measurement*, 2012, 45(1):74-78.

[10] P. Castellini, M. Martarelli, E. P. Tomasini. Laser Doppler Vibrometry: Development of advanced solutions answering to technology's needs. *Mechanical Systems and Signal Processing*, 2006, 20(6): 1265-1285.

IOP Conf. Series: Journal of Physics: Conf. Series **1149** (2018) 012018 doi:10.1088/1742-6596/1149/1/012018

A simple method for amplitude/phase calibration of tri-axial accelerometers in the low frequency range

G. D'Emilia*, A. Gaspari, E. Natale

Department of Industrial and Information Engineering and Economics, Via Gronchi, 18, 67100 L'Aquila, ITALY

E-mail: giulio.demilia@univaq.it

Abstract. A simple method for amplitude/phase calibration of tri-axial accelerometers in the low-frequency range is proposed, based on a linear slide, used to excite all the axes of the accelerometer at the same time, and a Laser Doppler Vibrometer (LDV) as a reference. For the phase evaluation a cross-correlation analysis has been carried out. The procedure includes many validation actions in order to verify the correctness of data processing and to reduce the calibration uncertainty.
Results show that the phase is a critical aspect to consider, more than the amplitude, and the comparison with the theoretical model is useful to verify the hypotheses. Different behaviors result depending on the elements of the measurement chain and not only on the type of accelerometer.

1. Introduction

Due to the wide use of accelerometers in the most varied applications, many kinds of accelerometers have been designed, using different physical principles, and having different characteristics, which are often conveniently integrated in practical applications.

Among all, the capacitive (and in particular MEMS capacitive) and the piezoelectric accelerometers are the most widespread kinds of accelerometers.

Micro-electro-mechanical system (MEMS) accelerometers have been receiving particular attention because of their low cost and small size [1]. MEMS capacitive accelerometers have the benefits of low power consumption, low sensitivity to temperature changes, low cost, and they are suitable for measuring low-frequency acceleration, but suffer from a limited bandwidth and a low signal-to-noise ratio.

Piezoelectric accelerometers are characterized by low noise output, and wide frequency range, but their cost is significantly higher and they are unable to measure static and quasi-static acceleration [4].

For this reason, they could be in crisis in the lower side of their frequency range, and the information of data sheets is not always comprehensive in this regard.

In literature, the integration of this kinds of accelerometers is proposed, to combine the best features of both [4].

In the field of the civil structures monitoring, the combination of these kinds of sensors could be useful. In fact, the need of covering wide areas through a multi-sensor network, in order to include different buildings, sets the requirement of low-cost solutions suitable in the low-frequency range [5-6], like capacitive MEMS accelerometers. Piezoelectric accelerometers are reliable and stable sensors and could be used as a reference for the MEMS sensors.

Knowing the behavior of these sensors in the low-frequency range, in terms of amplitude and phase, could be a useful information for the design of a distributed sensors network.

However, many other applications arise in different fields of engineering, where a single or a network of accelerometers are used, to monitor the system of interest in the low frequency range as, for instance, energy production [7], geotechnical applications [8], human vibration and bio-dynamics [9]. For this reason, micro accelerometers [10-11] and calibration methods [12-13] have been specifically designed for the low frequency range, but they are limited to the amplitude evaluation.

Furthermore, it should be considered that for piezoelectric accelerometers, the low-frequency response of the measuring set-up depends primarily on the amplifier used in the measuring system. The output of a piezoelectric accelerometer should be corrected by eliminating gain and phase shift caused by the amplifier [14].

Alternatively, the behavior of the entire measurement chain could be experimentally investigated, taking into account the contribution of all its elements.

Based on the above considerations MEMS accelerometers, in comparison to high-performance piezo-electric transducers, exhibit lower accuracy and, for this reason, a calibration of both amplitude and phase is particularly important for a reliable vibration analysis [2, 3].

The goal of this paper is to investigate the behavior of tri-axial piezoelectric and capacitive MEMS accelerometers in the lower limit of the operating frequency range, from the point of view of sensitivities and phases, considering the whole measuring chains.

For this purpose, a calibration method is proposed, based on a linear slide, used to excite the accelerometer being tested in the low frequency range, and a Laser Doppler Vibrometer (LDV), used as a reference. The low frequency behavior of the LDV is also a key-aspect to be considered and it is studied in [15]. The proposed method promises to be interesting, being simpler than the standard one, to be used for in-line and on-line calibration [12-13].

2. Materials and methods

This paper aims to investigate the behavior, in terms of sensitivity and phase, of capacitive MEMS and IEPE (Integrated Electronics Piezo-electric) accelerometers in the frequency range 0.5 – 12 Hz, considering the whole measuring chains.

For this purpose, a calibration bench is designed, based on a LDV VS 100, by Ometron, as a reference (Figure 1), and a simple procedure is developed, requiring a single set of tests to excite all the axes at the same time, and based on a data processing algorithm implemented in Matlab environment.

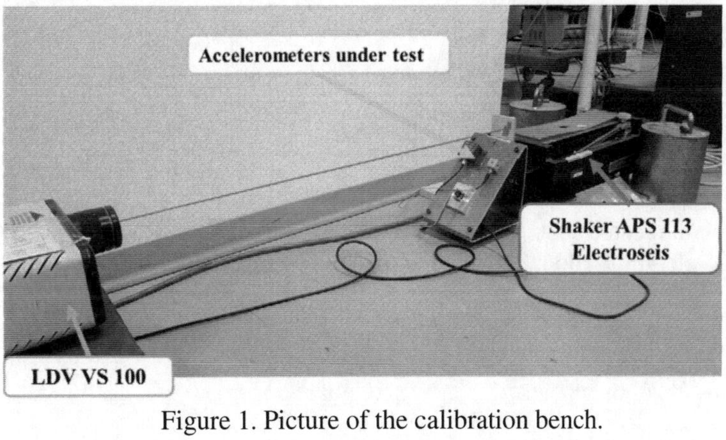

Figure 1. Picture of the calibration bench.

The test bench used is a vibrating table with a horizontal linear slide, the APS 113 ELECTRO-SEIS shaker. It is a long-stroke, electro-dynamic force generator specifically suitable for low-frequency vibration testing. The slide is moved according to a sinusoidal law.

The data acquisition system (DAQ) is the Compact RIO 9004 by National Instruments, together with the module NI 9234 for both IEPE and non-IEPE sensors, and the analog input module NI 9215.

The accelerometers tested and their measuring chains are shown in Table 1. The first two accelerometers are of the same type, but they have different sensitivities.

It must be noticed that the LDV output is acquired by the same module of the accelerometer under test to guarantee the synchronism of the acquisitions.

The accelerometer being tested is mounted on an inclined plate at an angle of 45° with respect to the horizontal plane on which the motion is realized. Furthermore, the accelerometer is rotated on the clamp surface with an angle of 45°, in order to simultaneously excite the three axes, with a single horizontal sinusoidal acceleration (Figure 2).

Table 1. Accelerometers under test and their measuring chains.

	Type of accelerometer	Acquisition module	Acquisition mode (accelerometer)	Acquisition mode (LDV)
1	Tri-axial IEPE accelerometer	IEPE module	IEPE AC coupling	AC coupling
2	Tri-axial IEPE accelerometer	IEPE module	IEPE AC coupling	AC coupling
3	Tri-axial MEMS capacitive accelerometer	Analog input module	DC coupling	DC coupling
		IEPE module	AC coupling	AC coupling
		IEPE module	DC coupling	DC coupling

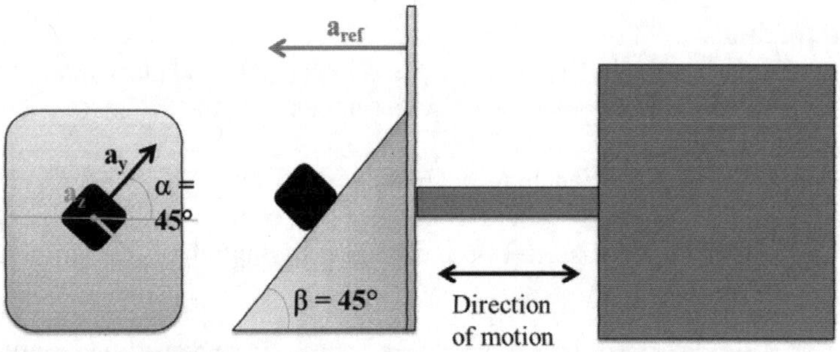

Figure 2. Scheme of the measuring set-up.

2.1. Sensitivity evaluation

Considering that the LDV output is proportional to the velocity, it must be differentiated to obtain the acceleration signal.

This acceleration, indicated as a_{ref} in Figure 3, has to be projected along the directions of the axes. So, the reference accelerations along the three axes are obtained as follows:

$$a_x = a_{ref} \cdot \cos(45°) \cdot \cos(45°) \quad (1)$$
$$a_y = -a_{ref} \cdot \cos(45°) \cdot \cos(45°) \quad (2)$$
$$a_z = a_{ref} \cdot \sin(45°) \quad (3)$$

The sensitivities are estimated by dividing the amplitudes of the x, y and z outputs, by the amplitudes of the reference signals, all evaluated by an FFT analysis:

$$S_{xx} = \frac{V_x}{a_x} \qquad S_{yy} = \frac{V_y}{a_y} \qquad S_{zz} = \frac{V_z}{a_z} \tag{4}$$

The transversal sensitivities are neglected in this analysis, in the hypothesis that they are less than 5%, but their effects are considered in the uncertainty budget.

2.2. *Phase evaluation*

With regard to the phase evaluation, the reference signals and the outputs of the accelerometer under test (Figure 3) have been compared by a cross correlation analysis, executed in Matlab environment by the "finddelay" function.

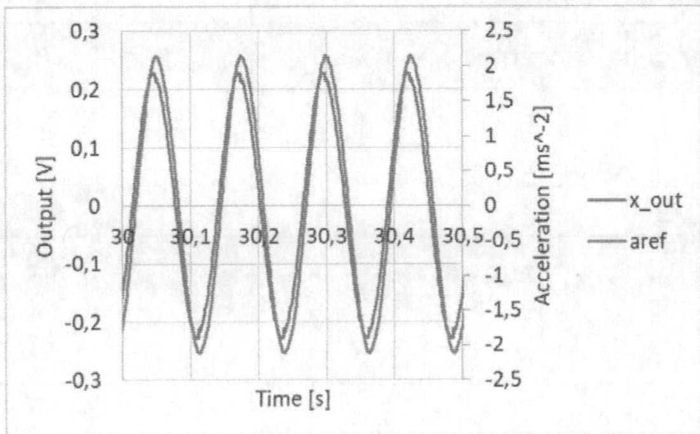

Figure 3. Example of delay between the reference acceleration and the output signal of the accelerometer under test.

In this way, the phase delay of the whole measuring chain with respect to the reference one is tested.

In the phase evaluation, the phase of the reference is a critical aspect to consider [15]. As a first approximation, the LDV phase behavior has been considered ideal (i.e. negligible phase effect of the frequency of vibration). That hypothesis will be discussed in the Result section.

2.3. *Validation*

Validation actions are necessary to confirm the correctness of the procedure and of data processing [16-17].

In particular, the following controls have been carried out:

- Check of the delay between the same signals acquired by different channels, both in DC (Direct Coupling) or AC (Capacitive Coupling) mode; in this case, the "finddelay" function provides a null delay.
- Check of the delays evaluated by the "finddelay" function, for several digitally built signals with different delays; the "finddelay" algorithm provides the digital set delay.
- Cross comparisons among different sensors and channels; the relative delays provide consistent results.
- Agreement of results with the indications of Lissajous figures.
- Systematic comparison between experimental results and theoretical models.

IOP Conf. Series: Journal of Physics: Conf. Series **1149** (2018) 012018 doi:10.1088/1742-6596/1149/1/012018

2.4. Uncertainty evaluation

In Table 2 the uncertainty budget for the sensitivity estimation is shown. It can be seen that the major contributions are represented by the uncertainty of the positioning angles of the accelerometers and the transversal sensitivities, whose uncertainty is in the order of 2 % in the hypothesis that they are less than 5% with respect to the main sensitivities. The method is not suitable for applications where transversal sensitivity effects could be relevant.

Table 2. Uncertainty budget for the sensitivity evaluation.

Main sources of uncertainty	**Relative standard uncertainty (%)**
Repeteability	0.2
Reproducibility	0.80
Reference uncertainty	0.30
Angle α	X 1.2 Y 1.2 Z 0.0
Angle β	X 0.21 Y 0.21 Z 0.42
Transversal sensitivities	2.0

The uncertainty budget for the phase evaluation is shown in Table 3. In this case, the positioning angles do not influence the results. The main contributions are the repeatability, the resolution linked to the sampling rate and the reference uncertainty.

Table 3. Uncertainty budget for the phase evaluation. f_c is the sampling rate, f is the oscillation frequency.

Main sources of uncertainty	**Standard uncertainty [°]**
Repeteability	2°
Resolution	$(f/f_c \cdot 360°)/(2 \cdot \text{radq}(3))$
Reference uncertainty	2°

3. Results

3.1. Sensitivity evaluation

The results for sensitivities will be represented as percentage differences with respect to the value at 12 Hz. For instance, for the x-axis:

$$DS_{xx}(f) = 100 \cdot [S_{xx}(f) - S_{xx}(12)]/S_{xx}(12) \qquad (5)$$

The results are similar, for both the IEPE accelerometers considered. The sensitivities for all axes decrease by 15% compared to the values obtained at 12 Hz, when the frequency is reduced to 0.5 Hz. (Figure 4).

The results are consistent with those obtained by a calibration bench of the National Institute of Metrological Research (INRiM).

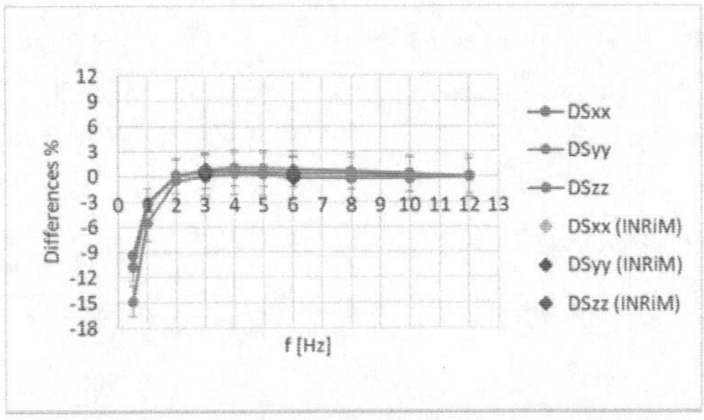

Figure 4. IEPE accelerometer, IEPE Module, IEPE AC coupled mode.

If the capacitive sensor is coupled DC mode, the differences with respect to 12 Hz remains within ± 3% for both Analog input module and IEPE module (Figure 5(a-b)). The sensitivities for all axes decrease by 6% compared to the values obtained at 12 Hz, when the frequency is reduced to 0.5 Hz.

If acquired in AC coupled mode (Figure 6(c)), for frequencies 0.5-1 Hz the sensitivities decrease by 0÷6% compared to the value obtained at 12 Hz.

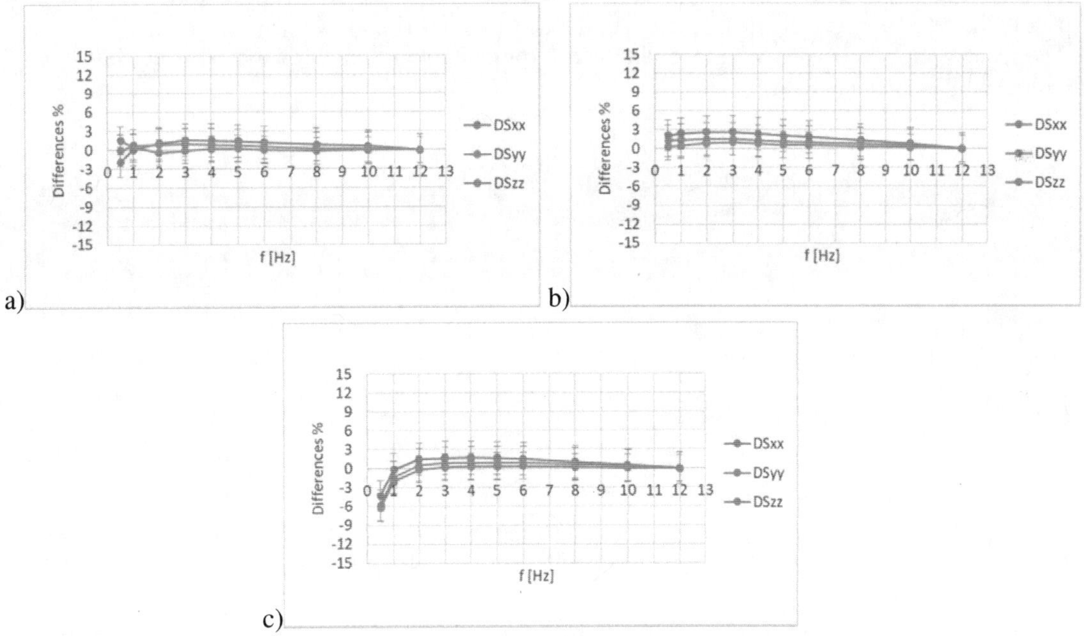

Figure 5. MEMS capacitive accelerometer. a) Analog input Module 9215; b) Module 9234, DC Coupled mode; c) Module 9234, AC Coupled mode.

3.2. *Phase evaluation*

As a first approximation, the LDV phase behavior has been considered ideal (i.e. negligible phase effect of the frequency of vibration due for example to internal high-pass filters).

A simple modeling of the whole measuring chain of the LDV allowed us to confirm this approximation, by comparing the LDV output when coupled AC either DC.

Differences appears due to only the different coupling mode (cut-off frequency of the AC mode being 0.5 Hz), suggesting that no more phase shift delay due to the LDV itself should be present, almost down to 0.5 Hz. This result is described in Figure (6)

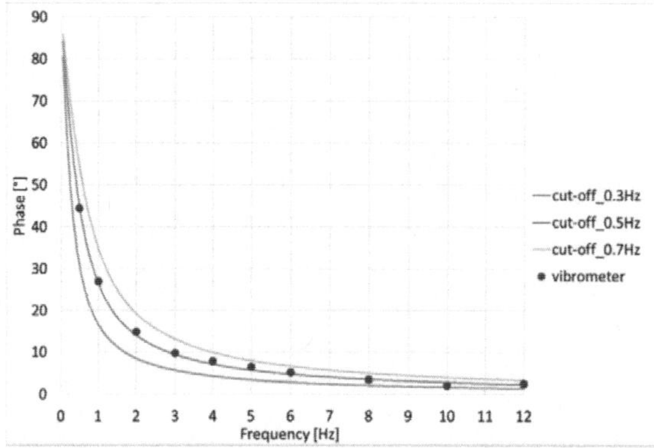

Figure 6. Comparison between theoretical curves of a first order high-pass filter with cut-off frequency in the range 0.3 – 0.7 Hz and experimental data.

In the following results it should be noticed that when the AC coupling is realized on both channels (reference LDV and accelerometer under test) the effect of AC coupling is compensated.

The results are similar, for both the IEPE accelerometers considered. The phase difference with respect to the reference, rapidly increases for frequencies < 2 Hz (Figure 7).

The results are consistent with those obtained by a calibration bench of the National Institute of Metrological Research (INRiM).

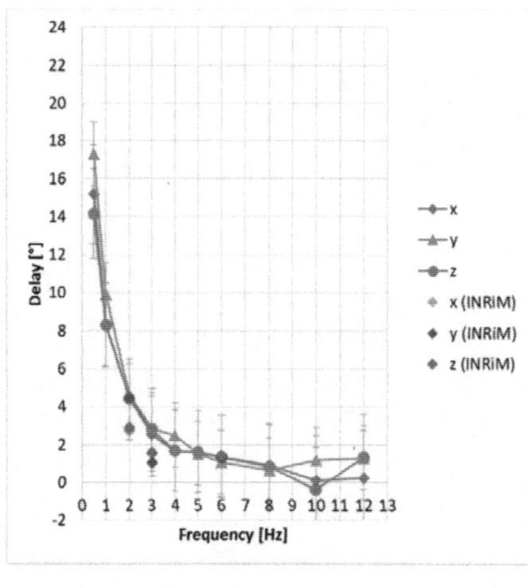

Figure 7. IEPE accelerometer, IEPE Module, IEPE AC coupled mode.

If the capacitive sensor is coupled DC mode, the differences with respect to LDV are in the range ± 4° up to 5 Hz for all configurations (Figure 8(a-b-c)).

It must be considered that the tested MEMS is a very good quality one. It would be interesting to examine the behavior of other types of MEMS accelerometers, in particular the very low-cost ones.

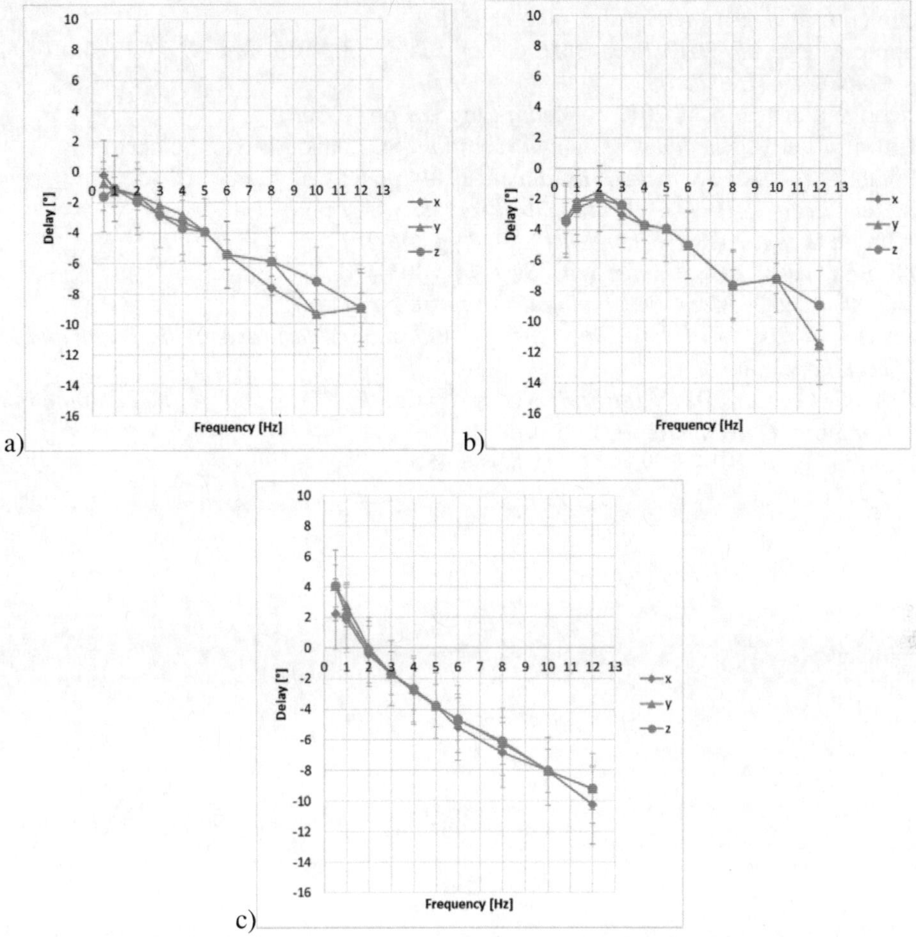

Figure 8. MEMS capacitive accelerometer. a) Analog input Module 9215; b) Module 9234, DC Coupled mode; c) Module 9234, AC Coupled mode.

4. Conclusions

A simple method to calibrate accelerometers in the low-frequency range has been presented.

The evaluation of sensitivity and phase of the whole measuring chains has been carried out by using a single set of tests, realized mounting the accelerometer on an inclined plate, rotated on the clamp surface by a known angle.

Validation actions have been carried out at every stage of data processing.

Results show that different behaviors, not completely predictable, can result depending on the elements of the measurement chain and not only on the type of accelerometer.

In particular, the phase is a critical aspect to consider, that can strongly affect the evaluations in some applications, so a quantitative evaluation could be useful.

References

[1] Son J D *et al* 2016 *Measurement* **94** pp 680–691.
[2] Albarba A *et al* 2009 *Measurement* **42** pp 790–795.
[3] Badri J K *et al* 2011 *Strain* **47** pp 242-257.
[4] Lu Y S *et al.* 2018 *Measurement* **125** pp 471–475.
[5] Sabato A *et al* 2017 *IEEE Sens. J.* **17** pp 226-35.

IOP Conf. Series: Journal of Physics: Conf. Series **1149** (2018) 012018 doi:10.1088/1742-6596/1149/1/012018

[6] Rainieri C *et al* 2013 *NDT E Int.* **60** pp 17-31.

[7] Ripper G P *et al* 2017 *Proc. IMEKO 23rd TC3, 13th TC5 and 4th TC22 Joint International Conference (Helsinki).*

[8] Czech K R and Gosk W 2017 *Procedia Eng.* **189** pp 45-50.

[9] Griffin M J 2014 *Handbook of human vibration* (St. Louis: Elsevier Science).

[10] Kavitha S *et al* 2012 *Sensors Transducers J.* **144** pp 62–75.

[11] Kavitha S *et al* 2016 *Mech Syst Signal Process.* **66-67** pp 410–424.

[12] D'Emilia G *et al* 2018 *J. Sens. Sens. Syst.* **7** pp 245-257.

[13] D'Emilia G *et al* 2018 *J. Sens. Sens. Syst.* **7** pp 403-410.

[14] Link A *et al* 2006 *Meas. Sci. Technol.* **17** pp 1888–1894.

[15] Sun H *et al* 2018 *13th Intl Conference on Vibration Measurements by Laser and Noncontact Techniques (Ancona).*

[16] D'Emilia G *et al* 2017 *Proc. IMEKO 23rd TC3, 13th TC5 and 4th TC22 Joint International Conference (Helsinki).*

[17] D'Emilia G *et al* 2014 *J. Phys.: Conf. Series* **658** 1.

Fitting algorithm of sine wave with partial period waveforms and non-uniform sampling based on least-square method

Zhiguo Liang[1], Dongmei Ren[1], Jingyu Sun[2] and Zhenyu Zhu[3]

[1]National Key Laboratory of Science and Technology on Metrology & Calibration, Changcheng Institute of Metrology and Measurement, Huanshan Village, Wenquan Town, Haidian District, Beijing 100095, China
[2]School of Instrumentation Science & Opto-Electronics Engineering, BUAA, No. 37 Xueyuan Road, Haidian District, Beijing 100191, China
[3]College of Precision Instrument and Opto-Electronics Engineering, Tianjin University, No. 92, Weijin Road, Nankai District, Tianjin 300072, China

Zhiguo Liang's e-mail address : Lzg304@sina.com
Dongmei Ren's e-mail address :Rendongmei@cimm.com.cn
Jingyu Sun's e-mail address: Saint@cimm.com.cn
Zhenyu Zhu's e-mail address: Zhuzhenyu@cimm.com.cn

Abstract. A novel four-parameter sine wave curve-fitting method for partial period and non-uniform sampling is presented. Based on the three-parameter sine wave curve-fitting method, the principle of this method is to turn the optimization of four parameters (amplitude, frequency, phase and offset) into the optimization of only one parameter (frequency). Compared to other curve-fitting methods, the characteristics of this method is that it does not need initial parameter estimation, and can be used to fitting partial period sinusoid, even with only 1/5 signal periods or less. It is shown that this method has wide convergence interval and excellent robustness. The validity and feasibility of this method have been proved by both the simulations and experiment. The four-parameter sine wave curve-fitting method can be applied to the parameter estimation of sinusoidal wave with non-uniform sampling series of partial period, especially it is very useful in the control of ultra lower frequency sine wave parameters.
Keywords: non-uniform sampling, partial period, sine wave, curve-fitting, parameter estimation, calibration
PACS: 07.05.Kf, 43.60.-c, 43.60.+d

1. Introduction

Sinusoid can be characterized by four parameters, the amplitude, frequency, phase and DC bias. The four-parameter curve-fitting of sinusoid is a basic algorithm in engineering measurements, and has been studied widely and deeply by many researchers [1-20]. Generally, initial parameter estimation is required in most curve-fitting methods. If the initial parameter is far enough from the target value, the final fitting results may not be always convergence to the target value.

In some applications, such as calibrations of ultra-low frequency vibration and ultra-low frequency sinusoid generator, the frequency may be very low, 10^{-4}Hz for example, that it takes a very long time to sample a whole signal period. So a method to estimate all the four sinusoidal parameters from partial signal period is required in the control, adjustment, test and calibration of sine wave parameters. Because the sampling series of signal is sometimes less than one period, and maybe it is only a little section of a period, it is difficult to estimate the initial parameters that many sinusoidal curve-fitting methods can't be used. Thus, it is an important and valuable work to estimate sinusoidal parameters from a partial period of signal.

In this paper, a partial period sine wave curve-fitting method with non-uniform sampling is presented. It can also be used to the uniform and random sampling sinusoidal series with both amplitude and time information known.

2. Three-parameter sinusoidal curve-fitting method

2.1. Principle of three-parameter sinusoidal fitting method [10]

Suppose an ideal sinusoidal wave can be expressed as

$$y(t) = E_1\cos(2\pi ft) + E_2\sin(2\pi ft) + Q = E\cos(2\pi ft + \Phi) + Q \quad (1)$$

Where, $y_1, y_2, \ldots, y_n$ are the sampling data at time points t_1, t_2, ..., t_n. The three-parameter sinusoidal curve-fitting process is, with known frequency f, to search for the A_1, B_1 and C that can minimize the ε in the following equation.

$$\varepsilon = \sum_{i=1}^{n}[y_i - A_1\cos(2\pi ft_i) - B_1\sin(2\pi ft_i) - C]^2 \quad (2)$$

Then a fitting equation is obtained, as shown below.

$$\hat{y}(i) = A_1\cos(2\pi ft_i) + B_1\sin(2\pi ft_i) + C = A\cos(2\pi ft_i + \theta) + C \quad (3)$$

$$A = \sqrt{A_1^2 + B_1^2} \quad (4)$$

$$\theta = \begin{cases} \arctan\left(\dfrac{-B_1}{A_1}\right); & A_1 \geq 0 \\ \arctan\left(\dfrac{-B_1}{A_1}\right) + \pi; & A_1 < 0 \end{cases} \quad (5)$$

The residual error of fitting is

$$\rho = \sqrt{\frac{\varepsilon}{n}}$$

$$\varepsilon = \sum_{i=1}^{n}(y_i - \hat{y}(i))^2$$

Since this is a closed arithmetic, the curve-fitting results can always be attained.

2.2. Discussion

The preceding three-parameter curve-fitting process is on the hypothesis that the frequency f is known. Especially, suppose t_{min}=min{t_i}=t_1 and t_{max}=max{t_i}=t_n (i=1,…,n), then the average sampling rate is $v = (n-1)/(t_{max} - t_{min})$, the average sampling period is $\Delta t=1/v$, and the digital angular frequency is $\omega=2\pi f/v$. In the case of uniform sampling, equation (1) can be expressed as

$$y(i) = E_1\cos(\omega\cdot i) + E_2\sin(\omega\cdot i) + Q = E\cos(\omega\cdot i + \Phi) + Q \quad (6)$$

The curve-fitting equation is

$$\hat{y}(i) = A_1\cos(\omega\cdot i) + B_1\sin(\omega\cdot i) + C = A\cos(\omega\cdot i + \theta) + C \quad (7)$$

Suppose the digital angular frequency is w instead of ω, by using a sine wave series of 1/3 periods, we can get the relationship between the normalized error ρ/E and the frequency ratio w/ω, as shown in figure 1, using the three-parameter curve-fitting method. Figure 2 shows the details of the beginning part where $\rho/E \leqslant 0.05$ of figure 1. Here, $\omega=2\pi/10000$, E=4, Q=0, and Φ are 0°, 35°, 70°, 105°, 140°, 175°, 210°, 245°, 280°, 315° and 350°, respectively. The three-parameter curve-fit is executed by replacing ω with w.

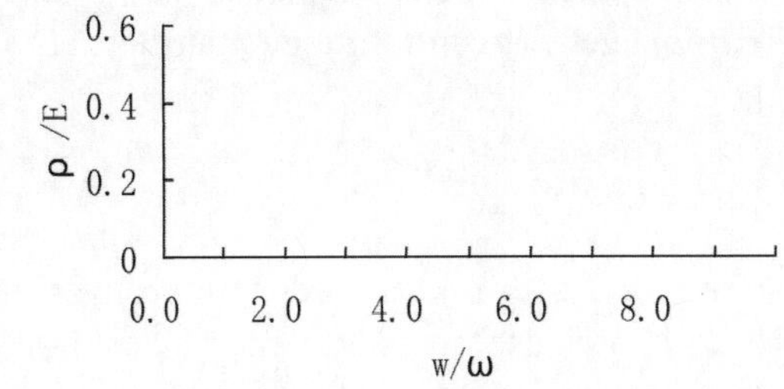

Figure 1. Variation of normalized error ρ/E with frequency ratio w/ω in curve-fitting of 1/3 period

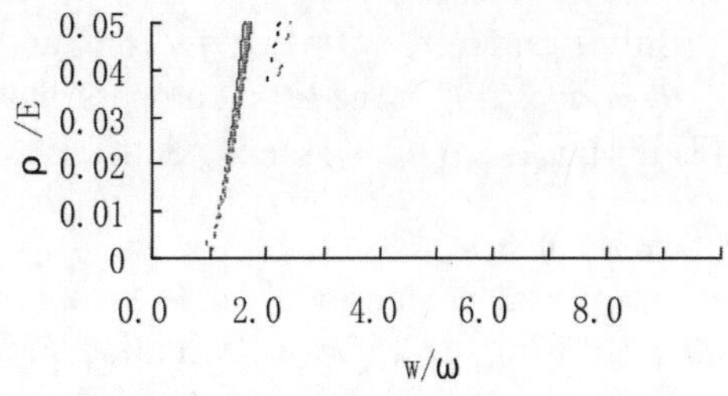

Figure 2. Details of Figure1 in the range of $\rho/E \leq 0.05$

It is shown in figure 1 and figure 2 that, for the curve-fitting of partial period sine wave with known frequency ω, the extremum of ρ/E of three-parameter curve-fitting exists and it is alone in the interval of $[0, 2\omega]$. The extremum is at the point where the frequency is ω. It is also shown that the waveform of ρ/E is not affected by variations of amplitude and phase, but the amplitude of ρ/E is affected by the variation of initial phase. When the fitting frequency w is bigger than 2ω, the values of ρ/E are bigger than that when w is smaller than 2ω. This can be used as the criterion to determine the upper boundary of convergence, and the lower boundary can be a small frequency that is close enough to zero. The point where ρ/E has the minimum value can be obtained by searching ω in the interval of $[0, 2\omega]$. The same conclusion can also be reached by using other parts of sinusoidal period waveform to make the fitting.

3. Four-parameter sine wave curve-fitting algorithm with partial period and non-uniform sampling

3.1. Principle and process

By rebuilding the preceding curve-fitting method, we can obtain a convergent four-parameter sinusoidal curve-fitting method, which can be used to partial period sinusoidal waveforms with non-uniform sampling. Suppose the average sampling rate is v, the true frequency to be estimated is f_0, the length of sampling series is p periods of sinusoidal waveform ($0<p\leq1$), and the time interval of p signal periods is τ. Then $f_0\leq1/\tau$. Select a factor q that is small enough ($q=10^{-5}$ for example) to make $f_0>q/\tau$. To each frequency f in the interval $[q/\tau, 2/\tau]$, we know that the extremum of square sum of residuals $\varepsilon(f)$ is in existence and alone. So, the four-dimension search for amplitude, frequency, phase and offset can be transformed into one-dimension search for frequency and $\varepsilon(f)$. In the interval of $[q/\tau, 2/\tau]$, the convergence of the four-parameter sinusoidal curve-fitting process, which is realized by using the three-parameter curve-fitting method, can be assured. The procedures of the four-parameter fitting are as following.

(1) Suppose the condition for iteration to stop is h_e.

(2) Obtain the time interval τ of sampling the partial signal period from the sampling series y_1, y_2, ..., y_n at known time points t_1, t_2, ..., t_n using points counting method, $\tau= t_n-t_1$. Calculate the average sampling rate, $v=(n-1)/(t_n-t_1)$. Then select a factor q that is small enough to make $f_0>q/\tau$. Determine the range in which the target frequency f_0 exists, $[q/\tau, 2/\tau]$.

(3) Determine the left boundary frequency, $f_L=q/\tau$, and the right boundary frequency, $f_R=2/\tau$, of the iteration.

(4) Let the median frequency be $f_M=f_L+0.618\times(f_R-f_L)$ and $f_T=f_R-0.618\times(f_R-f_L)$.

(5) Execute the three-parameter curve fit at f_L, f_R, f_M and f_T, respectively, to get A_L, θ_L, C_L, ρ_L, A_R, θ_R, C_R, ρ_R, A_M, θ_M, C_M, ρ_M, A_T, θ_T, C_T and ρ_T.

(6) If $\rho_M<\rho_T$, then $f_0\in[f_T, f_R]$. Let $\rho=\rho_M$, $f_L=f_T$, $f_T=f_M$, $f_M=f_L+0.618\times(f_R-f_L)$.

If $\rho_M\geq\rho_T$, then $f_0\in[f_L, f_M]$. Let $\rho=\rho_T$, , $f_R=f_M$, $f_M=f_T$, $f_T=f_R-0.618\times(f_R-f_L)$.

(7) Judge whether $|(\rho_{M(k)}-\rho_{T(k)})/\rho_{T(k)}|<h_e$ is true. If it is true, then stop the iteration. If $\rho=\rho_T$, get the curve-fitting parameters $A=A_T$, $f=f_T$, $\theta=\theta_T$ and $C=C_T$. If $\rho=\rho_M$, get the curve-fitting parameters $A=A_M$, $f=f_M$, $\theta=\theta_M$ and $C=C_M$. The fitting process ends.

(8) If $|(\rho_{M(k)}-\rho_{T(k)})/\rho_{T(k)}|<h_e$ is not true, repeat (5)–(7).

3.2. Convergence

The four-parameter sinusoidal curve-fitting is an iteration process, so it also needs to be convergent. As denoted in figure 1, for the sampling series of p ($0<p\leq 1$) periods, the convergence interval is (0, $2f_0$], where, f_0 is the true value of signal frequency.

4. Simulations

Assuming that the measurement range is -5V~+5V, the amplitude of sine wave signal is 4V, the DC bias is 0V, the initial phase is 100°, the frequency is 6Hz, the uniform sampling rate is 8kSa/s, the sampling interval is 0.125ms, the sampling number is 400, and the sampling series covers about 0.3 waveform periods, we can get the four-parameter sine wave curve-fitting results, as shown in table 1, by using the method discussed above. The fitting curve is shown in figure 3, along with the uniform sampling curve. Figure 4 shows the difference between the sampling series and its curve-fitting.

Table 1. Results of partial period sine wave curve-fitting

Parameter	Nominal value	Figure 3 Uniform sampling ($p\approx 0.3$)	Figure 5 Non-uniform sampling ($p\approx 0.4$)	Figure 7 Non-uniform sampling ($p\approx 1.2$)
A	4 V	4.003 V	4.000000 V	4.000000 V
ΔA	0 V	2.6 mV	0.24 μV	0 V
$\Delta A/A$	0	6.5×10^{-4}	6×10^{-8}	0
f	6 Hz	5.996 Hz	6.000001 Hz	6.000000 Hz
Δf	0 Hz	-4 mHz	1 μHz	0 Hz
$\Delta f/f$	0	-6.7×10^{-4}	1.7×10^{-7}	0
θ	100°	100.289°	99.999981°	100.0000038°
$\Delta\theta$	0°	0.289°	-0.00002°	0.0000038°
$\Delta\theta/\theta$	0	2.9×10^{-3}	-2×10^{-7}	4×10^{-6}
C	0 V	-2.082 mV	-0.476 μV	42.7 nV
ΔC	0 V	-2.082 mV	-0.476 μV	42.7 nV
$\Delta C/A$	0	-5.2×10^{-4}	1.2×10^{-7}	1.1×10^{-9}
ρ	0 V	95.7 μV	1.61 μV	2.46 μV

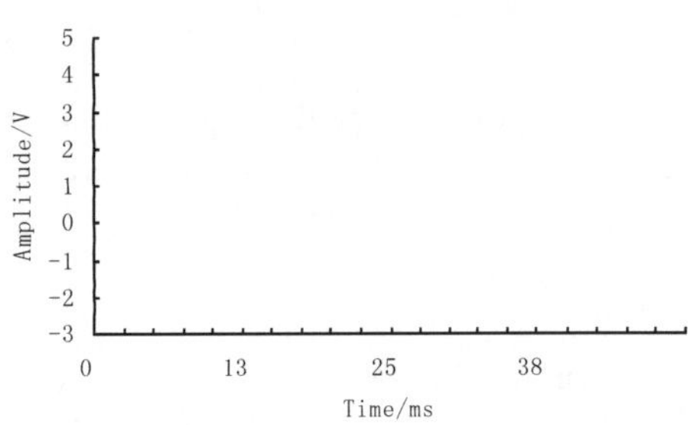

Figure 3. Uniform sampling points y_i and curve-fitting points $\hat{y}(i)$ ($p\approx 0.3$)

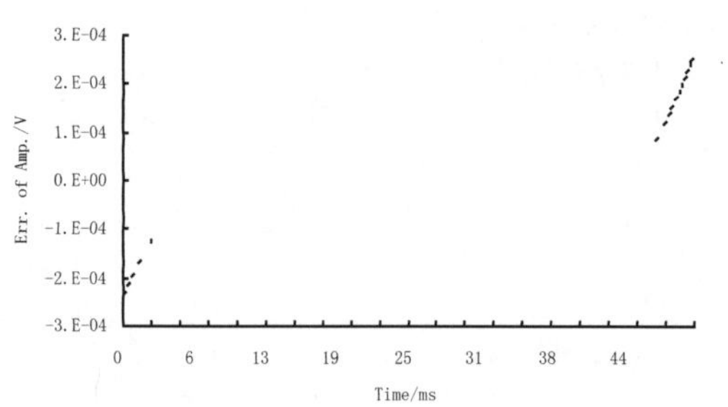

Figure 4. Difference between uniform sampling series and its fitting curve y_i- $\hat{y}(i)$

IOP Conf. Series: Journal of Physics: Conf. Series **1149** (2018) 012019 doi:10.1088/1742-6596/1149/1/012019

Keeping the parameters of sine wave model fixed and executing the random rate sampling, we can get the random sampling series and their fitting-curves by using the method in this paper, which are shown in figure 5 to figure 8. Figure 5 shows the non-uniform sampling points y_i and their curve-fitting points $\hat{y}(i)$ in about 0.4 periods. The sampling interval is (RND-0.4)×20τ_0, RDN is a random number with uniform distribution in [0,1]. It can be found that, since the sampling intervals are non-uniform and vary randomly, the form of sine wave curve drawn in the mode of uniform sampling in figure 5 varies evidently. Figure 6 is the difference between the sampling series and its curve-fitting. It can be seen that the sampling data are fitted well. Figure 7 shows the non-uniform sampling points y_i and their curve-fitting points $\hat{y}(i)$ in about 1.2 periods. The sampling interval is (RND-0.3)×20τ_0. Figure 8 is the difference between the sampling series and its curve-fitting. It can be found from the figure that the data are also fitted well. The curve-fitting parameters of non-uniform sampling are also listed in table 1. The data in table 1 indicates that the curve-fitting error of non-uniform sampling is smaller than that of uniform sampling. The difference is distinct enough. The reason for this needs to be further studied. Maybe it is because the residual error is more like the white noise in non-uniform series curve-fitting, as shown in figure 6 and figure 8, but it is like system error in uniform sampling series curve-fitting, as shown in figure 4.

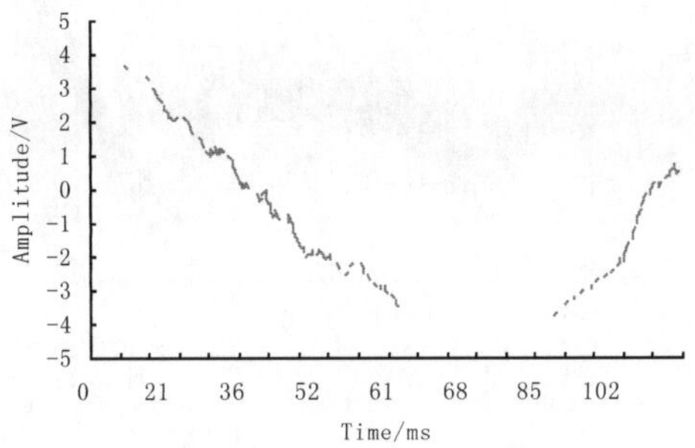

Figure 5. Non-uniform sampling points y_i and curve-fitting points $\hat{y}(i)$ ($p\approx0.4$)

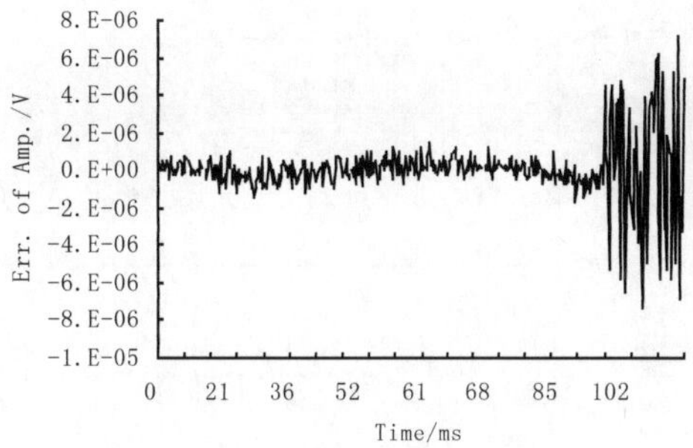

Figure 6. Difference between non-uniform sampling series and its fitting curve y_i- $\hat{y}(i)$

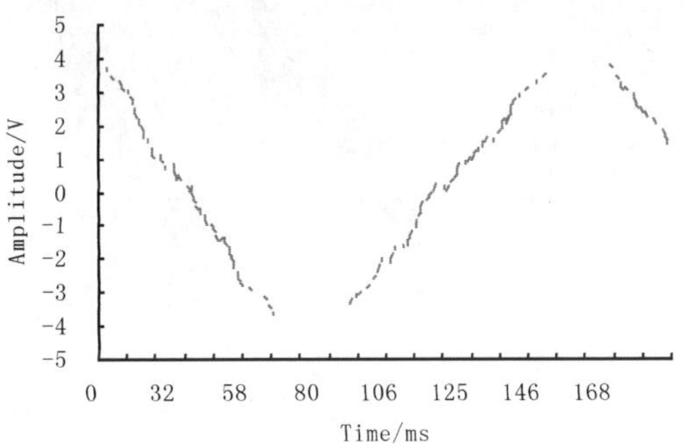

Figure 7. Non-uniform sampling points y_i and curve-fitting points $\hat{y}(i)$ ($p\approx1.2$)

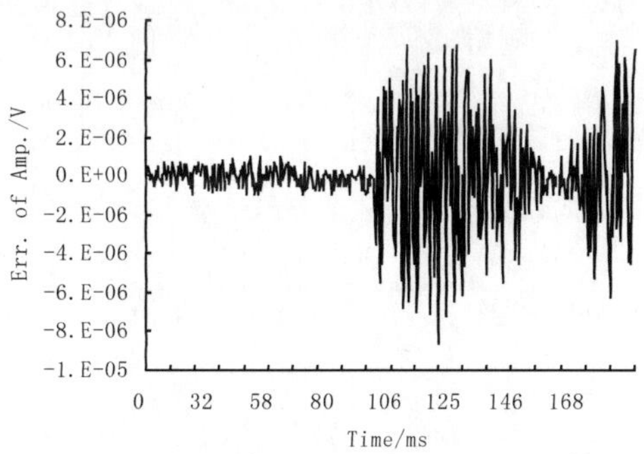

Figure 8. Difference between non-uniform sampling series and its fitting curve y_i- $\hat{y}(i)$

5. Experiment and result

By using an ultra-low frequency vibration standard as inspiriting source, a sinusoidal vibration signal is generated. The displacement amplitude and frequency of the sinusoid are 36.32 mm and 0.040000

Hz, respectively. The signal is measured using a displacement transducer of model ASQ-1CA. The data are collected using a NI PXI-6281 data acquisition system, the A/D of which is 18 Bits, the measurement range is ±2.5V and the sampling rate is 200 Samples/s. The sampling series contains 5000 data, which distribute in approximate 1 period of sinusoidal waveform.

The four-parameter sinusoidal curve-fitting are carried out using the method discussed above and the following results are obtained. The amplitude is 1771.02mV, frequency is 0.04056Hz, phase is 230.098°, offset is 120.53mV, the root mean square of residual error ρ is 30.157mV and the total waveform distortion is 2.4%. Both the fitting curve and sampling data are shown in figure 9, and their difference is shown in figure 10.

By cutting out two parts of sampling series in figure 9, the uniform sampling series is turned into a non-uniform sampling series, as shown in figure11. The results of four-parameter sinusoidal curve-fitting in this case are: amplitude is 1746.93mV, frequency is 0.04042Hz, phase is 231.354°, offset is 123.94mV, root mean square of residual error ρ is 34.154mV and the total waveform distortion is 2.8%. Both the fitting curve and the sampling data are shown in figure 11, and their difference is shown in figure 12.

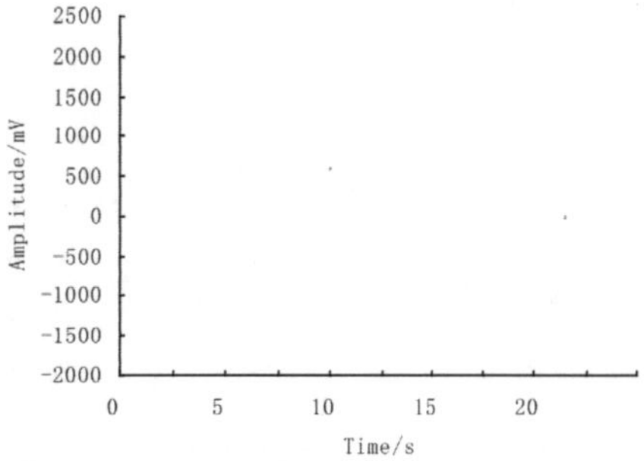

Figure 9. Uniform sampling points y_i and curve-fitting points $\hat{y}(i)$ $(p \approx 1)$

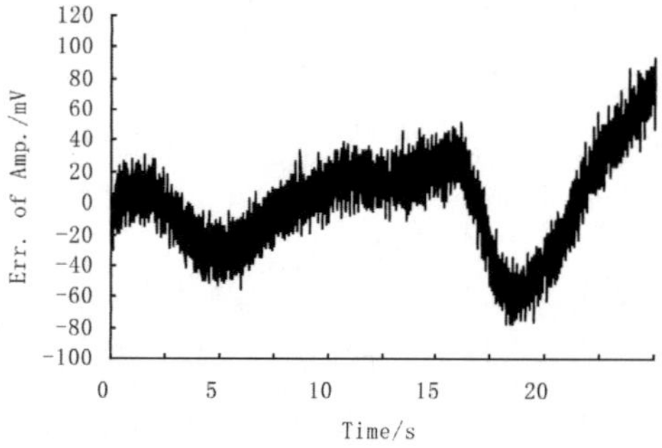

Figure 10. Difference between uniform sampling series and its fitting curve y_i- $\hat{y}(i)$

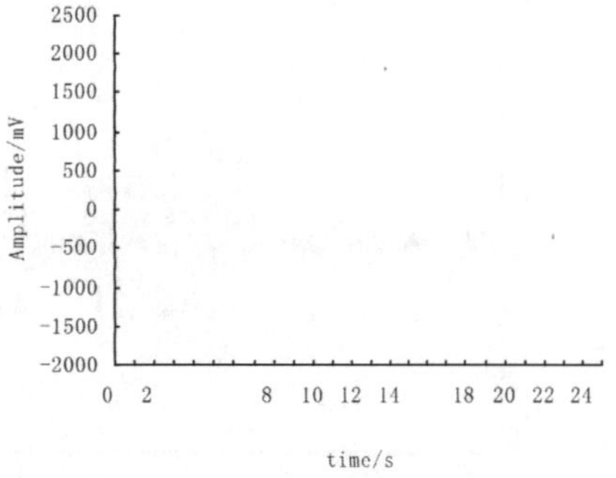

Figure 11. Non-uniform sampling points y_i and curve-fitting points $\hat{y}(i)$ $(p \approx 0.6)$

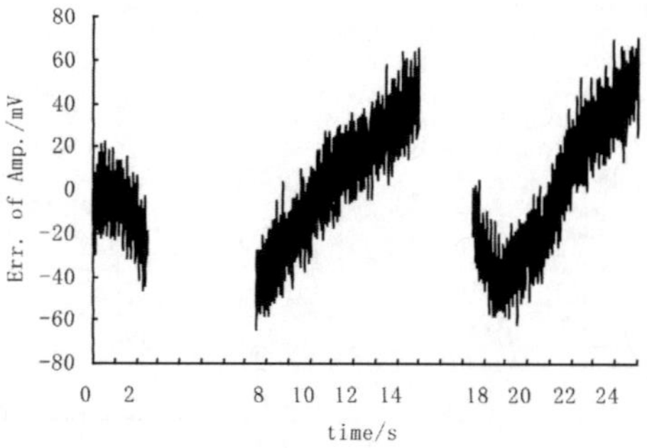

Figure 12. Difference between non-uniform sampling series and its fitting curve y_i- $\hat{y}(i)$

6. Discussion

Through the simulation and experiment above, we can see that the four-parameter sinusoidal curve-fitting method can be used to partial period sine wave fitting with non-uniform sampling. The amplitude, frequency, phase and DC offset of a sinusoidal wave can be obtained using this method. Although the original purpose of this method is to solve the partial period sine wave fitting problem, in fact, it can be used to any sinusoidal sampling series less than 2 periods directly. A distinguished feature of this method is that it fits for any sine wave sampling series, whether it is uniform sampling or not. Especially, it is very useful in the case that some abnormal parts are removed from a uniform sampling series. A uniform sampling is often turned into a non-uniform sampling for some reasons, such as the period is too long or the experiment condition is varied during the data acquisition. In random sampling, there are even some cases in which the time sequence reversal may occur. The method discussed above is valid in all these cases. It has shown excellent properties of robustness and convergence, and it can be used to solve various engineering problems above.

It is shown from the simulation that, compared with the uniform sampling, the sinusoidal curve-fitting error of non-uniform sampling is much smaller. In the simulation of non-uniform sampling above, the amplitude difference is 6×10^{-8}, frequency difference is 1.7×10^{-7}, phase difference is -0.00002° and the DC offset difference is 1.2×10^{-7} (refer to amplitude). The simulation results are proved by the experiment. The simulation experiment also shows that, compared with the multi-period series curve-fitting, the error of partial period curve-fitting is a little larger, especially when the DC bias is large. The result of parameter estimation is affected by many factors, such as information losing, waveform distortion, sampling jitter, series number and period width, etc. Normally, correct results can be obtained with a sampling length over 1/5 sine wave period by use of the partial period curve-fitting method presented above.

7. Conclusions

A four-parameter curve-fitting method for partial period sinusoidal waveform with non-uniform sampling is presented in this paper. The process of four-parameter curve-fitting is introduced and the convergence interval is given. The characteristics of this method is that it does not need initial parameter estimation, and can be used to fitting partial period sinusoid. Knowing the sampling series and sampling moment is enough for this method. Simulations and experiment are carried out to prove the validity of this method, and the curve-fitting results in random sampling and non-random sampling are compared. It can be concluded that the four-parameter sine wave curve-fitting method has excellent properties of robustness and convergence. It can be used to partial period sinusoidal curve-fitting with non-uniform sampling. The method has a special value to ultra-low frequency parameter estimation.

References

[1] Kuffel J, McComb T R and Malewski R 1987 Comparative evaluation of computer methods for calculating the best-fit sinusoid to the digital record of a high-purity sine wave *IEEE Transactions on Instrumentation and Measurement* 36(3) 418–22

[2] McComb T R, Kuffel J and Le Roux B C 1989 A comparative evaluation of some practical algorithms used in the effective bits test of waveform recorders *IEEE Transactions on Instrumentation and Measurement,* 38(1) 37–42

[3] Jeng Y C 1987 High precision sinusoidal frequency estimator based on weighted least square method *IEEE Transactions on Instrumentation and Measurement* 36(2) 124–27

[4] Jeng Y C and Crosby P B 1988 Sinewave parameter estimation algorithm with application to waveform digitizer effective bits measurement *IEEE Transactions on Instrumentation and Measurement* 37(4) 529–32

[5] Cennamo F, Daponte P and Savastano 1992 Dynamic testing and diagnostics of signal analyzers *IEEE Transactions on Instrumentation and Measurement* 41(6) 840–44

[6] Giaquninto N and Trotta A 1997 Fast and accurate ADC testing via an enhanced sine wave fitting algorithm, *IEEE transactions on Instrumentation and Measurement* 46(4) 1020–25

[7] Zhang J, Zhao X, Xiao H and Sun J 1997 Sine wave fit algorithm based on total least-squares method with application to ADC effective bits measurement *IEEE Transactions on Instrumentation and Measurement,* 46(4) 1026–30

[8] Handel P 2000 Properties of the IEEE-STD-1057 Four-Parameter Sine Wave Fit Algorithm

IEEE Transactions on Instrumentation and Measurement 49(6) 1189–93
[9] Tian S, Wang J, Yan D, et al. 2005 Base on inherit algorithmic sinusoidal parameter estimation,*Measurement Technique* 369 3–5
[10] IEEE Std 1057-1994, IEEE standard for digitizing waveform recorders, pp 79, 1994
[11] Kollar I and Blair J J 2005 Improved determination of the best fitting sine wave in ADC testing *IEEE Transactions on Instrumentation and Measurement* 54(5) 1978–83
[12] Tamás Zoltán Bilau, Tamás Megyeri, Attila Sárhegyi, János Márkus and István Kollár 2004 Four-parameter fitting of sine wave testing result: Iteration and convergence *Computer Standards & Interfaces* 26(1) 51–6
[13] Palfi V and Kollar I 2013 Acceleration of the ADC test with sine-wave fit *IEEE Transactions on Instrumentation and Measurement* 62(5) 880–8
[14] Belega D, Dallet D and Petri D 2011 A high-performance procedure for effective number of bits estimation in analog-to-digital converters *IEEE Transactions on Instrumentation and Measurement* 60(5) 1522–32
[15] Belega D and Petri D 2013 Statistical performance of the effective-number-of-bit estimators provided by the sine-fitting algorithms *IEEE Transactions on Instrumentation and Measurement* 62(3) 633–40
[16] Chen K F 2010 Estimating parameters of a sine wave by separable nonlinear least squares fitting *IEEE Transactions on Instrumentation and Measurement* 59(12) 3214
[17] Vucijak N M and Saranovac, L V 2010 A simple algorithm for the estimation of phase difference between two sinusoidal voltages *IEEE Transactions on Instrumentation and Measurement* 59(12) 3152
[18] Belega D, Dallet D and Petri D 2011 Performance comparison of the three-parameter and the four-parameter sine-fit algorithms *Instrumentation and Measurement Technology Conference (I2MTC), IEEE,* 1–4
[19] Sárhegyi Attila and István Kollár 2006 Robust sine wave fitting in ADC testing *IMTC 2006 - Instrumentation and Measurement Technology Conference, Sorrento, Italy* 24–7
[20] Ramos P M, Janeiro F M and Radil T 2010 Performance comparison of three algorithms for two-channel sinewave parameter estimation *Metrology and Measurement Systems* XVII (2) 255-70

IOP Conf. Series: Journal of Physics: Conf. Series **1149** (2018) 012020 doi:10.1088/1742-6596/1149/1/012020

Measuring the speed of gravity with laser Doppler vibrometers and silicon mechanical resonators

I Cagnani

EPS Gravitational Physics Division, 6 rue des Frères Lumière, 68200 Mulhouse, France

ivan.cagnani@gmail.com

Abstract. The old question of the speed of propagation of gravitational interactions still needs a laboratory answer. An experimental setup for direct measurement of the propagation delay in the gravitational interaction between resonating gravitationally coupled oscillators is proposed. The oscillators are designed to act like transmitter and receiver in an AM radio communication. The signal, a dynamic Newtonian field, is generated by the precise vibrations of a massive tungsten disc, in resonance with its suspension system like a trampoline oscillator. The disc is driven by a shaker locked to a GPSDO frequency reference, while its movement is tracked with a laser Doppler vibrometer and a lock-in amplifier, locked to the same frequency reference. The gravitational signal is received by an antenna, a gold-coated silicon oscillator with resonance frequency of 2 kHz and a Q of 200 000 in thermal vacuum. The vibrations of the antenna are tracked by a second laser Doppler vibrometer with a noise level of 30 fm/√Hz and a lock-in amplifier using the output of the first lock-in amplifier as reference. High-vacuum chambers and multi-stage vibration and acoustic isolation systems are required for both the generator and the antenna. Cross-correlating the data of the two oscillators, their phase difference is then measured and, using the frequency of the signal and the distance between oscillators, the speed of gravity will be calculated. Numerous measurements at various distances spanning a 300 mm range are collected and used for the determination of other parameters, including G.

1. Introduction

The speed of propagation of Newtonian gravitational fields has been questioned by physicists for centuries. Newton's Universal Gravitation implicitly assumes an instantaneous propagation of the gravitational interaction, and using that assumption astronomers could correctly predict orbits, yet philosophically, the concept of instantaneous action-at-distance was unacceptable to many and Newton himself was notoriously skeptical. Laplace attempted a solution proposing a highly superluminar speed of gravity, but that would imply a faster-than-light transfer of information, which for modern physics is again unacceptable. We now known that stable orbits are like an undamped harmonic oscillator, or particles within a potential well: no transfer of information is needed to keep the periodic motion going steadily. A gravitational perturbation is instead an information, and it cannot propagate faster than light according to General Relativity. This last affirmation can now be tested in a room-size laboratory, using only commercial technology and without even the need for modern physics.

IOP Conf. Series: Journal of Physics: Conf. Series **1149** (2018) 012020 doi:10.1088/1742-6596/1149/1/012020

2. Description of the experiment

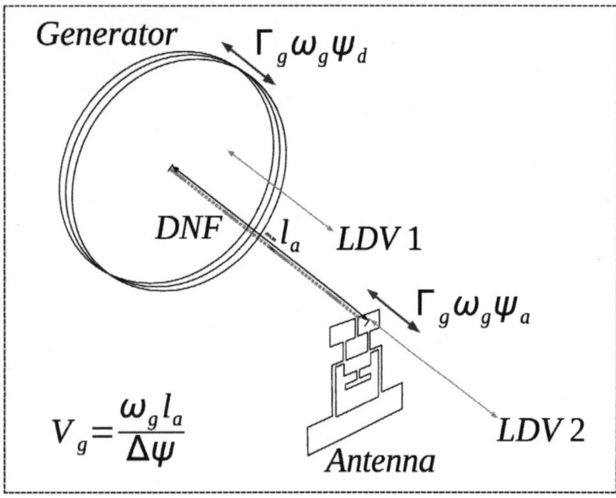

Figure 1. Simple representation of the basic schematics of the experiment. The disc vibrates along its radial axis. Aligned with the centre of the disc, one vibrating pad of the antenna is vibrating at the same frequency of the disc, stimulated only gravitationally by the dynamic gravitational attraction of the disc. One laser Doppler vibrometer is used to track the oscillations of the disc while the other is used to track the vibrations of the antenna. The value of the speed of gravity can be calculated from the phase difference between oscillators, their distance, and the frequency of their coupling gravitational signal.

As illustrated in Figure 1, the periodic displacement of a large tungsten disc, made to oscillate along its axial direction, can generate a gravitational signal able to excite a high-Q mechanical oscillator aligned with the disc. The maximum transmission efficiency of the gravitational information is reached when the frequency of the gravitational signal is exactly equal to the resonance frequency of the antenna. Since the frequency of the gravitational signal must be equal to the frequency of the motion of the object generating the gravitational field, the frequency of the signal generator and of the antenna should be perfectly matched or at least always lie within the bandwidth of the antenna. For a high-Q antenna, such a bandwidth is very restricted.

The antenna selected for the experiment is a Phonon Transfer Resonator [1]. It is made of monocrystalline silicon and characterized by an effective mass of $m_a = 0.3\ g$, a resonance frequency of $f_a = 2.0 \cdot 10^3\ Hz$ in the highest Q flexural mode in the out of plane direction (aligned with the disc), and a Q factor of $2 \cdot 10^5$.

The resonance bandwidth of the antenna is then

$$\Delta f_a = \frac{f_a}{Q_a} = 10\ mHz \tag{1}$$

while its damping ratio is

$$\zeta_a = \frac{1}{2\,Q_a} = 2.5 \cdot 10^{-6} \tag{2}$$

and its mechanical impedance is

$$Z_a = \left(\frac{\omega_a^2}{Q_a^2} + \frac{\left(\omega_a^2 - \omega_g^2\right)^2}{\omega_g^2}\right)^{\frac{1}{2}} \tag{3}$$

where of course $\omega_a = 2\,\pi\,f_a$ and ω_g is the angular frequency of the gravitational force $F_g = \Gamma_g\,m_a$ driving it. When the gravitational signal driving the antenna matches perfectly the frequency of the antenna, the impedance of the antenna reaches its minimum value

$$\omega_g = \omega_a \Rightarrow Z_a = \left(\frac{\omega_a^2}{Q_a^2}\right)^{\frac{1}{2}} = 3.95 \cdot 10^{-3}\ \Omega \tag{4}$$

and this will result in the maximum amplitude of vibration when stimulated by the gravitational signal

$$\omega_g = \omega_a \Rightarrow A_a = \frac{\Delta\Gamma_g}{Z_a\,\omega_g} \tag{5}$$

IOP Conf. Series: Journal of Physics: Conf. Series **1149** (2018) 012020 doi:10.1088/1742-6596/1149/1/012020

where Γ_g is the gravitational field strength of the generator, the intensity of the gravitational signal at the position of the antenna, then at distance l_a from the geometrical center of the disc, seen as a cylinder, along the direction of its axis. Apart from the distance l_a between the generator and the antenna, Γ_g depends also on the geometry and on the density ρ_d of the disc:

$$\Gamma_g = 2\,\pi\,G\,\rho_d\left\{h_d - \left[(h_d + l_a)^2 + r_d^2\right]^{\frac{1}{2}} + \left(l_a^2 + r_d^2\right)^{\frac{1}{2}}\right\} \tag{6}$$

where h_d is the thickness of the disc and r_d is its radius. The formula is valid only along the axis of the disc and considers the gravitational field as originating at a point in the geometrical center of the disc seen as a cylinder with height h_d and radius r_d [2].

The disc of the generator is made of 99.95% pure tungsten, it has a radius r_d of $175 \pm 0.1\ mm$, a height h_d of $10 \pm 0.01\ mm$ and polished surfaces. It has a uniform density ρ_d of $19020\ kg/m^3$.

Defined the disc, the value of the gravitational field strength Γ_g depends only on the distance l_a, anyway, l_a is not constant, as the disc vibrates along the same direction. When the disc is closer to the antenna the field will be slightly more intense than when the disc is farther away. What really drives the antenna into vibration is not the field itself, which acts like the carrier of an AM radio signal, but its periodic temporal variation, acting like the envelope of an AM signal.

If we call the amplitude of the oscillation of the disc A_d, we can define $l_{amax} = l_a + A_d$ and $l_{amin} = l_a - A_d$ the maximum and minimum distance, then we can define $\Gamma_{gmin} = \Gamma_g(l_{amax})$ and $\Gamma_{gmax} = \Gamma_g(l_{amin})$ the minimum and maximum of the intensity of the gravitational signal. To complete Equation 5 we can then define

$$\Delta\Gamma_g = \Gamma_{gmax} - \Gamma_{gmin} \tag{7}$$

For a distance of $l_a = 250\ mm$ and an amplitude of $A_d = 1 \cdot 10^{-5}\ m$ for the oscillations of the disc, Equation 5 leads to an amplitude $A_a = 3.3 \cdot 10^{-14}\ m$ of the antenna. Being a linear system, increasing the oscillations of the disc to $A_d = 1 \cdot 10^{-4}\ m$ results in an amplitude of $A_a = 3.3 \cdot 10^{-13}\ m$ in the driven motion of the antenna, but it is not necessary. An instrument like the OFV-5000 Xtra, a single-point laser Doppler vibrometer by Polytec, can resolve a minimum displacement of $A_{min} = 2 \cdot 10^{-14}\ m$ for a bandwidth of $\Delta f = 1 \cdot 10^{-2} Hz$ so it can be able to track every oscillation of the antenna. Considering only the intrinsic noise of the LDV, the measurement would have a $SNR = 10$. Of course, the motion of the antenna due to thermal noise far exceeds the steady oscillation due to the driving from the gravitational signal. Because of that, a lock-in amplifier with a large effective dynamic reserve, like model SR860 from SRS, must be used to recover the signal from the overwhelming noise. The reference signal used by the lock-in amplifier of the LDV of the antenna is the output signal of the lock-in amplifier of the LDV of the generator. There is indeed no better reference frequency than the actual movement of the generator [Figure 2].

The outputs of both lock-in amplifiers must then be cross-correlated in data analysis to measure the phase difference of the antenna with respect to the generator. The measured quantity has two components

$$\psi_{tot} = \psi_{res} + \psi_g \tag{8}$$

one of them is given by theory: the phase of a harmonic resonator at resonance is always $\psi_{res} = \frac{\pi}{2}$, lagging 90° from the driving force, the other component is the aim of the quest: if gravity propagated at infinite speed, ψ_g would be zero independently of the distance l_a or the frequency ω_g of the signal, but assuming a speed of propagation equal to the speed of light c

$$v_g = c \Rightarrow \psi_g = \frac{l_a\,\omega_g}{c} \tag{9}$$

In the case of $l_a = 250\ mm$, the component $\psi_g = 1 \cdot 10^{-5}\ s = 6 \cdot 10^{-4}$°.

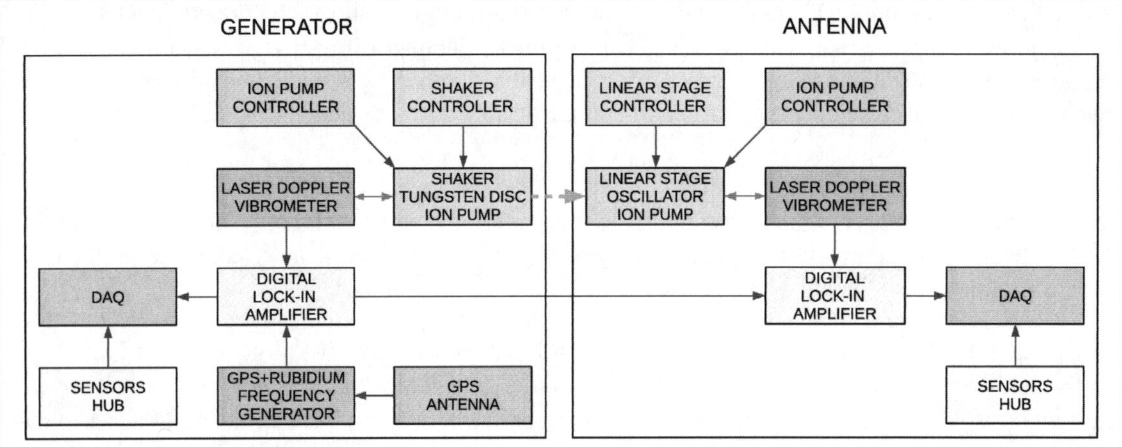

Figure 2. General schematics of the experiments, with connections. The DNF gravitational signal is represented in green, the laser of the LDV in red, connections in blue are electrical (analog or digital).

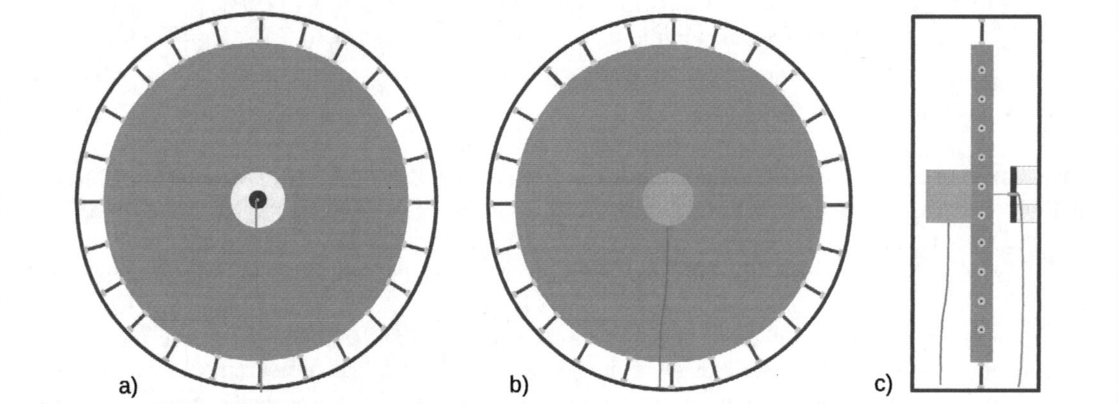

Figure 3. The generator is a tungsten disc connected like a trampoline, with wires, to the wall of its own vacuum chamber. The dimensions of the wires are exaggerated in the drawing for visibility. In 3a, the front view, an optical fibre connects the head of the LDV (not illustrated) to the lens, allowing the measurement of the vibrations of the disc in high vacuum. The lens is held in place by a special vibration damping mount. In 3b, the back view, an electrical cable connects the shaker, in orange, with the its controller external to the chamber (not illustrated). In 3c, the side view, the attachment points of the wires are indicatively represented. The thickness of the disc is exaggerated for visibility.

3. Signal generation and noise management

As illustrated in Figure 2, a GPS-locked and rubidium-locked frequency reference, like the model FS740 by Stanford Research Systems, generates a signal with an Allan Deviation of $1 \cdot 10^{-14}$ over an integration time of 3.5 days, and a phase noise of -150 dBc/Hz for the same period. The signal is set to the desired frequency of the gravitational signal, after a precise measurement of the selected resonant mode of the antenna in the actual experimental conditions. The instrument allows a frequency resolution of 1 μHz.

The signal of the frequency generator will be amplified by the controller of the mechanical transducer. The duty of the transducer is sustaining a fixed level of kinetic energy in the highest Q vibration traversal mode of the mechanical system consisting of the tungsten disc, the suspension wires connecting the disc to the wall of the vacuum chamber [Figure 3] and the transducer itself. The transducer can be a calibration-grade shaker, a high-power vibration speaker, or a device using a different actuation technology, as long as the conversion of the electric signal into mechanical force, imparted toward the center of the disc, can be well controlled in intensity, frequency and phase.

The suspension wires of the disc act like parallel springs in a mass-spring system: their total spring constant is the sum of the individual ones. The total spring constant required is:

$$k_{tot} = md\ \omega_g^2 = 2.89 \cdot 10^9 \frac{N}{m} \tag{10}$$

The required length of each wire, l_w, is given by E, the Young's modulus of its material, A, its cross-section area, and n, the number of wires:

$$l_w = \frac{E\,A\,n}{k_{tot}} \tag{11}$$

As illustrated in Figure 3, a total of 24 wires are used for suspending the disc, so $n = 24$. Considering it is easier to weld tungsten with tungsten, tungsten wire is the obvious choice. Its Young's modulus is 411 GPa. For a wire diameter of 1.5 mm, the length of each wire should be

$$l_w = 7.68 \cdot 10^{-3}\ m \tag{12}$$

The resulting natural frequency of the generator is 2000.008 Hz. Cutting the wires with 0.5 micron of tolerance, the frequency of the generator can be fine-tuned to be precisely at the center of the measured resonance bandwidth of the antenna.

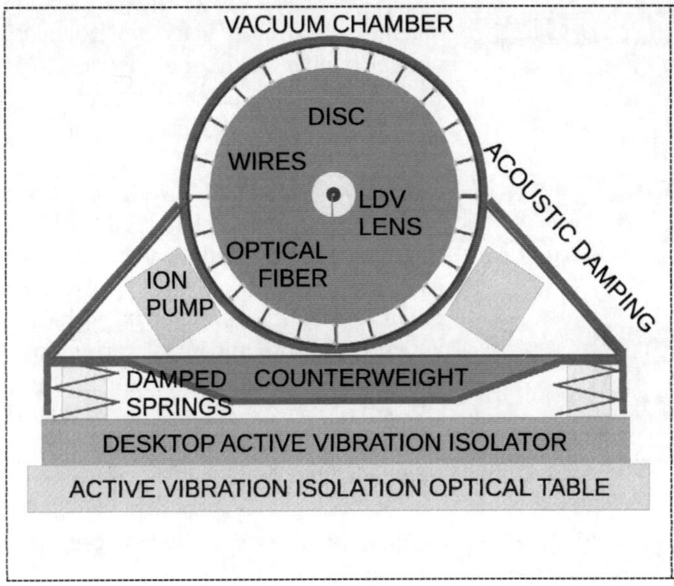

Figure 4. The two main challenges in the design of the generator are keeping the oscillation within a very narrow band and preventing the transmission of non-gravitational signals. The main solutions are fine-tuning the mechanical and control systems and adding plenty of damping and isolation. The third challenge is delivering the minimum quantity of energy to drive the maximum amplitude of oscillation: a side effect of the thick covering used for obtaining a high grade of acoustic damping is indeed thermal isolation. The main thermal dissipation mechanism is conduction from the shaker to the disc, radiation to the vacuum chamber, then conduction to the counterweight.

The vacuum chamber of generator is a cylindrical chamber made of copper for emf radiation shielding and thermal conductivity. It has a base hosting two ion pumps and a counterweight under the base. The whole system is suspended over damped springs on top of a table provided with active vibration isolation. Active isolation has the advantage of being able to isolate vibration sources on top of the table in addition to ground vibration. All the surfaces of the generator that can transfer acoustic energy to the surrounding air are covered with layers different highly effective acoustic damping products like Moongel® made by RTOM Corporation and Sorbothane® made by Sorbothane, Inc.

The same approach to acoustic and vibration isolation used for the generator applies to the antenna, with the fundamental difference that the antenna does not generate energy of any kind, but it is extremely receptive to external energy inputs.

The antenna is a silicon oscillator clamped vertically to a pure copper structure specially designed for thermal and vibration management of the highest Q mode of the oscillator. Its high-vacuum

IOP Conf. Series: Journal of Physics: Conf. Series **1149** (2018) 012020 doi:10.1088/1742-6596/1149/1/012020

chamber is also made of copper for electromagnetic isolation (alternatively, a chamber of mu-metal can be used). An optical flange allows the connection of the optical fibre of the LDV, while a lens holder is fixed inside to point the lens of the LDV to one of the two pads of the silicon oscillator. The chamber is also surrounded by layers of different acoustic isolation materials like Moongel® and Sorbothane®.

The vacuum chamber is bolted on top of its ion pump and both are displaced by a motorized delay line used as a linear stage positioner like model 8MT160-300 by Standa Ltd. The main characteristic of the positioner must be the straightness, and the rigidity and stability when not in motion. The delay line is bolted to a desktop active vibration isolator like model 1TS-140 by Standa Ltd. placed on top an optical table with active vibration isolation like model T1020CK by Thorlabs, Inc.

On top of the table of the antenna, several sensors are also hosted to monitor the noise levels along with the controllers for the linear positioner, the controller of the ion pump, the controller of the LDV, the lock-in amplifier of the antenna and a DAQ with SSD storage.

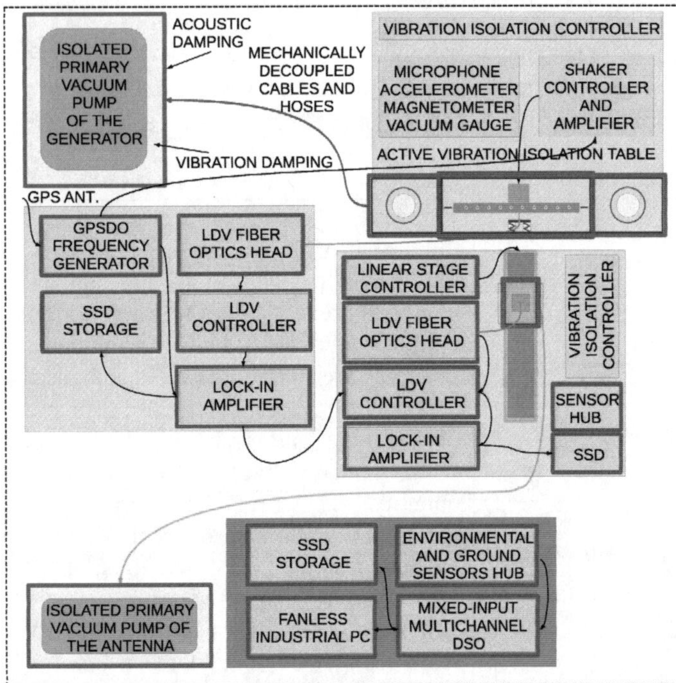

Figure 5. The experiment relies on a high level of control of spurious signal generation and management of internal and external noise sources. The generator and the antenna are located on optical tables separated by a small distance. A third actively isolated optical table can be used for instruments of the generator that need special isolation. Other vibration-resistant devices can be located on a passively isolated table, while primary vacuum pumps must be fitted with a high level of acoustic and vibration damping. Special attention must be given to coupling through the cables, which can also be potential carriers of sound and vibration. Electromagnetic emissions can be managed reducing power levels and using emf shielding materials like copper.

The need for vibration and acoustic isolation can be easily quantified: as specified before, an oscillation of $A_d = 1 \cdot 10^{-5}\ m$ in the generator corresponds to an oscillation of $A_a = 3.3 \cdot 10^{-14}\ m$ of the antenna, then the gravitational transmission has an attenuation of $Loss_g = -170\ dBm$ (where m indicates metres not millivolts). If we desire a SNR=10 the loss of the vibrational and acoustic counterparts of the gravitational transmission must be $Loss_v = -190\ dBm$ and $Loss_s = -190\ dBm$. To realize this, it is necessary to use a system of multiple isolation layers. For example, to minimize the transmission of vibrations, it is possible to apply at least four active mechanical filters providing each a $Loss_{active} = -40\ dBm$ and rely on passive filters, including springs, dampers and natural loss in the ground propagation between tables, for the remaining attenuation $Loss_{passive} = -30\ dBm$, which is within reason. In practice, both the generator and the antenna must be placed on top their own desktop active vibration isolator and each active isolator must be placed on top its own optical table with active isolation.

The required acoustic isolation of the antenna implicates a similar split of the attenuation budget between the generator and the antenna. Acoustic absorption and deflection technologies are very mature anyway, and high frequency sound, like high frequency vibration, is easier to attenuate, so in this case the only minor challenge is how to keep the layers of sound damping materials from becoming unacceptably thick. The solution anyway is focusing the damping on the required frequency, even using, if necessary, materials with a non-linear response at 2000 Hz.

The procedures for calibration, data taking, data analysis and the determination of gravitational parameters including the value of the universal gravitational constant will be soon the subject of a further publication by the author, while a third paper could be focused on results of numerical multi-physics simulations and preparatory tests.

4. Conclusions

The current level of commercial LDV technology has already enabled gravitational measurements that were impossible until recently without designing and assembling custom laser measurement devices. The prospective of an improving dimensional resolution in laser Doppler vibrometry [3] will allow new measurement scenarios in experimental gravity, advancing basic research on fundamental constants and forces.

Acknowledgments

I deeply thank all the collaborators, professors, industry experts and friends who have supported me in this project with technical and scientific advice, review and encouragement.

References

[1] Serra E Bonaldi M Borrielli A Conti L Pandraud G Sarro P M 2014 Selective coating deposition on high-Q single-crystal silicon resonators for the investigation of thermal noise statistical properties *Procedia Engineering* **87** pp 1485-1488 (10.1016/j.proeng.2014.11.579)

[2] Tatum J B 2018 Classical Mechanics **5** p 13 *University of Victoria's website* (Lecture notes http://astrowww.phys.uvic.ca/~tatum/celmechs.html)

[3] Rembe C Kadner L Giesen M 2016 Approaching attometer laser vibrometry *Review of Scientific Instruments* **87** 102503 (https://doi.org/10.1063/1.4964625)

[4] Chen Y T Cook A 1993 Gravitational experiments in the laboratory, *Cambridge University Press*

[5] Astone P et Al 1998 Experimental study of the dynamic Newtonian field with a cryogenic gravitational wave antenna *The European Physical Journal C* **5** 651-664 (https://doi.org/10.1007/s100529800987)

[6] Ferretti B 2003 On the gravitational field (A suggestion about a possible experimental research) (arXiv:gr-qc/0307030)

[7] Ferretti B 2005 On the gravitational field. A suggestion about a possible experiment (arXiv:gr-qc/0507120)

IOP Conf. Series: Journal of Physics: Conf. Series **1149** (2018) 012021 doi:10.1088/1742-6596/1149/1/012021

Gravity acceleration measurement based on Atom interferometry

Dong Hu[1], Yu Wang[1*], Huijuan Ma[1], Ning Ru[1] and Li Zhang[1]

1National Key Laboratory of Science and Technology on Metrology & Calibration, Changcheng Institute of Metrology and Measurement, Huanshan Village, Wenquan Town, Haidian District, Beijing 100095, China

Corresponding e-mail address : wangyu@cimm.com.cn

Abstract. Cold atom interferometer based on two-photon stimulated Raman transition has been proved to be a promising way to measure g with higher precision. In this work, we first introduce the principle of atomic interference gravimetry, including the interaction between the atoms and light, the interferometry with the Raman beams and the measure of the g. Then our gravity acceleration measurement system will be introduced, including atomic fountain, Raman system and detection system. With this device, we can get the value of the g, with the uncertainty in the order of ten minus seven.

Keywords: Gravity acceleration, Atom interferometry, Atomic fountain, Raman transition
PACS: 04.80.Cc, 03.75.Dg, 37.10.De

1. Introduction

In 1991, Steven Chu and Mark Kasevich proposed the atom interferometry with three Raman pulses[1]. They measured the value of gravity with that kind of atom interferometry. After that, there are more and more works which concentrate on the gravity measured by atom interferometry[2-4]. What's more, it can be used to measure the gravitational constant[5], the inclination[6]. There are also some gravimeters based on coherent Bragg diffraction of freely falling cold atoms[7]. The compact Gravimetric Atom Interferometer is also developed to make the gravimeter portable and movable[8]. The gravimeter also can be used to test the universality of free fall with atoms in different spin orientations[9].

In this article, we first introduce the basic theory of the Gravity acceleration measurement based on Atom interferometry. Then we introduce the experiment setup. At last, the results of this experiment will be showed.

2. Theory

The principle of gravity acceleration measurement based on Atom interferometry has been described before[10]. Figure 1 is the sketch map of the interferometry. Through the figure we can find the atom internal state evolution through two ways after the first half π pulse. Then finally combined together with a π pulse and another half π pulse. We can get the interference fringe by detecting the population of the atom internal state. In this kind of atom interferometry, there are two mechanisms which contributes to the phase of the interferometry's fringe: the phase shift contributed by the gravity and the phase shift contributed by frequency sweeping of the Raman light.

$$\phi = \mathrm{k_{eff}} g T^2 - \alpha T^2 \quad (1)$$

g is the local gravitational acceleration, T is the interval time in figure 1,$\mathrm{k_{eff}} = k_1 - k_2$ is the effective wave vector,α is the chirp rate of the Raman beams.

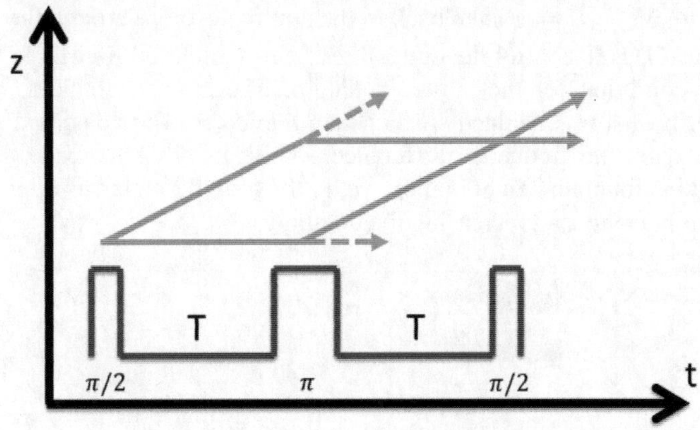

Figure 1. The sketch map of the interferometry.

We can get fringe by sweeping the α. In the atom's interference fringe, the minimum in the fringe in relation with $\phi = \mathrm{k_{eff}} g T^2 - \alpha T^2 = 2n\pi, n = 0,1 \ldots$ The value of g is determined by the case of n = 0. For n=0, the fringe will always be the trough for arbitrary interval time. This can help us to deduce the value of the gravity acceleration.

3. Experiment setup

Our system consists of several parts(see figure 2). The left side is the vacuum system where the atomic fountain and atom interference happen. The right side is the optical system with two layers correspond to magnetic optical trap(MOT) system and Raman system individually.

Figure 2. The photo of the gravimeter system.

IOP Conf. Series: Journal of Physics: Conf. Series **1149** (2018) 012021 doi:10.1088/1742-6596/1149/1/012021

The cold atoms are captured in the MOT which combines the optical part(see figure 3(a)) and the magnetic part which generated by a pair of anti-helmholtz coils. In the experiment, we locked the cooling laser on the cross of $5S_{1/2}, F = 2 \rightarrow 5P_{3/2}, F' = 2$ and $5S_{1/2}, F = 2 \rightarrow 5P_{3/2}, F' = 3$ by the Dichroic Atomic Vapor Laser Lock(DAVLL). Then acoustic optical modulators(AOMs) are used to make the light slightly red detuned to cooling the atoms. Besides that, we also have a repumping light to get the atoms which dropped to $5S_{1/2}, F = 1$ state back to the cooling cycle. Through the figure 3(a), we can find that there are two AOMs to control the upper three beams and the lower three beams individually which is a necessary condition for the atomic fountain. The atoms will be thrown upward when frequency of the upper beams was detuned smaller than the lower. The velocity and the height of the fountain are determined by the frequency difference. While in order to make the atomic fountain temperature lower and the fountain height higher, we do the moving molasses after launching stage by ramping down the light intensity and increasing the detuning.

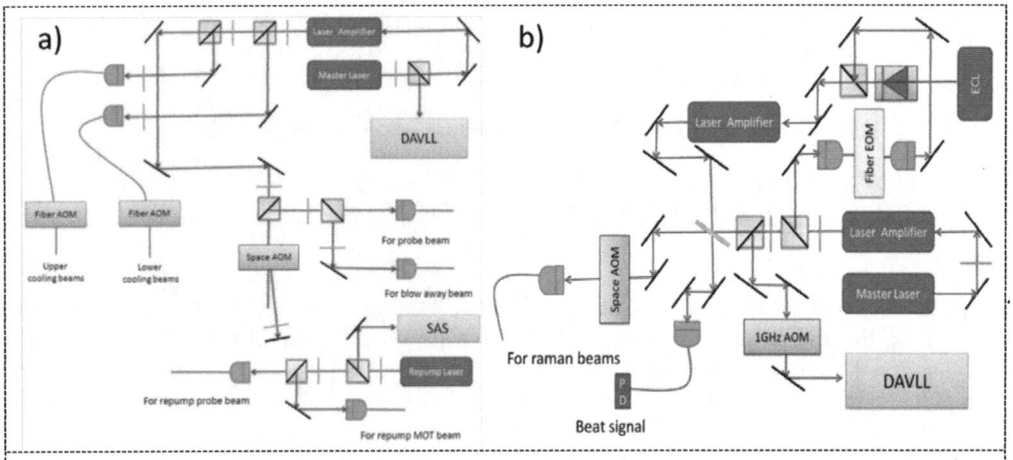

Figure 3. The optical system for the MOT(a) and the Raman(b).

The sketch map of the atomic fountain is shown in figure 4. Besides the atomic fountain, we still need the Raman system to get the interference fringe. The Raman light is locked by the electrooptic modulation phase lock, the main problem for this method is the 0 order and the ±1 order after the EOM cannot be divided. The combined light will has obviously power fluctuation which has an adverse effect on the atom interferometer. To overcome this problem, we use the grating as the frequency selection device. We inject the modulated light to a homemade external cavity diode laser(ECDL), which can filter the 0 order light that has a 6.8GHz shift with the mode which we chose. What's more, this light can be used as the seed for the taper amplifier. The detuning of 1GHz is produced by an AOM, the main Raman laser's frequency was frequency shifted before locked on the transition line.

Detection system(as figure 5 shows) is also needed in the experiment which can help us to deduce the atom's temperature and the population of the states. We use a thin light as the detect light. With the difference of the atom's velocity, the time it gets to the detection area is different. The photodiode's signal is changed with time. So it denotes the velocity attribution of the atoms which can be used to do the Gauss fitting to get the temperature. For the state detection, we have two layers of detection system with a repumping light in the middle. After the atoms in $F = 2$ state be detected, the repumping light will get the atoms in $F = 1$ state to $F = 2$ state. So we can judge the number of the atoms in $F = 1$ state by the signal which detected later.

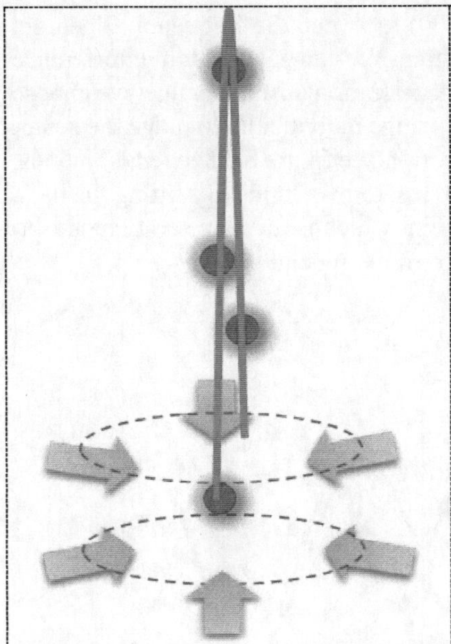

Figure 4. The sketch map of atomic fountain.

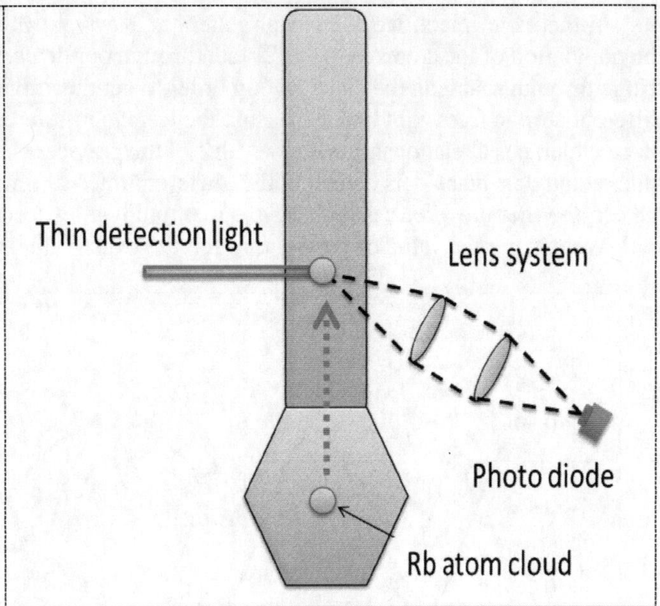

Figure 5. The diagram of the detection system.

4. Experiment results

We first measure the atomic fountain's temperature to judge whether the fountain is OK or not. Through the fitting in figure 6, we can get the temperature is about 10 μK, which can satisfy our need in the gravimeter.

For the Raman process, the very important thing is to choose the proper time of the Raman pulse. By changing the time of the Raman duration we can get a Rabi oscillation, through which we can got the proper Raman duration for the subsequent experiment as shown in figure 7.

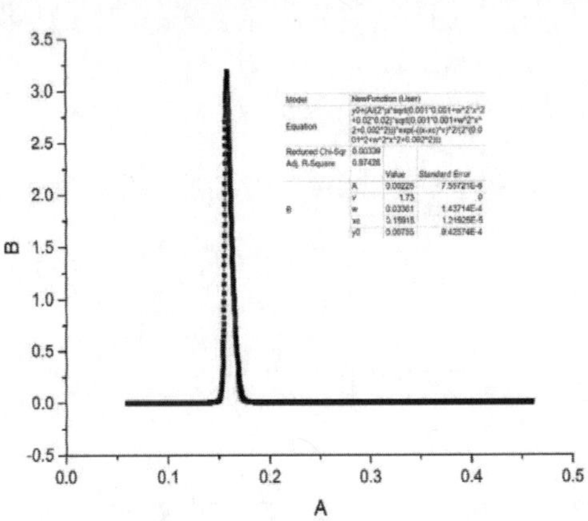

Figure 6. The fitting of the temperature of atomic fountain.

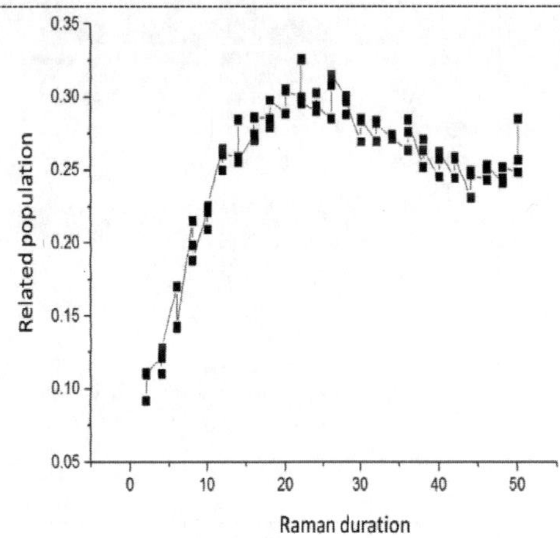

Figure 7. The Rabi oscillation with the different Raman durations.

In the experiment, we use signal generator(Keysight E8257D) to sweep the frequency. Then get the population of the atoms in the F=2 state through our detect system. We can get the atom interference fringe through scanning the chirp rate. In order to get the gravity acceleration absolute value, we choose different atomic free flight times, measure the Raman interference fringe individually to judge the center place which has a relationship with the value of the gravity acceleration(see figure 8). The red diamonds, blue rectangles, black dots correspond to different interval time 10 ms,15 ms, 20 ms. By fitting the data, we can see that the three cases all have a maximum at a special point which is the place corresponds to n=0. We can get the value of gravity through the exact value of α in the special point.

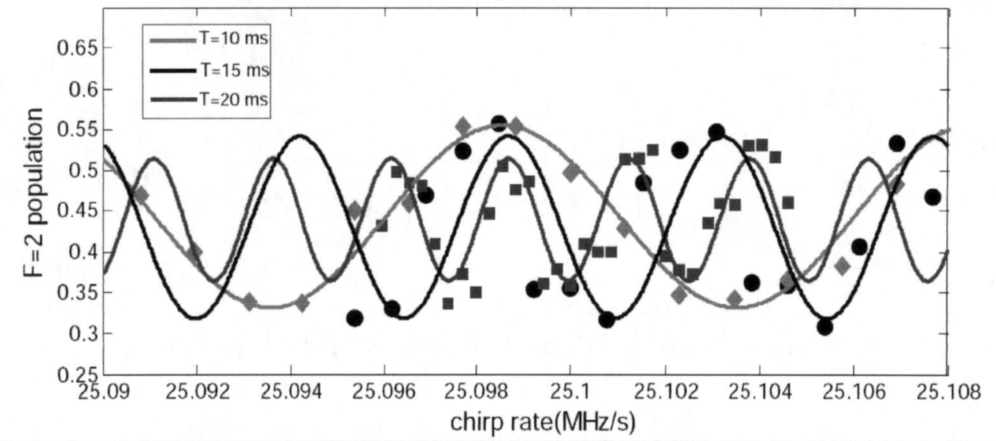

Figure 8. The atoms population change with different chirp rate.

In order to investigate the stability of gravity acceleration measurement, we continuous measure the value and got the Allan deviation as the figure 9 shows.

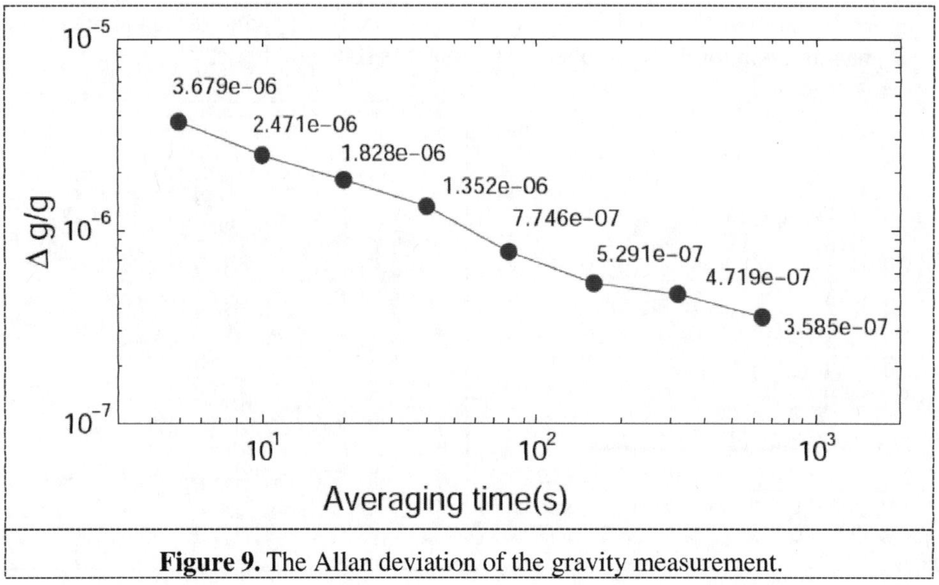

Figure 9. The Allan deviation of the gravity measurement.

5. Conclusions

In this article, we proposed our gravimeter system and some results. The gravimeter is based on atom interferometry. The atomic fountain is realized by applying a frequency shift between the upper and lower light. The Raman system is accomplished by injection-locking of a master laser to a slave laser. The interference fringe and Allan deviation have been showed. We will try to make the atomic fountain temperature lower next. With a higher fountain, we can prolong the interval time. The longer interference time will lead to a more precise value of the gravity.

References

[1] M Kasevich, S Chu 1991 atomic interferometry using stimulated raman transitions Phys.Rev.Lett.67,181

[2] M De Angelis, A Bertoldi, L Cacciapuoti, A Giorgini, G Lamporesi, M Prevedelli, G M Tino, 2009 Precision gravimetry with atomic sensors. Measurement Science and Technology, 20(2).

[3] Z K Hu, B L Sun, B, X C Duan, M K Zhou, L L Chen, S Zhan, J Luo. 2013 Demonstration of an ultrahigh-sensitivity atom-interferometry absolute gravimeter, 043610, 1–6.

[4] M K Zhou, Z K Hu, X C Duan, B L Sun, L L Chen, Q Z Zhang and J Luo, 2012 Performance of a cold-atom gravimeter with an active vibration isolator, 043630, 5–9.

[5] F Sorrentino, Y H Lien, G Rosi, L Cacciapuoti, M Prevedelli, and G M Tino, 2010 Sensitive gravity-gradiometry with atom interferometry: Progress towards an improved determination of the gravitational constant. New Journal of Physics, 12.

[6] W J Xu, M K Zhou, M M Zhao, K Zhang, and Z K Hu 2017 Quantum tiltmeter with atom interferometry, PHYSICAL REVIEW A 96, 063606.

[7] P A Altin, M T Johnsson, V Negnevitsky, G R Dennis, R P Anderson, J E Debs, S S Szigeti, K S Hardman, s Bennetts, G D McDonald, L D Turner, J D Close and N P Robins. 2013 Precision atomic gravimeter based on Bragg diffraction, , New J. Phys. 15, 023009.

[8] M Hauth, C Freier, V Schkolnik, A Senger, M Schmidt, and A Peters. 2013 First gravity measurements using the mobile atom interferometer GAIN. Applied Physics B: Lasers and Optics, 113(1), 49–55.

[9] X C Duan, X B Deng, M K Zhou, K Zhang, W J Xu, F Xiong, Z K Hu. 2016 Test of the Universality of Free Fall with Atoms in Different Spin Orientations. Physical Review Letters,117(2), 1–6.

[10] K Moler, D S Weiss, M Kasevich and S Chu. 1992 Theoretical analysis of velocity-selective Raman transitions. Physical Review A, 45(1), 342–348.

IOP Conf. Series: Journal of Physics: Conf. Series **1149** (2018) 012022 doi:10.1088/1742-6596/1149/1/012022

Towards laser Doppler vibrometry from unmanned aerial vehicles

Dr Ben J Halkon[1] & Prof. Steve J Rothberg[2]

[1]Centre for Audio, Acoustics & Vibration, Faculty of Engineering and IT, University of Technology Sydney, Ultimo, NSW 2007, Australia
[2]Wolfson School: Mechanical, Electrical and Manufacturing Engineering, Loughborough University, Loughborough, Leics., LE11 3TU, U.K.

Benjamin.Halkon@uts.edu.au

Abstract. Laser Doppler vibrometers are technically well suited to general application but they offer special benefits in a variety of challenging measurement scenarios which are now well documented and accepted. An interesting and potentially powerful example of such a challenging measurement scenario is one where the laser vibrometer is mounted on/in an unmanned aerial vehicle in order that autonomous measurement campaigns can be undertaken in remote and/or harsh environments. One important challenge to overcome in such a scenario is the measurement sensitivity to vibration of the instrument itself or indeed of any steering optics used to point the probe laser beam toward the target of interest. In this paper, recently reported means by which this measurement sensitivity can be rectified by simultaneously obtained correction measurements will be developed. Specifically, this development is intended to lead towards laser Doppler vibrometry from unmanned aerial vehicles (UAVs) with correction of instrument motion being presented herein for the first time from a single, rather than a pair of, uniaxial accelerometers.

1. Introduction

Laser Doppler vibrometers (LDVs) measure target surface vibration velocity and are technically well-suited to general application with benefits over traditional contacting transducers in a range of challenging measurement scenarios [1]. Until quite recently [2]-[4], however, little attention had been given to i) a quite fundamental aspect of operation which is that the measurement is of <u>velocity relative to the instrument itself</u> and ii) measurement scenarios in which interpretation of the data must consider instrument or laser beam steering optic motion. Instead, practical application of LDV has typically either involved mounting of the instrument on a stable platform, i.e. on a tripod, or, in the case of significant levels of ambient ("base motion") vibration, the use of passive or active anti-vibration mounting arrangements [5]. Often such measures will be sufficient to yield high quality outcomes but an alternative, in which the instrument vibration can be fully compensated for across a broad frequency range, would be clearly preferable. Such compensation would be particularly beneficial where the measurement campaign involves the mounting of the instrument onto a platform in which the vibration level might be comparable with that of the target of interest, for example when making measurements from unmanned or other aerial vehicles.

IOP Conf. Series: Journal of Physics: Conf. Series **1149** (2018) 012022 doi:10.1088/1742-6596/1149/1/012022

Previously, comprehensive vector-based mathematical treatments of the velocity measured for both scenarios in which significant levels of i) base motion vibration [2] and ii) steering optic vibration [3] have been presented. Substantial experimental investigations have conclusively verified both scenarios providing practical compensation schemes based on traditional contacting vibration measurement transducers which can be conveniently located on the instrument or steering optic itself. For realistic vibration levels for both target and instrument/steering optic vibration, reductions of at least 37/47 dB, respectively, were demonstrated. Ultimately in both cases, real-world measurements were presented, in accordance with conventional noise and vibration engineering paradigms, showing that the required accelerometer compensation measurements yield a substantially enhanced understanding of the dynamic characteristics of the systems under test.

More recently, these earlier contributions have been reviewed in such a way that the practising vibration engineer might readily realise the measurement error reduction benefits [4]. In the case of instrument vibration correction, which is the primary focus of *this* article, the recommendation up to now, given the need to determine the instrument vibration in the direction of the probe laser beam, was to mount a pair of accelerometers equidistant around the probe laser beam axis. A single compensation accelerometer was shown to be viable in the case of only translational vibration, with angular vibration components present resulting in reduced correction performance. A single correction accelerometer would, of course, be optimal, indeed certain commercially available systems come complete with a mounting location for just such a single "compensation sensor". In this paper, therefore, a revised scheme for complete correction of LDV measurements, taken in the presence of substantial instrument vibration, based on the use of a single accelerometer is presented for the first time. A further, novel contribution of this paper is to make use of variable capacitance "DC response" accelerometers, rather than the previously utilized traditional general-purpose piezoelectric equivalent, for correction.

2. Accelerometer relative calibration

When embarking on a new instrument vibration correction measurement campaign, the first step is to "calibrate" the *correction* transducer(s). Additionally, where a "true" target vibration reference measurement is required (to prove the principle of the instrument vibration correction technique [4]) and this measurement is to be from a contacting transducer (rather than another, "Fixed LDV" [2]), this also applies to the *target* transducer. Indeed, the full measurement chain must be calibrated and this initial transducer "accelerometer relative calibration" stage takes two distinct forms: amplitude and phase relative (to the LDV) calibration.

Firstly, and consistently with all traditional measurement campaigns, the actual sensitivity of the accelerometer must be determined. In traditional measurement campaigns, this is achieved using an accelerometer calibrator such as the B&K Type 4294. Such devices generate a controlled amplitude single frequency component, e.g. a sinusoid at 159.2 Hz (1000 rad/s), against which the output of the accelerometer(s) can be verified, and the sensitivity(ies) accordingly adjusted. In the case of instrument base motion correction, however, the LDV will be used as the reference vibration (velocity) signal with the vibration being (arbitrarily) generated using a broadband (white) noise signal amplified to drive an electrodynamic shaker on which the accelerometers are mounted. Figure 1 shows the typical experimental arrangement in which the LDV probe laser beam is directly in line with the accelerometer sensitive axis (and, of course, positioned at the optimal stand-off distance).

Using a frequency domain integration approach [2],[3] to convert the accelerometer signal into velocity for direct comparison with that of the LDV, the RMS of the difference between the two signals within the frequency range of interest (2.5 – 100 Hz in this case) can be determined. The ratio of these two values for the accelerometer and the LDV can clearly be used to revise the accelerometer sensitivity such that the ratio ultimately becomes 1. Typically, this calculation is performed for a minimum of five averages and for two distinct levels of vibration amplitude. Figure 2 shows typical example accelerometers vs. LDV data *after* amplitude calibration. The curves may be challenging to differentiate from one another when reproduced in greyscale; that, however, is the very point – that, for each of the two vibration levels shown, the (integrated) accelerometer signals are indistinguishable from that of the LDV.

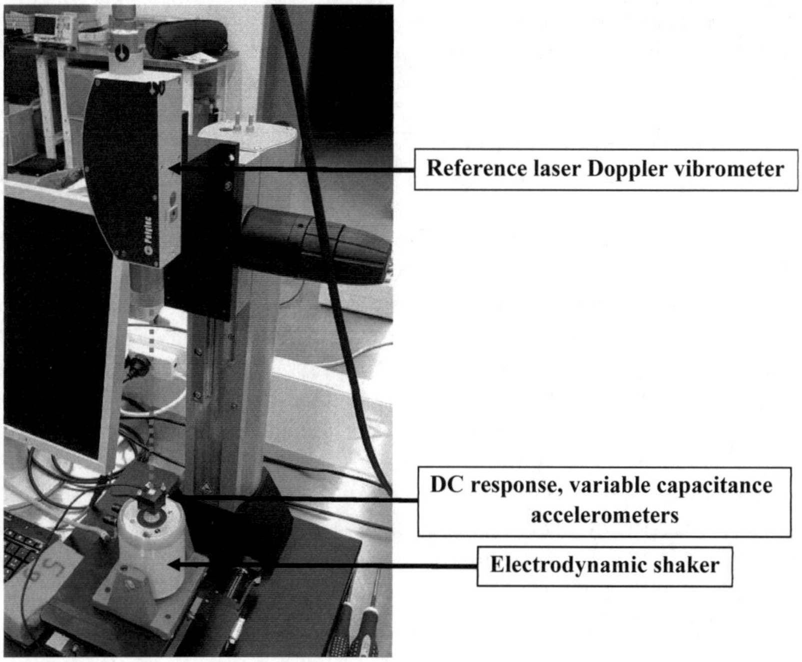

Figure 1. Typical accelerometer calibration experimental arrangement (with laser beam path highlighted)

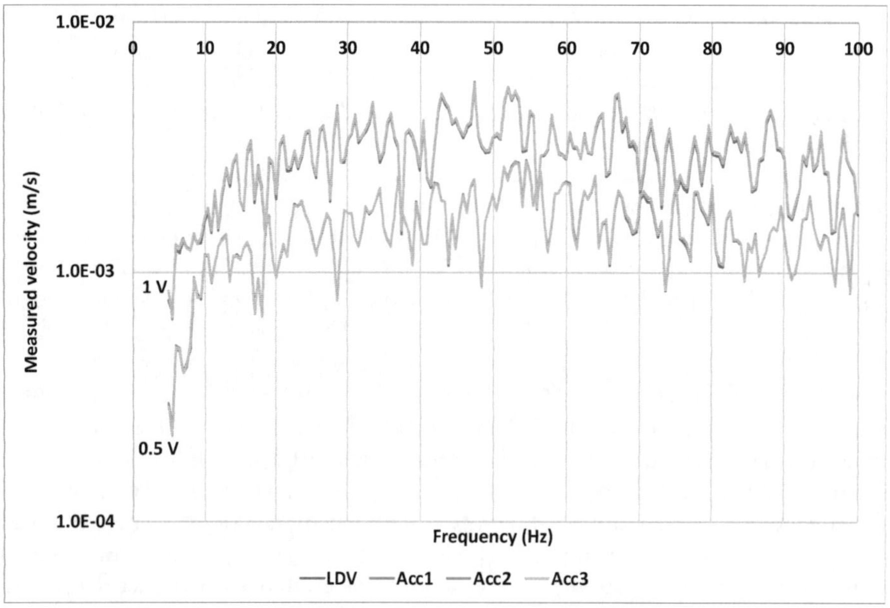

Figure 2. Typical *amplitude* difference comparisons LDV vs. accelerometers

In addition to *amplitude* calibration, *phase* differences between the accelerometer and LDV, mostly attributed to relative time delays between the signal conditioning electronics in each, also need to be accounted for. Broadband calculations are again performed in frequency domain processing. In this case, the absolute unwrapped phase differences between the accelerometer and LDV signals are averaged on a frequency-by-frequency basis for a minimum of five averages. The ratio of the mean of these values to the mean angular frequency of the frequency range of interest (2.5 – 100 Hz) is the effective time delay between the two signals. Figure 3 shows typical phase difference data *prior to* time delay correction. Again, in greyscale these various signals will be challenging to differentiate from one another but, again, this is indeed the very point – the three similar accelerometers should be expected to have nominally the same characteristics with respect to the LDV. The effect of the finite time delay can be seen in these curves as the underlying upward slope. In essence, the algorithm determines the gradient of the slope from which the time delay is derived (τ = slope gradient / 2π); clearly some noise immunity is necessary. Taking the mean for two different vibration levels yields values per accelerometer as shown in Table 1 below. These values are then readily incorporated into the subsequent frequency domain data processing when using the accelerometers to correct the measured, vibrating LDV signal. This will be described in the following subsection.

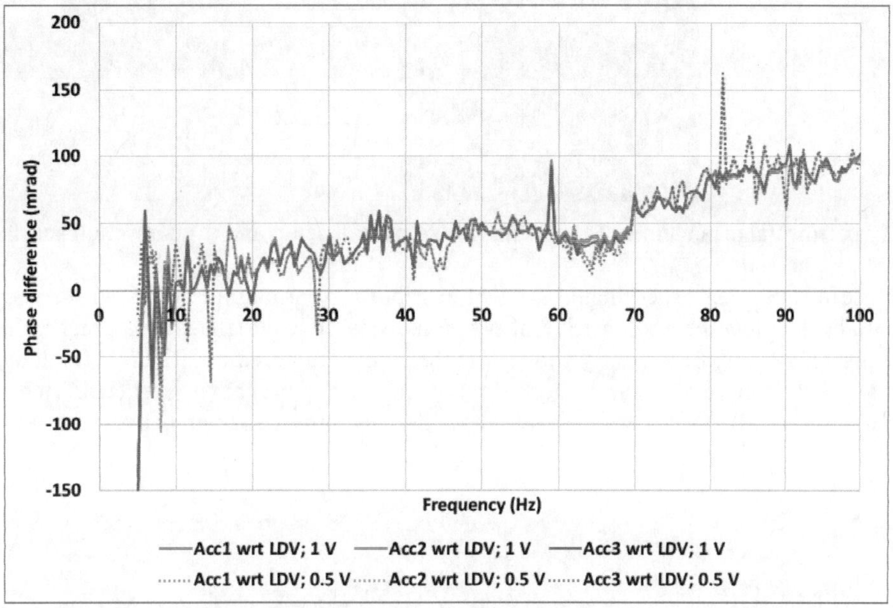

Figure 3. Typical *phase* difference comparisons LDV vs. accelerometers

Table 1. Example accelerometer vs. LDV time delays

Acc. model	Acc. s/n	Time delay (ms)		
		1V	0.5V	Mean
770F-10-U-120	10046	0.137	0.139	0.138
770F-10-U-120	10047	0.136	0.139	0.138
770F-10-U-120	10048	0.139	0.143	0.141
3225F	15763	-0.044	-0.046	-0.045

3. Instrument vibration correction

Sensitivity to instrument vibration (in the direction of the laser beam) was readily confirmed by a straightforward experiment, with an arrangement as set out in figure 4. Here, a "Vibrating LDV" sensor head was mounted on an electrodynamic shaker and excited with a broadband signal. The speaker cone, being relatively lightweight and of small radius of curvature, was fitted with a Dytran 3225F "teardrop" piezoelectric accelerometer to provide the 'true' "Vibrating target" measurement. With nominally zero target vibration, the true accelerometer reference measurement is clearly reflective, as shown in figure 5, of the target vibration whereas the LDV measurement is not. The LDV measurement is 100% sensitive to the vibration of the instrument itself as expected and previously shown in significant detail [2],[4].

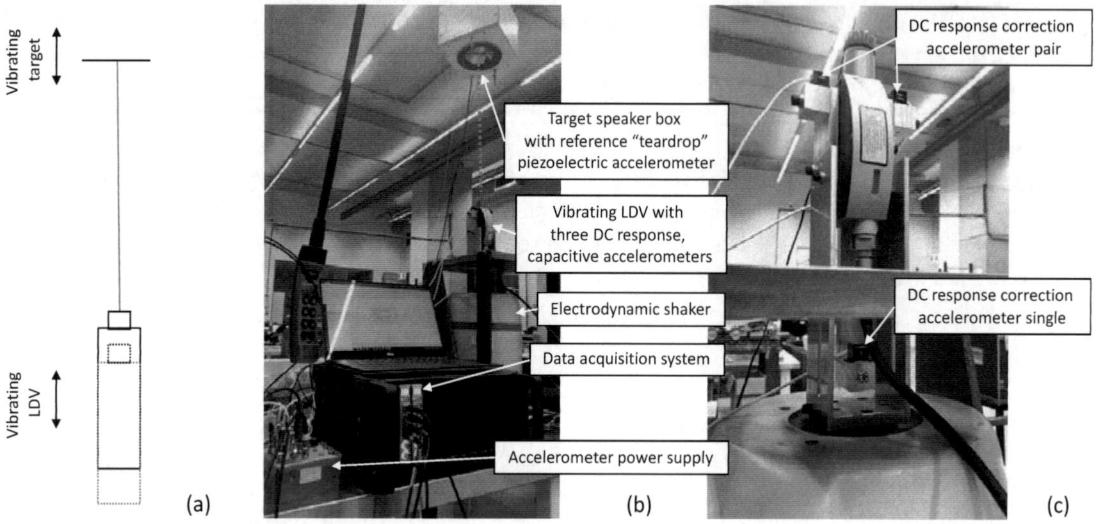

Figure 4. Experimental arrangement (a) schematic, (b) overall setup and (c) correction accelerometers

Making use of the same experimental arrangement but now arranging both instrument *and* target vibration (of absolute levels representative of those likely to be experienced in a practical real-world measurement scenario), figure 6 clearly shows that the signal measured by the LDV (long dash curve) is substantially different to the true target vibration (short dot curve) as expected. Here the most significant, approximately one order of magnitude, differences are seen to be between c5 and c20 Hz.

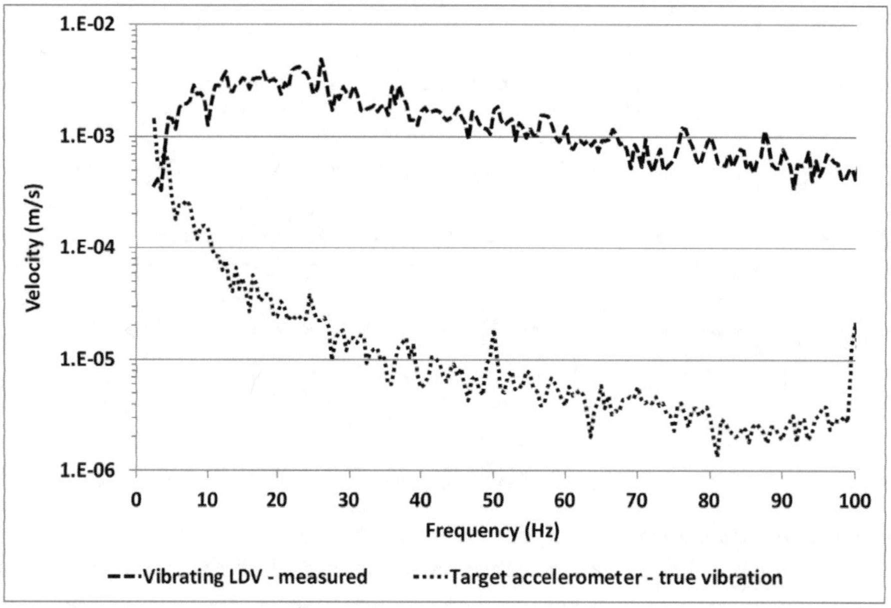

Figure 5. LDV vs. true target vibration velocity: instrument vibration, no target vibration

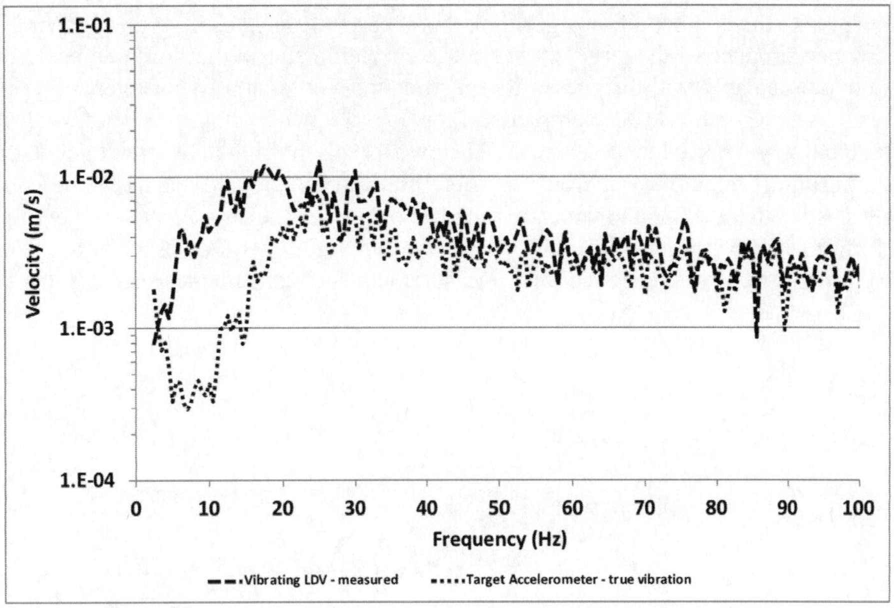

Figure 6. LDV vs. true target vibration velocity: instrument vibration and target vibration

As set out in significant theoretical and experimental detail previously [2], for complete correction of the erroneous LDV measurement, a simultaneous measurement of the instrument velocity at some point along the laser beam axis is required. This can be straightforwardly shown as:

$$U_{corr} = U_m - U_0 \quad (1)$$

where U_m is the measured velocity, U_{corr} the corrected velocity and U_0 the required correction measurement. Given that positioning a single transducer *forward* of the instrument would clearly result in a blocking of the laser beam, the required correction measurement was previously obtained by a pair of accelerometers positioned equidistant about, with their sensitivity axes parallel to, the laser beam axis.

Figure 4c shows a practical example of this accelerometer pair positioning strategy where a simple bracket and a pair of mounting blocks are used to provide accelerometer mounting locations. In other commercially available instruments, mounting is often even more straightforward since the laser beam optical axis is typically normal to a flat surface which is the front of the sensor head body. There is such a surface on the Polytec NLV-2500 Compact Laser Vibrometer used in this study but it is not sufficiently expansive for the relatively large footprint Endevco 770F-10-U-120 variable capacitance accelerometers. Figure 4c also shows an alternative instrument vibration compensation sensor arrangement based on a single accelerometer mounted on the laser beam axis but *behind* the instrument sensor head. This alternative option is presented and compared with the previously proven accelerometer pair here for the first time.

Integrating these accelerometer measurements in the frequency domain, incorporating the finite time delays as set out in Table 1 and subtracting the appropriate correction measurement in accordance with equation (1) leads to *corrected* LDV signals as shown in figure 7. Here, as expected, complete correction of the LDV measurement is shown for both cases of the accelerometer pair (bold solid curve with +) and the single accelerometer (bold solid curve); indeed, they are indistinguishable from one another. The dB reduction in the error is 17.5 dB in both cases. Phase error, as shown in figure 8, is comfortably within 100 mrad between 20 and 100 Hz; below 20 Hz phase error is larger as would be expected.

There is a small exception to the excellent agreement between the corrected LDV signal and the true vibration reference measurement and that is in the low frequency region below c10 Hz. Rather than this being a problem with the compensation algorithm, this is certainly as a result of the comparatively poor low frequency performance of the target piezoelectric accelerometer in that low frequency range (it is performing quite acceptably with respect to its specification – it is simply not intended to provide as reliable a result as a DC response accelerometer in this low frequency range). When the accelerometer signal is integrated, any error in the measured data below 20 Hz is multiplied by a relatively large number ($1/\omega$) and the resulting velocity can be seen to "tick up" whereas the corrected LDV measurement is more reliable since both the LDV and the correcting DC response accelerometers have very reliable low frequency performance. Future studies will make use of an additional DC response accelerometer or another LDV for the "true" reference vibration measurement such that a better agreement is achieved.

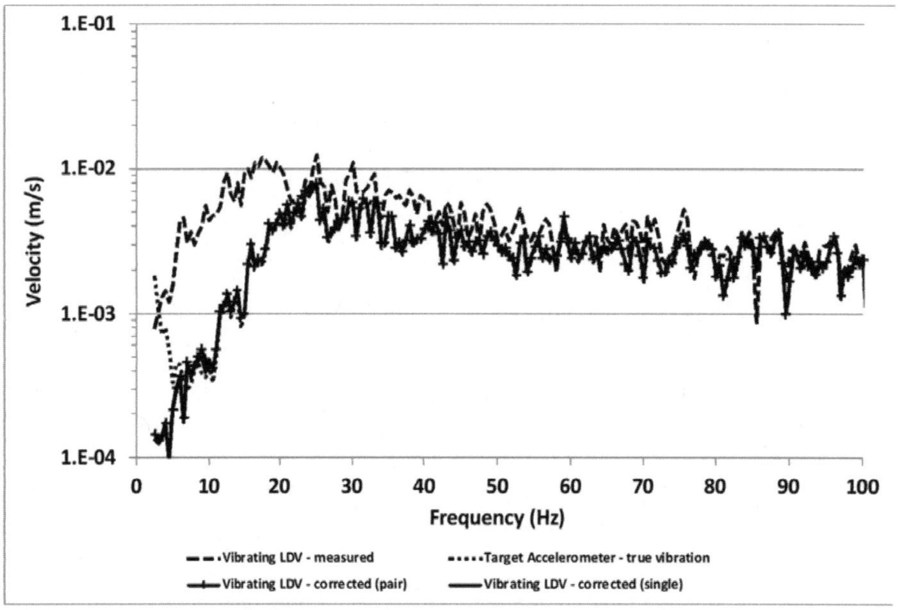

Figure 7. Corrected LDV measurement for a pair and a single correction accelerometer

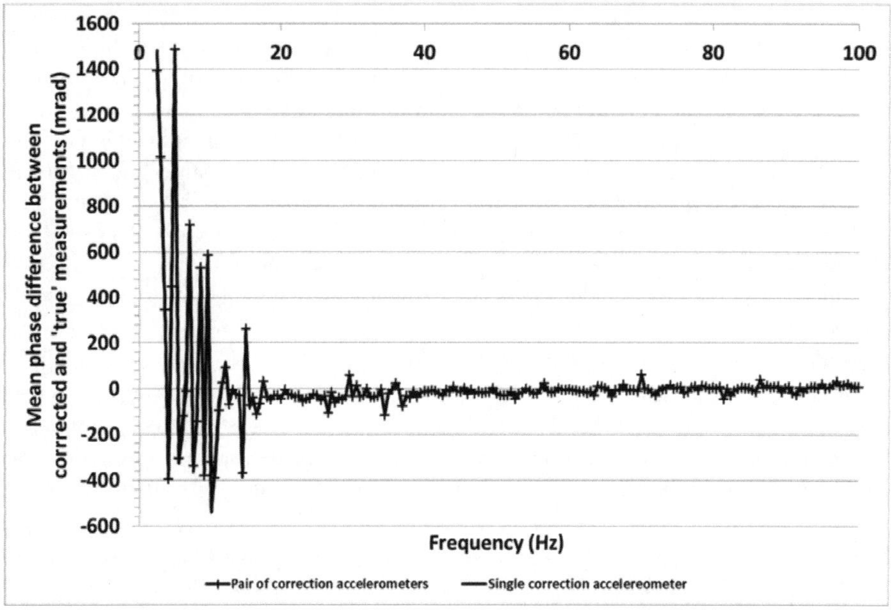

Figure 8. Mean phase difference between corrected LDV and true vibration measurements

4. Conclusions

This paper has reviewed the practical means by which an erroneous LDV measurement, due to significant levels of instrument vibration, can be corrected by subtracting a simultaneous measurement of the instrument vibration at some point along the axis of the laser beam. The means by which the required correction accelerometers should be calibrated both in terms of amplitude *and* phase (finite time delay) has been presented with newly obtained measurement data included herein for the first time. A measurement scenario, with typical real-world levels of instrument and target vibration, was presented, again incorporating data presented herein for the first time. Recommendations for the practical application of this, readily deployable by the practising noise and vibration engineer, are included. Novelties in this article include the use of DC response accelerometers to provide the correction measurement and, the use of a single accelerometer, rather than a pair, mounted behind the instrument sensor head. For the vibration levels used, error reduction of 17.5 dB was achieved between uncorrected and corrected measurements with phase agreement within 100 mrad between 20 and 100 Hz.

In order to achieve better correction at frequencies down to DC, which is important for applications such as seismic measurements and handheld application, correction measurement transducers with excellent low frequency performance should be utilised. Similarly, comparison measurements of the true vibration, for the sake of confirming the validity of the technique, should have equally excellent low frequency performance. In future, a DC response accelerometer or another LDV will be used to determine the true target vibration. The piezoelectric accelerometer used in this study was chosen for its small footprint and light mass which made for easy and reliable attachment to the light speaker cone domed surface. However, it has the problem of relatively unreliable low frequency performance after integration to velocity (for comparison with the LDV measurement) which causes the signal to "tick-up" as shown in figures 6 and 7 at frequencies below 5-6 Hz. In order to move "towards LDV from unmanned aerial vehicles", future research will also investigate the extension of the technique presented herein to the measurement of *transient*, rather than only stationary, vibration signals where time, rather than frequency, domain processing will clearly need to be developed and deployed.

5. References

[1] Rothberg S J et al. 2017 An international review of laser Doppler vibrometry: making light work of vibration measurement *Opt. and Lasers in Eng.* **99** 11–22
[2] Halkon B J and Rothberg S J 2017 Taking laser Doppler vibrometry off the tripod: correction of measurements affected by instrument vibration *Opt. and Lasers in Eng.* **91** 16–23
[3] Halkon B J and Rothberg S J Restoring high accuracy to laser Doppler vibrometry measurements affected by vibration of beam steering optics *J. of Sound and Vib.* **405** 144-157
[4] Halkon B J and Rothberg S J 2018 Taking laser Doppler vibrometry off the tripod. *Int. Cong. on Sound and Vib. (Hiroshima)*
[5] G.M. Revel, P. Castellini, P. Chiariotti, E.P. Tomasini, F. Cenedese and A. Perazzolo 2011 Laser vibrometry vibration measurements on vehicle cabins in running conditions: helicopter mock-up application *Opt. Eng.* **50** 101502-1-14
[6] Polytec GmbH 2017 *RSV-150 Remote Sensing Vibrometer product brochure* [Online] available: https://www.polytec.com/fileadmin/d/Vibrometrie/OM_DS_RSV-150_E_42319.pdf (last accessed 2-9-18)

Acknowledgements

The authors wish to acknowledge the invaluable contributions of Mr Chris Chapman, Scientific Officer, Dynamics & Mechanics of Solids Laboratory and Workshop colleagues in the Faculty of Engineering and IT, University of Technology Sydney.

IOP Conf. Series: Journal of Physics: Conf. Series **1149** (2018) 012023 doi:10.1088/1742-6596/1149/1/012023

A novel approach to reduce the influence of external conditions on the laser Doppler vibrometer signal acting in the nanometer range

L K Yarovoi[1,] and I V Stepakhno[2]

[1] Mechanics and Mathematics Faculty of Taras Shevchenko National University of Kyiv, Glushkova st., 4d, Kyiv, 03127, Ukraine.

[2] Department of Higher and Applied Mathematics of National University of Life and Environmental Sciences of Ukraine, Heroyiv Oborony str.15, Kyiv - 03041, Ukraine.

Address for correspondence: PO Box 85, Kyiv, 03191, Ukraine.

yarovoi@univ.kiev.ua, uos@gluk.org

Abstract. In this work, the operation of laser Doppler vibrometer (LDV) studies in the presence of a third coherent light wave that interferes in detector in addition to the two waves of classical LDV. If amplitudes of vibration are small, presence of the third wave may give a practical advantage. Specifically, such "tree-wave interference" (TWI) vibrometer with the probing wave and additional waves propagating along the same path through the media with unstable refractive index (such as fiber or turbulent air) would have a higher resistance to the external influence on the propagating light in comparison to the classical LDV. The challenge here is to provide the optimal relations between phases and amplitudes of the interfering waves. We propose a simple approach to check optimal phase difference ($\Delta\phi = \pi$) by monitoring the high frequency component of LDV photo-detector signal, particularly spectrum of its amplitude. Our experiments have demonstrated that LDV modified by the additional coherent wave have increased signal-to-noise ratio (SNR) from 1dB to 18 dB in the presence of wideband noise. Therefore, TWI vibrometry can be used in noisy environmental conditions where classical LDV shows poor SNR.

1. Introduction

Generally, the presence of additional waves that interferes with the two common waves of the classical laser Doppler vibrometer may cause critical distortions of the output velocity signal in form of spurious ripple and spikes [1, 2]. In most instances this is a result of unwanted light reflection.

The attention to the negative impact of the interference in LDV was drawn in 1998 on the 7th AIVELA conference, where an effective method to reduce the distortion of output vibrometer signal was proposed [1]. Later, in 2004, a phenomenon of three-wave interference (TWI) in LDV was explored in detail [2], especially for cases when the vibration amplitudes exceed the operational wavelength of the laser. At about the same time, another study describing competition of the LDV signals from various sources was published [3].

The mentioned above as well as some other studies [4, 5] regarded the phenomena resulting from unwanted interference in vibrometers as a source of the signal distortion. However, in certain conditions, it may be used to attain a positive result. As an example, the appearance of distinctive spikes in the velocity signal might be exploited for incremental acquisition of object displacement as an alternative measurement method [2, 6].

Moreover, when the amplitude of the oscillations is less than a half of the laser wavelength, ripples disappear in the output signal. In certain conditions of phases and amplitudes of interfering waves, under which the deformation of waveform in output vibrometer signal disappears, a significant raise of output signal amplitude may be present [7]. It can be used to increase the SNR in the presence of spurious influence of an environment on phase of the probing wave [8, 9]. The environmental influence may reveal itself as a movement and/or vibration of any optical element outside the vibrometer, and/or as an impact on the phase of propagating waves by atmosphere turbulence, fiber optic, *etc*. For example, the study [8] demonstrated an increase of SNR in the TWI vibrometer with long fiber probe.

The difficulty of this approach underlies the need to check and maintain the required conditions, especially the zero phase difference between probing and additional waves that should be close to $\Delta\phi = \pi$ [7]. The matter is that a deviation $\Delta\phi$ from this optimal value may lead to considerable mistakes in the velocity measurements [10].

This study comprises a detailed analysis of the approach to reduce the influence of external conditions on the laser Doppler vibrometer signal by usage in vibrometer setup an additional coherent wave. We also propose a novel method to support the optimal phase conditions for this approach.

2. Mathematical analysis of laser vibrometer signals under three-wave interference

Following [2, 6], we evaluated the result of the three coherent waves interference on the surface of the photo-detector. There are interfering waves: $e_0 = E_0 \exp\left(-i\cdot\left(\omega_{opt}t + \Phi_0(t)\right)\right)$ is the reference wave, $e_1 = E_1 \exp\left(-i\cdot\left(\omega_{opt}t + \Phi_1(t)\right)\right)$ is the probe wave, and $e_2 = E_2 \exp\left(-i\cdot\left(\omega_{opt}t + \Phi_2(t)\right)\right)$ is the third wave, where $\Phi_0(t), \Phi_1(t), \Phi_2(t)$ are the whole phases of the waves that may have arbitrary dependence on time, and ω_{opt} is the angular frequency of the laser radiation.

2.1. Base equations

The current at an output of the detector is proportional to the total intensity of laser radiation $\left(|e_0 + e_1 + e_2|\right)^2$. The photo-detector current i_{detHF} at the heterodyne carrier f_0 ("heterodyne signal") is $i_{\det HF} = I\cos(2\pi f_0 t - \phi)$ [2], where:

$$I(t) = 2E_0\cdot\left(2E_2E_1\cos\left(\Phi_2(t) - \Phi_1(t)\right) + E_1^2 + E_2^2\right)^{1/2}, \tag{1}$$

is the amplitude of the "heterodyne signal" and

$$\phi = \arctan\left[\frac{E_1\sin\left(\Phi_1(t)\right) + E_2\sin\left(\Phi_2(t)\right)}{E_1\cos\left(\Phi_1(t)\right) + E_2\cos\left(\Phi_2(t)\right)}\right] \tag{2}$$

the resulting phase angle.

The instantaneous frequency is determined as the time derivative of the modulated phase:

$$\omega(t) = \frac{1}{2}\left[\frac{d}{dt}\left(\Phi_1(t) + \Phi_2(t)\right)\right] + \frac{\frac{1}{2}\left(\eta^2 - 1\right)\cdot\left[\frac{d}{dt}\left(\Phi_1(t) - \Phi_2(t)\right)\right]}{2\eta\cdot\cos\left(\Phi_1(t) - \Phi_2(t)\right) + \eta^2 + 1}, \tag{3}$$

where $\eta = E_1 / E_2$ denotes the ratio between the electrical field amplitudes of the probe and the third waves.

In general, the phase of probing wave is represented by $\Phi_1(t) = 2kX_1\cos(\Omega t) + \phi_R(t) + \phi_1$, were the first term is associated with the vibration signal $x_1(t) = X_1\cos(\Omega t)$, X_1 – is a vibration amplitude, Ω –

IOP Conf. Series: Journal of Physics: Conf. Series **1149** (2018) 012023 doi:10.1088/1742-6596/1149/1/012023

is a circular frequency of vibration, k – is the wave number of the laser radiation, $\phi_R(t)$ – is a volatile phase component that may include both a determine fraction and a random one caused by impact of environment, ϕ_1 – zero phase.

By supposing that $E_2 = 0$, the complex formula (3) may be easily reduced to the classical case of two interfering waves: $\omega \to \omega_D$, where $\omega_D(t) = -2kX_1\Omega\sin(\Omega t)$, or $\omega_D(t) = 2k \cdot v(t)$ – is true Doppler frequency associated with velocity $v(t)$.

2.2. The extra signal gain

Let's compare two layouts of laser vibrometer measuring vibrations with small amplitudes: $X_1 << \lambda/2$.

1. The additional interfering wave is absent ($E_2 = 0$), the random phase $\phi_R(t)$, and the modulating vibration influence on output vibrometer signal. Substituting $\Phi_1(t)$ into (3) we get:

$$\omega_{clas,R}(t) = 2k \cdot v(t) + \frac{d}{dt}\phi_R(t) = \omega_D(t) + \frac{d}{dt}\phi_R(t) \tag{4}$$

2. In the case of the TWI vibrometer that is depicted in the figure 1, both the probing and the additional waves, travel along the same path and encounter the same influence of environment associated with $\phi_R(t)$. The phase of the additional wave is: $\Phi_2(t) = +\phi_R(t) + \phi_2$.

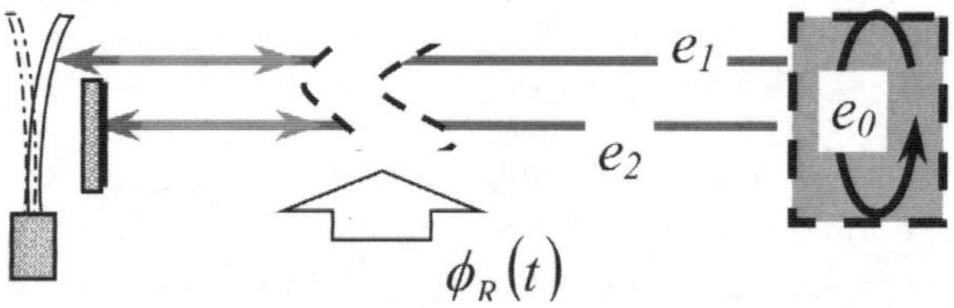

Figure 1. The TWI vibrometer layout.

When phase difference $\Delta\phi = \phi_2 - \phi_1$ is close to an "optimal" condition $\Delta\phi = \pi$, equation (3) may be transformed to:

$$\omega_{TWI,R}(t) = 2kv(t) \cdot \frac{\eta}{\eta - 1} + \frac{d}{dt}\phi_R(t) = \omega_D \frac{\eta}{\eta - 1} + \frac{d}{dt}\phi_R(t) \tag{5}$$

Comparing (4) with (5), it can be seen that this formula differs from the two-waves case by a factor $\alpha = \frac{\eta}{\eta - 1}$. Parameter α shows an extra gain of the velocity signal in the output TWI vibrometer in comparison to the classical LDV. In order to get true instantaneous velocity signal $v = \omega_D / 2k$, the instantaneous frequency with noise component $\omega_{TWI,R}$ (5) in TWI vibrometer needs to be divided by factor $2k\alpha$:

$$v(t) = \frac{\omega_{TWI,R}(t)}{2k\alpha} \tag{6}$$

Accordingly, the noise component $\frac{d}{dt}\phi_R(t)$ for TWI vibrometer will be in α times less in comparison to classical LDV (4) and signal-to-noise ratio for TWI vibrometer will be α times greater It is illustrated by dependence in figure 2. For instance, the SNR of the TWI vibrometer may be improved up to 20 dB for parameter $\eta = 1.1$ by the optimal phases setting.

The simulation in figure 3 (for a target amplitude of 0.3 nm and frequency of 1 kHz) demonstrates an increase SNR when using the TWI vibrometer with the optimal phase setting .

The true Doppler frequency signal from a vibrating object measured by classical two-wave LDV is represented by a dashed sinusoid. The solid magenta line represents the same signal, but under the influence of phase noise (4) with the RMS = 8.12e-005 radians. The cyan line simulates the instantaneous frequency signal of TWI vibrometer (η=1.05) under optimal phase condition for the same noise and vibration.

The simulation demonstrates an evident increase of the SNR (namely, about 40 dB) in the TWI vibrometer in comparison with the classical one.

2.3. The methods to control the optimal initial parameters.

In order to effectively apply the opportunity given by the TWI layout, one needs to control and maintain both, the ratio η on a selected level and the phase difference as close as possible to the optimal value $\Delta\phi = \pi$.

The control of the parameter η may be achieved by applying the means to manage intensity of interfering waves together with monitoring the backscattered intensity in each channel.

To solve the problem of independent zero phase difference control, we propose to observe amplitude $I(t)$ of the photo-detector signal at the heterodyne carrier (1) and analyze its spectrum.

In the approximation of small displacement X_1, expression (1) may be expanded in a series:

$$I(t) = E_0 E_1 \cdot \left[A - I_I \cos(\Omega t) - I_{II} \cos(2\Omega t)\right] \tag{7}$$

where

$$I_I(t) = \frac{2kX_1\eta \sin(\Delta\phi)}{D}, \qquad I_{II}(t) = \frac{(2kX_1)^2 \eta \cos(\Delta\phi)}{4D}, \tag{8}$$

$$D = \left(1 + \eta^2 + 2\eta\left(1 - (kX_1)^2\right) \cdot \cos(\Delta\phi)\right)^{1/2}$$

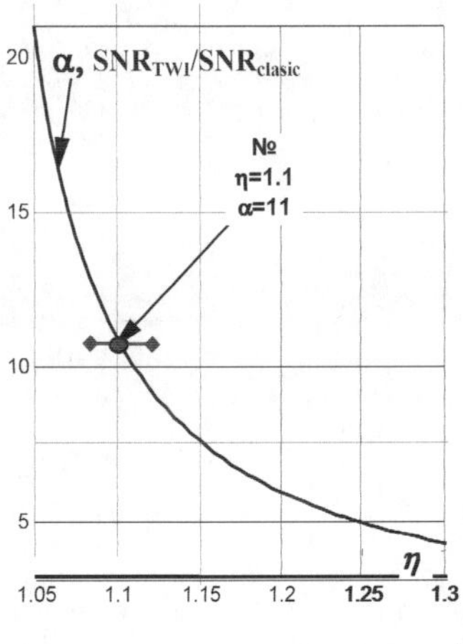

Figure 2. The dependence $\alpha(\eta)$ shows the SNR of TWI vibrometer with $\Delta\phi = 0$ in comparison to SNR = 1 in classical LDV

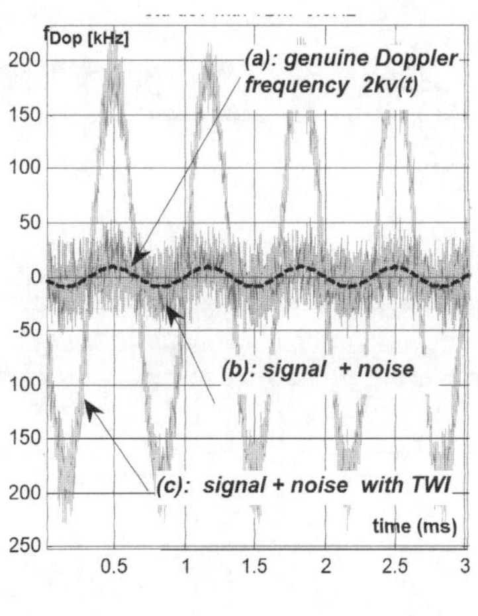

Figure 3. Simulation of instant frequency signals. (a): classical LDV; (b): classical LDV + the noise ($\omega_{clas,R}$); (c): TWI LDV + noise ($\omega_{TWI,R}$), $\Delta\phi = 0$

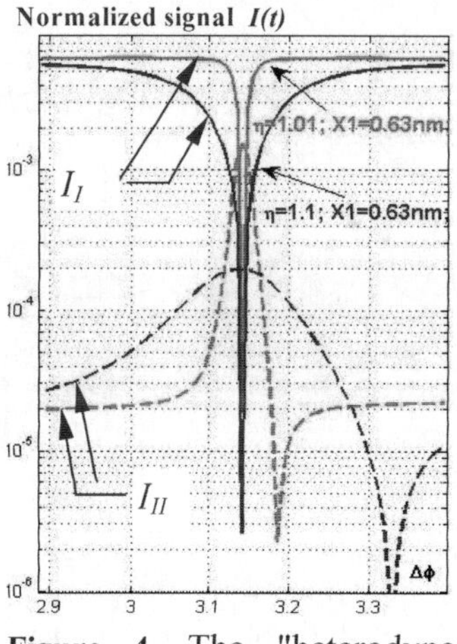

Figure 4. The "heterodyne signal" amplitude $I(t)$. $X_1 = 0.63\mu$; I_1, I_{II} are the fundamental and the second harmonic accordingly.

Analysis of the (7) shows that with the phase $\delta\phi \to 0$ the main harmonic of $I(t)$ approaches zero and the second harmonic reaches its maximum.

The simulation in figure 4 demonstrates behavior of the fundamental harmonic $I_I(\Delta\phi)$ and the second one $I_{II}(\Delta\phi)$ of "heterodyne signal" amplitude $I(t)$ as a function of zero phase difference. The dependence shows an evident minimum of $I_I(\Delta\phi)$.

The calculation of the formulas (8) shows that for $\eta = 1.1$ ($\alpha = 11$) and $I_I(\Delta\phi)/I_{II}(\Delta\phi) \leq 2$ the maximal deviation of $\Delta\phi$ from π is only 0.04 radians for all vibration with amplitude less than 1 nm.

Therefore, to control the current state of the phase difference we propose to monitor the spectrum of $I(t)$ signal. In the case of narrowband vibration of a target with a frequency *F*, appearance of a component at the frequency *2F* is the substantial evidence that $\Delta\phi$ is close to the optimal $\Delta\phi = \pi$. After this, we should change the phase difference and support such value of $\Delta\phi$, that the second harmonic of the probe will be present in the $I(t)$ signal spectra and further.

3. Experimental set-up

To verify experimentally the upper mathematical simulations we used the set-up depicted schematically in figure 5. A compact laser vibrometer of the CLV series (1), manufactured by POLYTEC GmbH [11] was a base devise of the set-up. It includes interferometer, where the referent wave $e_0(t)$ circulates, photo-detector, Bragg cell and other elements.

At the distance 135 cm from the vibrometer output a splitter (2) was located to divide the primary probing beam into two waves. One of them $e_1(t)$, which was used as the probe, directed onto vibrating target (3) covered with retro-reflective foil. A generator (6) controlled the target vibration.

The other wave $e_2(t)$ was reflected back from the surface of piezo (4) that was used to tune phase difference $\Delta\phi$ by changing optical path of wave $e_2(t)$ by a slowly varying voltage from a driver (7).

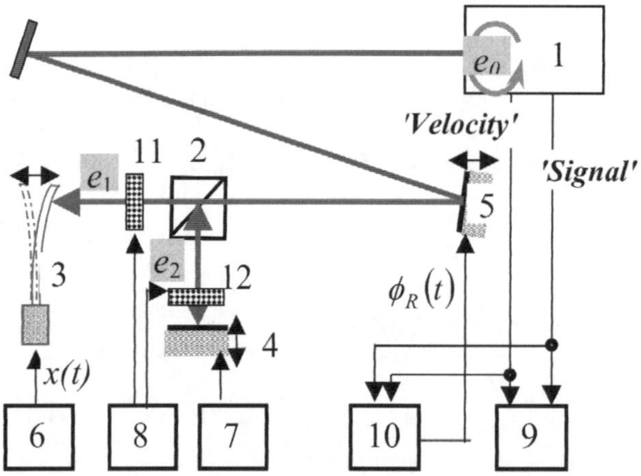

Figure 5. Experimental set-up.

The other piezo with a mirror (5) controlled common optical path of waves $e_1(t)$ and $e_2(t)$. It was used to simulate external impact on the phase $\phi_R(t)$ of laser channel and was powered by signal arranged by PC (10).

A vibrometer (1) controller provided an ac voltage output proportional to the vibration velocity "Velocity" and an additional voltage dc output (marked as "Signal Level"). It is proportional to the logarithm of the signal of amplitude $I(t)$. Both of this signals were observed with oscilloscope (9) and recorded for further processing in the PC-based data acquisition system (10)

Electrically controlled neutral density filters (11) and (12) were used to change the parameter η. The driver (8) managed manually the filters density.

The backscattered signals in each of the channels were checked by shutting the other one by appropriate filter. The filter (11) was also used to quickly switch between the classical and TWI vibrometer layouts

4. The experimental results

Figure 6 comprises waveforms of the "velocity" output and "Signal Level" signals of the classical two–wave vibrometer (a) and TWI laser vibrometer (b) for the same parameters of the experiment: X_1= 6.7 nm, vibration frequency is 2.5 kHz, the ratio of light amplitudes for TWI LDV is η≈1.2, and phase difference $\Delta\phi$ is changing from approximately 3.3 values to π and little bit farther.

(a) 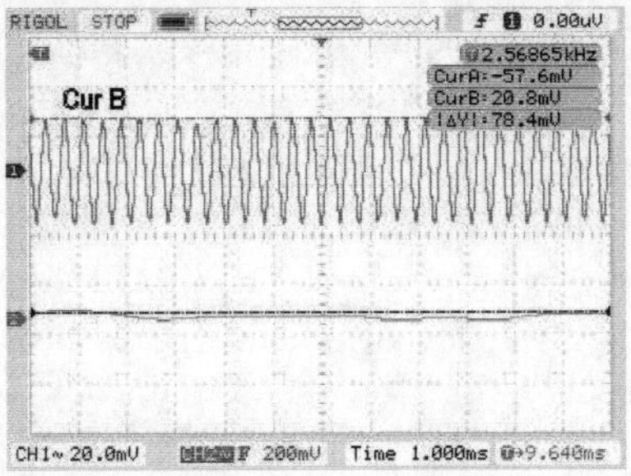

(b) 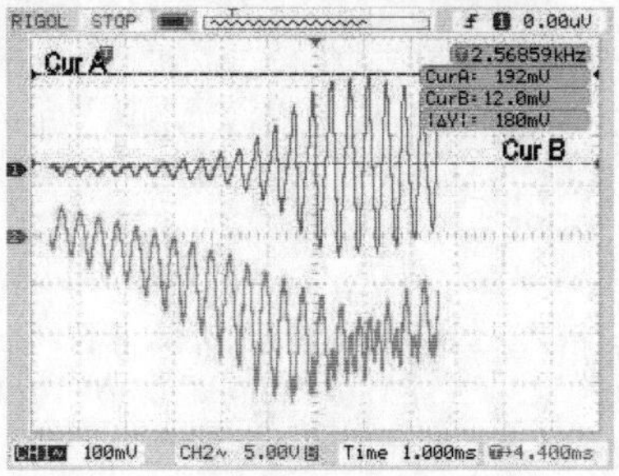

Figure 6. Oscilloscope diagrams of vibrometer outputs: up is velocity signal; down is the "Signal" vibrometer output that is $\log(I(t))$. Vibration: F_s=2.5 kHz; X_1= 6.7 nm. *(a)*: Classical LDV (E_2 = 0); *(b)*: TWI LDV; $\Delta\phi \approx \pi$; $\eta \approx 1.2$; $\alpha \approx 8.5$.

Comparing the traces from both diagrams, first, we can see that the amplitude of "heterodyne signal" $I(t)$ in TWI vibrometer 6*(b)* oscillates with the frequency of studied vibration at 2.5 kHz, unlike the classical LDV 6*(a)*.

Second, the signal $I(t)$ decreases and its visual frequency is doubled, while the velocity signal is reaching its maximum. As we have shown in section 2.3, the disappearance of the main harmonic in the spectrum of $I(t)$ signal is the evidence of $\Delta\phi = \pi$.

(a) 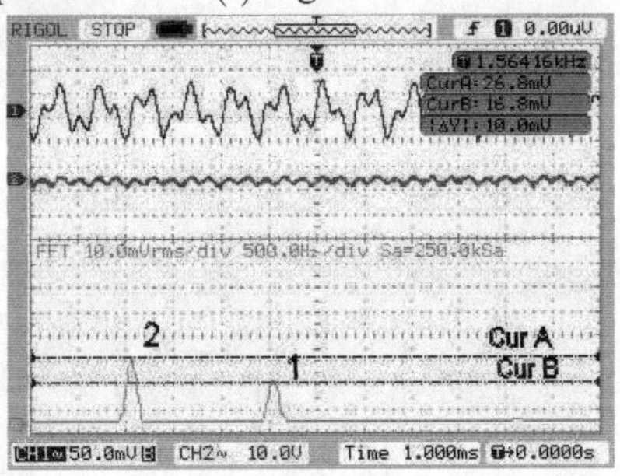

(b) 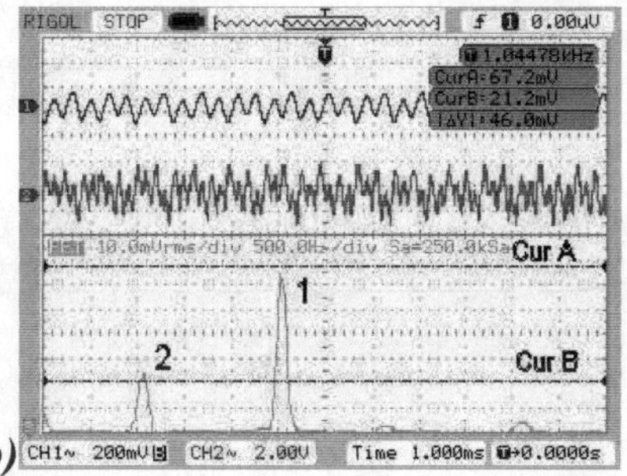

Figure 7. Experimental data for TWI vibrometer with η≈ 1.3…1.4. A studied signal (1) at F_s = 2.5 kHz; X_1=10 nm; a narrowband noise (2) at F_R=21.0 kHz, X_R= 45 nm. Up traces are velocity signal; middle (red) traces is the "Signal" vibrometer output ($\log(I(t))$; down (cyan) line is the spectra of velocity signals. *(a)*: $\Delta\phi \neq \pi$; *(b)*: $\Delta\phi \approx \pi$.

In another experiment (see figure 7) we analyzed the signals of TWI vibrometers ($\eta \approx 1.3$) in presence of a vibration that simulates an undesired impact at frequency 1.0 kHz on both the probing and the additional waves. Amplitude of the tested vibration at 2.5 kHz was 6.7 nm and the amplitude of the noise was 40 nm. When the phase $\Delta\phi$ was far from the optimal value $\Delta\phi = \pi$, the noise velocity signal (2) exceeded the tested signal (1) by factor 1.5, as you can see in figure 7(a).

In contrary, the output signal (1) exceeds the noise (2) by factor 3, when the phase difference is $\Delta\phi = \pi$ (figure 7(b)). The calculation also shows that the extra gain α= 4.6 accused in the experiment is close to value α= 4.3 that is evaluated for $\eta \approx 1.3$.

Note, that the noise doesn't affect the amplitude $I(t)$ of photo-detector signal at heterodyne carrier for both the detuned and the optimally tuned cases at all. The visible ripple in the trace (red color) of the logarithm of $I(t)$ is an oscillation in frequency of tested signal at 1 kHz (a) or at 2 kHz (b). The latter is the second harmonic of the tested signal. Therefore, the channel of the signal $I(t)$ is protected from environmental noise.

The figures 8(a)–(c) show an influence of a wideband noise ($X_{RMS} = 0.5$ *nm*, the bandwidth *0...5 kHz*) on the signals of both LDV types: classical LDV (figures 8(b)) and TWI LDV (figures 8(c)). The signals and their spectra in figure 8(a) correspond to the referent case of classical LDV with a tested signal (1) from a vibrating target with amplitude $X_1 = 0.5$ *nm* and frequency *2.5 kHz* in absence of noise. The parameter η was $\eta \approx 1.1$ for the TWI layout.

Analyzing the experimental results we can note that in the presence of noise (figures 8(b)) the response of the test signal is undetectable in both time (upper oscilloscope trace) and frequency (upper spectrum) domains.

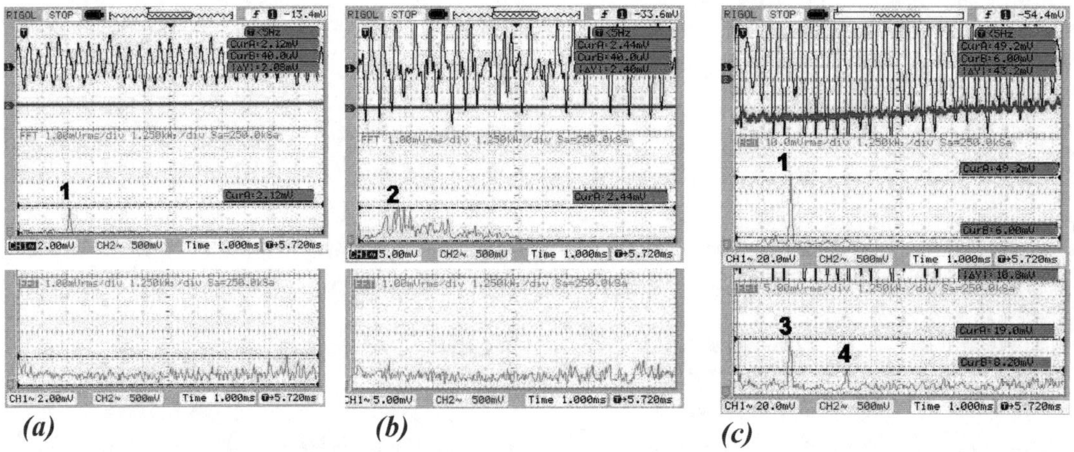

Figure 8. Experimental data. Classical LDV (a), (b); TWI vibrometer (c) with $\eta \approx 1.1$. Tested signal: $F_s = 2.5$ kHz; $X_1 = 0.5$ nm; a noise: $X_{RMS} = 0.5$ *nm*, the noise bandwidth *is 0...5 kHz;* the dark blue lines (up) are velocity signals and the cyan lines (up) are their spectra; the red traces are the signals $\log(I(t))$, the down cyan lines are their spectra. (a): the classical LDV without noise, (1) is velocity spectrum; (b): the classical LDV with noise, (2) is noise spectrum; (c): TWI vibrometer close to optimal state $\Delta\phi \approx \pi$, (3) and (4) are the spectral responds of the fundamental and the second harmonics of $\log(I(t))$ accordingly.

For the same vibration and noise TWI layout provides SNR $\approx$ 18 dB (upper spectrum trace in figures 8(c)) when it is optimally tuned. The moment of the optimal phase is indicated by appearance of the second harmonics in logarithm of $I(t)$ signal, as you can see in the bottom spectrum of figures 8(c).

5. Conclusion

The presence of a third interfering wave in the laser Doppler vibrometer may provide an extra increase of the output signal, but only for vibrations with the amplitudes less than a half of the laser wavelength. This phenomenon arising from the three-wave interference (TWI) appears in the case of a close equality of the amplitudes of backscattered probing and additional waves, and when zero phase difference is opposite ($\Delta\phi = \pi$).

Furthermore, when environment affects the phases of the probing and additional waves synchronically, the TWI vibrometer layout demonstrates a noticeable suppression of the environmental impact.

We propose an effective approach to control the optimal zero phase difference by monitoring the spectral components of the amplitude $I(t)$ of photo-detector signal that oscillates with frequency of a vibrating target. When the TWI vibrometer phase setting is close to the optimal value $\Delta\phi = \pi$, fundamental harmonic of $I(t)$ decreases and reaches zero at $\Delta\phi = \pi$. In contrast to this, the second harmonic of $I(t)$ arises and increases. Therefore, any appearance of the second harmonic in spectrum of $I(t)$ signal is the substantial evidence that $\Delta\phi$ is close to the optimal value $\Delta\phi = \pi$.

These propositions were verified experimentally for narrowband vibration in the interval 0.5…5 kHz and vibration amplitudes of nanometer range in the presence of environmental noise impact on the phase of propagating waves. The experiment demonstrated the rise of SNR up to 18 dB for TWI vibrometer in comparison to SNR $\approx$1 for classical two waves LDV.

Although the approach is developed and verified only for narrowband vibration signal, it is likely to be applicable to signals with complicated vibration spectrum by introducing a probe phase modulation that is far distant from the spectrum of vibration being tested.

Acknowledgements
We express our deep gratitude to SIC "INFOZAHIST" for funding the author's visit to the 13 AIVELA conference to present this study.

References

[1] Dussarrat O J, Clark D F and Moir T J 1998 *Vibration Measurements by Laser Techniques, Proc. SPIE* **3411** pp 2–13
[2] Yarovoi L K and Siegmund G 2004 *Meas. Sci. Technol* **15** 2150–56
[3] Rochefoucauld O L, Shyam M, Khana S M and Olson E S2005 *J. Accoust. Soc. Am.* **117** 1267–84
[4] Siegmund G 2008 *Vibration Measurements by Laser Techniques, Proc. SPI*E **7098** pp 70980Y-1–Y-13
[5] Yarovoi L K, Rozumnuk V I and Stepakhno I V 2013 *Proc. of XXIII National Scientific Symp. with Int. Participation Metrology and Metrology Assurance September 09-13th 2016 Sozopol, Bulgaria* (Sofia: (ISSN 1313-9126)) pp 176–182
[6] Yarovoi L K 2006 *Bulletin of Kiev University, Series: Mathematics and mechanics* **15-16** 53–7
[7] Yarovoi L K 2011 *Bulletin of Kiev University, Series: Mathematics and mechanics* **26** 39–43
[8] Ivashchenko D O, Yarovoi L K and Rozumnuk V I 2012 *Bulletin of Kyiv University. Series: Phys.- Math. Scuiience* **4** 47–51
[9] Yarovoi L K, Rozumnuk V I, Stepakhno I V and Chunihina T V 2015 *Information processing systems. Proc.* Kharkiv **4** 69–72
[10] Yarovoi L K 2016 *Proc. of XXV National Scientific Symp. with Int. Participation Metrology and Metrology Assurance September 07-17th* 2016 *Sozopol Bulgaria* (Sofia: (ISSN 1313-9126)) pp 241–7
[11] Lewin A 1998 *Vibration Measurements by Laser Techniques: Advances and Applications, Proc. SPIE* **3411** pp 61–67

IOP Conf. Series: Journal of Physics: Conf. Series **1149** (2018) 012024 doi:10.1088/1742-6596/1149/1/012024

Using an Interferometric radar to assess post-earthquake damage status of an urban building: a case study

Guido Luzi[1], Ramon Gonzalez-Drigo[2], Lluis Pujades Beneit[3], Esteban Cabrera Velez [2]
Yeudy Vargas[3], Michele Crosetto[1], Eulàlia Parés[1]

[1] *Centre Tecnològic de Telecomunicacions de Catalunya (CTTC/CERCA), Av. Gauss 7, 08860 Castelldefels, Spain*
[2] *Department of Structural Engineering, UPC, Barcelona C/ Eduard Maristany 10-14, 08019, Barcelona, Spain*
[3] *Department of Civil and Environmental Engineering, UPC, Barcelona, J.Girona 1-3, D2, 08034 Barcelona,Spain*

Corresponding author: gluzi@cttc.es

Abstract. This paper reports the use of an interferometric radar to support the detection of the main vibrating frequencies of a building damaged after the Mw 5.1, May 11th 2011, Lorca earthquake (Southeast Spain). The surveyed building, a reinforced concrete structure, was located in the San Fernando neighbourhood of the city of Lorca, where the earthquake caused severe damages. The radar survey was carried out some months after the earthquake. To assess the building capacity and fragility, a pushover analysis has been carried out. The building has been modelled by using the SAP2000 software. The goal of the study is to detect changes of the main natural frequencies of a specific building after the earthquake and, through a comparison of the building frequencies before and after the earthquake, to estimate the effect of the earthquake on the monitored building.

1. Introduction

The capability of a radar interferometer to detect displacements with sub-millimeter accuracy, main frequencies, and modal shapes of civil structures, is well-consolidated and a wide literature is available about different applications [1]. Although this technique is able to provide only an estimate of a component of the actual displacement, i.e. the radar Line-Of-Sight component, it can consistently contribute to check its structural health conditions providing estimates of their vibrating characteristics and its main vibration modes. In the last decade, literature showed the successful monitoring of different structures and facilities (bridges, buildings, wind towers). In the case of urban building, the small extent of the vibrating amplitude under micro tremor, makes the detection more difficult to tackle [2][3]. Anyway, this technique can be particularly useful when the use of in situ sensors needs hazardous installations, as in the emergency case, e.g. after an earthquake, and when a fast and safe survey is demanded. This paper reports the use of an interferometric radar to support the detection of the main vibrating frequencies of a building damaged after the Mw 5.1, May 11th 2011, Lorca earthquake that caused localized damage in the Region of Murcia, Spain. The surveyed building was located in the San Fernando district neighbourhood of the city of Lorca (Murcia, Spain), where the earthquake caused severe damages in 15 buildings which were finally demolished three years after the earthquake. The radar survey was carried out some months after the earthquake and before final demolition. The building has a reinforced concrete structure with unidirectional floors, having a thickness of 20 cm, and closed with infill-unreinforced masonry walls. The goal of the study is to detect changes of the main natural frequencies of a specific building after the earthquake. Because the vibrational behaviour of the undamaged building remains unknown, the frequencies of the undamaged building are calculated through a pushover analysis of the building in the pre-event status, while an interferometric radar is used, during the post-event, for the estimate of the vibrating frequencies of the potentially damaged building. To assess the building capacity and fragility, a pushover analysis has been carried out. The building has been modelled by using the SAP2000 software [4].

The fundamental periods measured on the damaged building using the interferometric radar, are in the range of the values obtained by means of the numeric procedures, and computational assessments are consistent with the observed damage. This agreement allows considering radar interferometry a complementary non-invasive prospecting tool to assess the structural health conditions in a post-earthquake scenario, thus contributing to estimate the damage index, and validating the simulation results. This approach could be also used for rapid screening, immediately after the disaster with alerting and supporting purposes. The paper is organized in the following sections. Starting from this introduction, the rationale of the study is introduced. A short description of the structural model used to simulate the response of the undamaged building is also given. The variation of the vibrating period expected from the model for the damaged building is analysed. The experimental results are then commented: radar outputs, combined with the results of the numerical analysis of the modelled structure, are used to estimate the damage condition. Some final observations conclude the paper.

2. The rationale

When a damage occurs to a building, due for example to a seismic event, from a simplified point of view, the stiffness of the structure reduces. Considering the vibrating behaviour of the structure, the main natural periods are expected to increase correspondingly. In fig. 1 we show a simple representation of this rationale: here the structure is represented by an elemental oscillating mass.

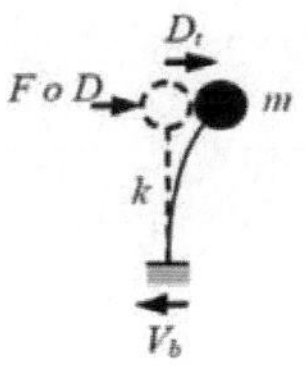

$$T = 2\pi\sqrt{\frac{m}{k}}$$

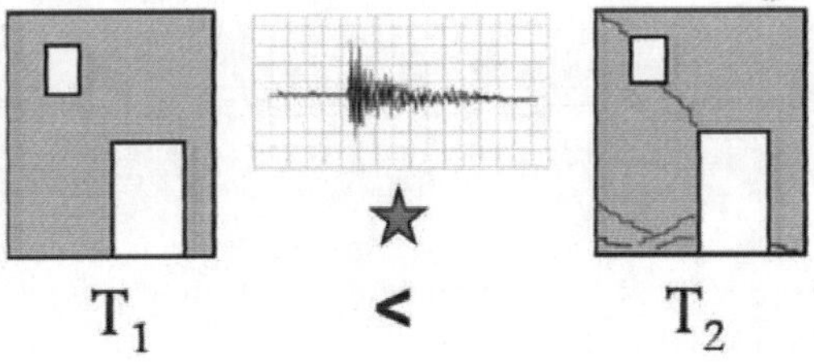

Figure 1. Simple scheme to explain the main rationale of the paper from left to right: the mechanical model, k stiffness, m mass, and T oscillating period. The expression of the vibrating period and the effect of an earthquake (the red star).

On this basis, if we are able to compare the period before and after the event, after evaluating with a sensitivity study the relationship between the effect of modifications of the structural elements of the building, and the main periods, we could estimate the damage. Definitely, because an earthquake is not predictable, an experimental measurement of the natural periods of the structures is usually not available, but we can calculate the main vibration modes of a building using a numeric approach. Starting from the geometry and the structural characteristics of the building, we can estimate, with a certain approximation, the behaviour of the structure in presence of an energizing force, as the acceleration induced by a seismic wave. The model can be also used to estimate the damage occurred to the building. Finally, comparing the periods of the undamaged building and those measured after the earthquake, we can estimate the damage degree. Note that here we consider a damage as changes introduced into the mechanical properties of a building, which affects the behaviour and performance of the building.

3. The event and the structural model results

3.1. The undamaged building

The investigated building is a reinforced concrete building, representative of a group of 15 housing buildings located in the neighborhood of San Fernando in the city of Lorca (Murcia, Spain) (Fig. 2). This building was severely damaged during the Lorca (Spain) May 11th 2011 earthquake [5]. The building, today demolished and substituted by a new construction, had 5 stories and six spans in the studied direction. The design for most buildings was based on a diaphanous first level as a solution in front of severe floods as those occurred in the region during the 1960-70 decade. This soft-storey configuration had influenced the building seismic performance due to the concentration of shear forces and displacements on the weaker frames induced by an abrupt difference on stiffness in height. The common failure mechanism in the damage report of the survey completed after the earthquake is ascribed to this effect. The described damage distribution mainly affected the first level. The few infill walls present in the entrance (first level) of most buildings collapsed, and the typical infill failure was described as induced by the movement of the surrounding frame. Regarding the columns of the first level, plastic hinges appeared at the bottom and top ends of almost all columns. Additionally, a reduced and non-severe distribution of the damage was observed in the upper levels. The undamaged building has been modelled and analyzed by using the SAP2000 software [4]. The obtained periods, listed in Table 1, are within the expected range of periods when similar buildings are considered, i.e. in the range from T=0.08.n to T=0.13.n seconds, where n is the number of levels. Table 1 lists also the mass participation and orientation of the first three modes.

Table 1. Periods obtained through the simulation for the undamaged building.

Mode	Period (s)	% Mass participation	Axis
Mode 1	0.570	84.1	Horizontal Y
Mode 2	0.567	84	Rotational Z
Mode 3	0.454	73	Horizontal X

3.2. The earthquake event and the damaged building modelling

The reinforced concrete building studied in this work was severely damaged during the Lorca (Spain) May 11th 2011 earthquake (Fig. 2b), which with an Mw=5.1 magnitude produced an anomalous high acceleration (PGA=0.37g) in the city [5].

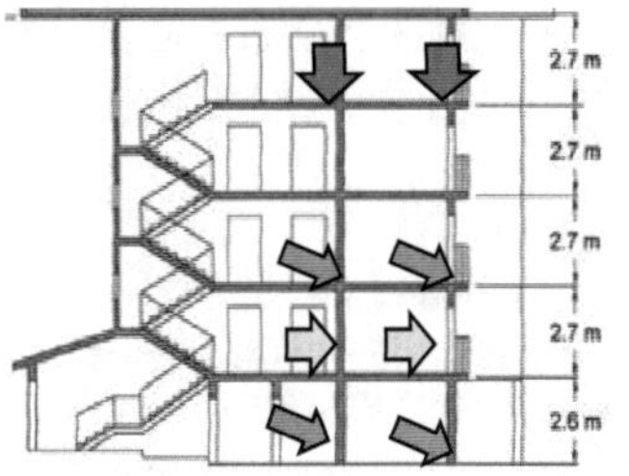

Figure 2. Shape and size of simulated building. Arrows indicate the elements where the damages have been simulated.

In fig. 2 a scheme of the building is shown with indicated the locations of the simulated damage as suggested by the damage report drawn after the earthquake [5].

In order to estimate the current damage state of the building a deterministic procedure, based on a numerical assessment of the building capacity and fragility, is carried out, with the object of linking the experimental results with a damage state of the building. The procedure provides a graphical representation of the capacity curve of the structure, and compares it to the specific seismic demand spectrum of the Lorca earthquake. According to the guidelines described in FEMA-440 [6], the capacity curve for the analyzed building is obtained by means of a pushover analysis. The capacity curve and its corresponding bilinear simplified representation, are converted to the Acceleration-Displacement Response Spectra (ADRS) format (or elastic acceleration–displacement response spectrum). For the present study, the iterative procedure A, described in FEMA-440 [6], has been utilized to calculate the performance point of the building. The performance point is the maximum expected displacement of a control node in the roof of the building obtained following an iterative process by intersecting the capacity of the building represented by the capacity spectrum and the demand of the earthquake, represented in this work by the response spectrum of the Lorca (Murcia, Spain) May 11th 2011 earthquake, shown in Fig. 4. The effective period corresponding to the building state reaching the performance point is 0.729s. According to the damage report of the analysed building, 3 types of damage have been recognized 1) breakage and detachment of walls of first floor, 2) rotation of the feet of pillars, crushing of the heads of the pillars. In fig. 2 the arrows indicated the locations of these damages.

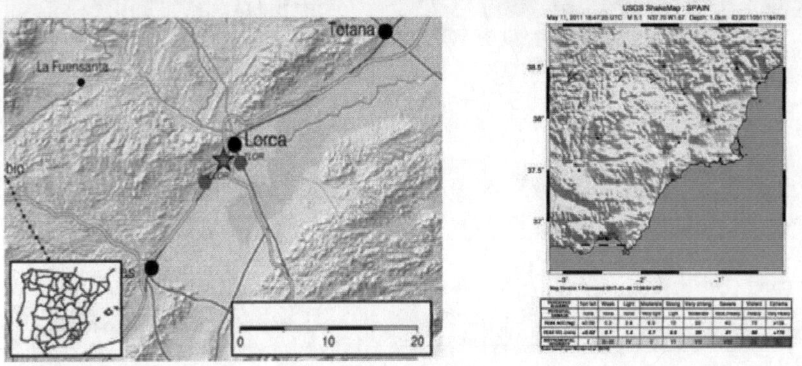

Figure 3. a) Location of the city of Lorca (Spain). b) USGS Shakemap of the event from https://earthquake.usgs.gov/earthquakes/eventpage/usp000j1en#shakemap.

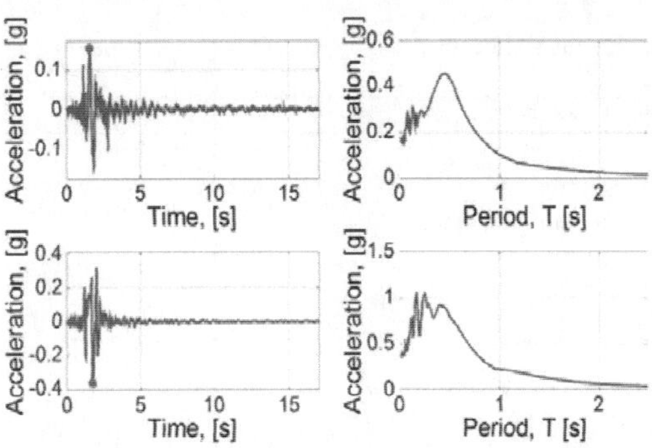

Figure 4. E30N and N30W accelerograms registered during the Lorca earthquake [7] and their response spectra.

IOP Conf. Series: Journal of Physics: Conf. Series **1149** (2018) 012024 doi:10.1088/1742-6596/1149/1/012024

In Table 2 the periods obtained through the simulation for the different worsening damage conditions of the building are reported; only the first mode is considered. The simulated damage degrees consist of three steps: D1, breakage and detachment of walls of first floor; D2 rotation of the feet of pillars: D3 crushing of the heads of the pillars.

Table 2. periods obtained through the simulation for the different worsening damage conditions of the building compared to that of the undamaged D0, for the first mode. Damage degree: D1>D2>D3

Period	D0 (No damage)	D1	D2	D3
T (s)	0.570	0.611	0.622	0.721

4. The results from the radar survey

To evaluate the variation actually occurred of the vibrating periods of the main modes of the building, a radar acquisition has been carried out some months later than the seismic event. The radar survey can give an estimate response of the damaged building performance.

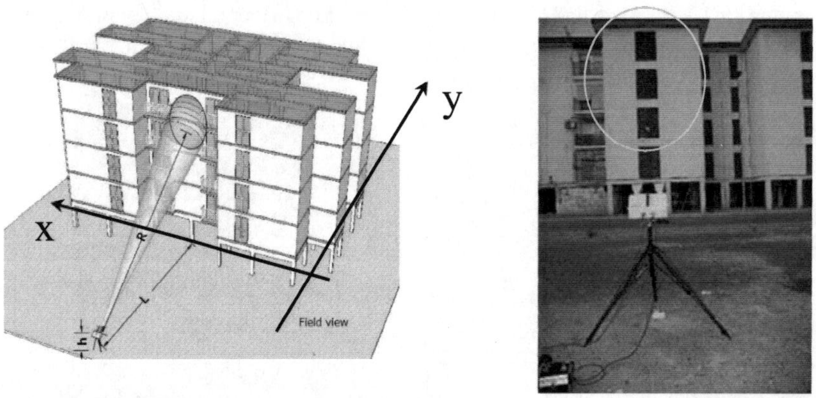

Figure 5. Left: Scheme of the radar survey. Right: picture of the sensor: the orange ellipse indicates approximately the area corresponding to the antenna field of view.

Table 3. Radar parameters

Parameter	
Operating frequency	17.2 GHz (Ku band)
Maximum Sampling frequency	200 Hz
Maximum operating distance	1000m
Range resolution	0.5 m
Battery Autonomy	5 hours
Nominal precision	0.00001 m

The radar system used in this study works at 17.2 GHz centre frequency (Ku band), i.e. 1.743 cm wavelength in vacuum, and it allows detecting equivalent sub-millimetre displacements [8]. It is of main concern to recall that the interferometric technique can estimate distance variations only along the line of sight (LOS) direction. The phase is linearly related to the variation of the distance between the sensor and the target. Interferometric capability: the IBIS-S, manufactured and marketed by IDS (Ingegneria dei Sistemi SpA). The system is designed for monitoring large structures and it has been largely tested by several research teams. The main operational characteristics of the system are summarized in Table 3. The system consists of a sensor module, a control PC and a power supply unit. The maximum acquisition rate is 200 Hz, but the available sampling frequency. The sensor unit is managed by control PC through a standard USB communication, which is provided with system management software, used to configure the acquisition parameters, store the measurement data and show the displacements in real time. A tripod equipped with a rotating head, used to adjust the bearing of the sensor towards the investigated structure, hold up the radar sensor. The radar is able to detect displacement components directed along the LOS and contained into the plane perpendicular to x direction (see fig.5) within the Field of view (FOV) of the antenna. Despite the difficulties of the measurements, due to the very low signal, whose peak-to-peak amplitude is of the order of a few tens of microns, the radar data allowed estimating the main vibrating frequencies. In this preliminary analysis we show the results only from bin35, the only where a good accuracy is expected using the estimated Signal to Noise Ratio [9], and where significant peaks of frequency outstand from its spectrum analysis, in the portion of the expected periods for building with similar characteristics. In fig. 6 the Power spectral density of bin35 is shown. Two main peaks are detected: 1.36 Hz (T=0.733) and 1.513 Hz (T=0.661 s). The corresponding periods are resumed in table 3. The periods measured in the experimental campaign are higher than those calculated from the numerical undamaged model indicating the presence of damage in the existing building. The ratio between the first period obtained in the experimental campaign and the corresponding first period in the modal analysis of the numerical undamaged model is about 1.28.

Table 4. Periods obtained through the radar measurements.

T (s)	0.731	0.661	0.42

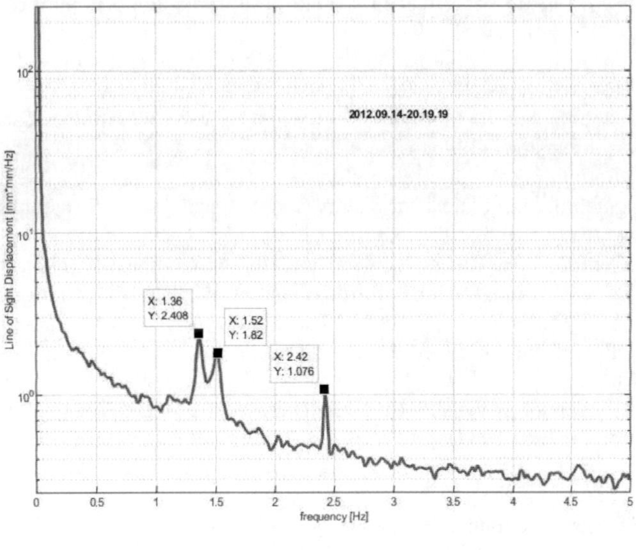

Figure 6. Power Spectral Density (PSD) calculated with the Welch method [10] applied to displacement samples obtained for bin 35. Sample duration=2160s, Overlapping 60%, Subsample=500 s.

5. Conclusions

Combining numeric modelling and experimental data acquired through an interferometric radar, the effect of an earthquake on the conditions of a building has been analysed. On the basis of the rationale that in a damaged structure the modification of the vibrating behaviour of the structure can be an indicator of a change of its stiffness, its main natural periods are expected to increase correspondingly. Numeric procedures and computational assessments allowed to evaluate the frequency expected from an undamaged building, and for the same structure when affected by different degrees of damage. The results of the radar survey are consistent with model results. This agreement allows considering radar interferometry a complementary non-invasive prospecting tool to assess the structural health conditions in a post-earthquake scenario, thus contributing to estimate the damage index, and validating the simulation results. This approach could be also used for rapid screening, immediately after the disaster with alerting and supporting purposes.

References

[1] M. Pieraccini, Monitoring of civil infrastructures by interferometric radar: A review. - The Scientific World Journal.Volume 2013, Article ID 786961, 8 pages http://dx.doi.org/10.1155/2013/786961, 2013

[2] G. Luzi, O. Monserrat, M. Crosetto, The Potential of Coherent Radar to Support the Monitoring of the Health State of Buildings, Research in Non-destructive Evaluation, Issue 23:3 (2012) 125-145. http://dx.doi.org/10.1080/09349847.2012.660241.

[3] C. Negulescu., G. Luzi, M. Crosetto, D. Raucoules, A. Roullé, D. Monfort, L. Pujades, B. Colas, T. Dewez, Comparison of seismometer and radar measurements for the modal identification of civil engineering structures. Engineering Structures V. 51 (2013), 10–22.

[4] CSI Analysis Reference Manual for SAP2000, ETABS SAFE. (2016), Berkeley-Ca. USA.

[5] J.M. Artés Carril, Informe sobre actuaciones realizadas y daños en el grupo de 232 viviendas sociales del barrio de San Fernando de Lorca (Murcia) como consecuencia del terremoto del día 11-05-2011. Instituto de vivienda y suelo. Consejería de obras públicas y ordenación del territorio (2011) [in Spanish].

[6] FEMA-440, Improvement of nonlinear static seismic analysis procedures. Applied Technology Council ATC-55 Project (2005). Redwood City, California, USA.

[7] IGN Spanish Geographic Institute http://www.ign.es/web/ign/portal/sis-catalogo-terremotos/-/catalogo-terremotos/open30Spain.

[8] C. Gentile, G. Bernardini, An interferometric radar for non-contact measurement of deflections on civil engineering structures: laboratory and full-scale tests. Structure and Infrastructure Engineering 6(5) (2010) 521-534. DOI: 10.1080/15732470903068557

[9] F. Coppi, C. Gentile, P.A. Ricci, Software tool for processing the displacement time series extracted from raw radar data. Proceedings of the 9th Int. Conference on Vibration Measurements by Laser and non-contact Techniques, Ancona, Italy; 22-25 June (2010). AIP Conference Proceedings 1253, ed. E.P. Tomasini.

[10] P.D. Welch, The use of fast Fourier transform for the estimation of power spectra: A method based on time averaging over short, modified periodograms. IEEE Trans. Audio Electroacoust, AU-15(2): 70-73, Jun. (1967).

Acknowledgments

This research has been partially funded by the Ministry of Economy and Competitiveness (Ministerio de Economía y Competitividad-MINECO) of the Spanish Government and by the European Regional Development Fund (FEDER) of the UE through projects referenced as: CGL2011-23621 and CGL2015-65913-P (MINECO/FEDER, UE). E. Cabrera work is sponsored by the University Of Azuay, through a scholarship from the Ministry of Higher Education, Science and Technology, SENESCYT, Government of Ecuador, for doctoral studies in the Earthquake Engineering program of the UPC-Barcelona Tech.

IOP Conf. Series: Journal of Physics: Conf. Series **1149** (2018) 012025 doi:10.1088/1742-6596/1149/1/012025

Uncertainty contribution of the laser-beam orientation for laser Doppler vibrometer measurements at the carotid artery

L Mignanelli[1] and C Rembe

Clausthal University of Technology,Leibnizstraße 28, Clausthal-Zellerfeld, Germany

[1]mignanelli@iei.tu-clausthal.de

Abstract. Displacement signals acquired at the carotid artery with laser Doppler vibrometry (LDV) present a characteristic waveform that is correlated to the blood pressure inside the vessel. There are different models in literature that relate the displacement amplitude to the pressure. Therefore, a wrong estimation of the displacement amplitude would lead to errors on the extraction of the pressure information. LDV measures the vibration component in the direction of the laser beam. If the laser beam is not parallel to the vibration direction of the target, only a projection of the real vibration amplitude is measured. With a 3D-LDV measurement it is possible to measure the real amplitude of the displacement. For 1D-LDV measurements it is necessary to calculate the contribution of the uncertainty of the laser beam orientation to the displacement. In this paper, we investigate the uncertainty contribution due to the laser beam direction on the displacement amplitude of vibration signals acquired at the carotid artery. We used a multipoint vibrometer for the investigation. Our aim is to quantify this uncertainty for 1D-LDV measurements at the carotid artery. Moreover, we quantify its influence on the detection of the pressure. We derive an uncertainty contribution due to the laser beam direction of 13% for the displacement amplitude.

1. Introduction

The employment of laser Doppler Vibrometers (LDV) for biomedical applications has several advantages due to the contactless nature of the technology itself, the sensitivity and its other beneficial metrological properties. Moreover, vibrational signals acquired on the human body contain information that allows the extraction of medical parameters [1-10]. This technique is named Optical Vibrocardiography (VCG) [1]. In particular, the measurement point on the carotid artery is often chosen for investigations because of its accessibility and its characteristic strong signal (Figure 1). The skin layer at this point is very thin and most of the mechanical energy generated from the blood flow is converted into skin deflections. Deflections of the diameter of the carotid artery relate to the internal blood pressure, which LDV can detect without contact. Usually, continuous pressure waves are monitored in contact or invasively with pressure sensors. To realize noncontact measurements, several studies [5-9] relate the amplitude of the displacement signal acquired with LDV to the internal pressure of the vessel using models that approximate the pressure-radius relationship. In these studies, ideal measuring conditions are assumed for the post-processing [7-9]. However, variations from the ideal situation are always present in practice. These lead to an uncertainty of the velocity and displacement signals and, therefore, of the derived cardiovascular parameters like the pressure wave amplitude. The LDV measures only the vibration component in the direction of the laser beam. If the vibration of the target is not parallel to the laser beam, the projection of the real vibration amplitude in the measuring direction is measured. This leads to a distortion of the characteristic waves of the signal [10]. Pointing the spot at the location of the carotid artery with the right beam orientation is more difficult if the LDV is adopted for tracking applications, like monitoring the health condition of a moving subject. Such conditions can generate severe measurement deviations. 1D measurements with a single point LDV are critical because it is difficult or even impossible to determine the exact angle of incidence. A single point LDV measurement is not appropriate if the real 3D-vibration amplitude of the signal needs to be detected. A 3D measurement represents an optimal solution to compensate these amplitude deviations. Alternatively, it will be helpful to know the uncertainty related to the laser-beam orientation for 1D measurements.

In this paper, we investigate the uncertainty contribution due to the laser beam direction on the displacement amplitude of vibration signals acquired on the carotid artery. Our aim is to quantify the uncertainty due to the laser beam orientation for 1D-VCG measurements at the carotid artery and evaluate its effects on the detection of cardiovascular parameters. We report our results on the blood pressure evaluation. The measurement point and its position deviation affect also the extraction of parameters from LDV signals, but they are not considered in this work. The real direction of the vibration was investigated with the Multipoint vibrometer (MPV)[11-13]. This allowed to quantify the uncertainty due to the laser beam orientation for 1D-LDV measurement at the carotid artery. With the MPV it is possible to measure from three different direction simultaneously on the same spot and thus, to reconstruct the 3D vector of the vibrations.

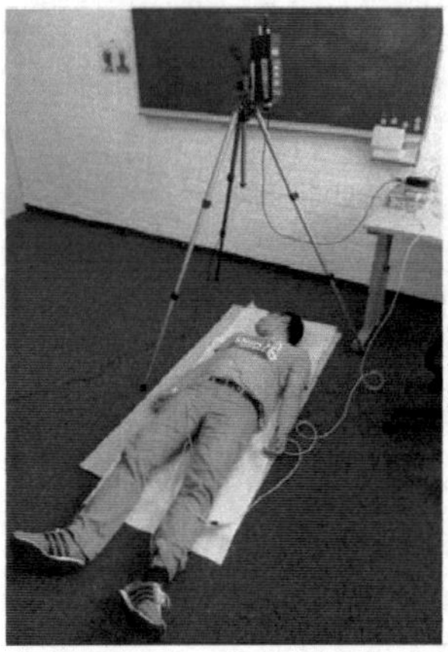

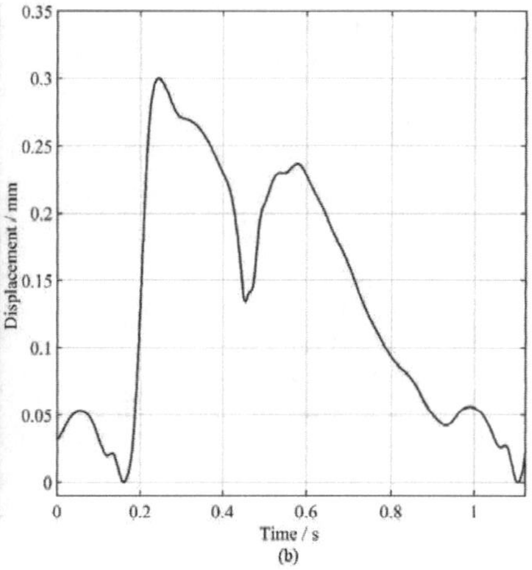

Figure.1 VCG typical set-up (a) and typical VCG signal at the carotid artery acquired with a single point LDV (b).

2. Material and Methods

2.1. Measurement set-up

The real direction of the vibration has to be known to quantify the uncertainty due to the laser beam orientation for 1D-LDV measurement at the carotid artery. We measured the time-dependent orientation of the heartbeat vector in the tridimensional space with the multipoint vibrometer MPV-800 (Polytec) in order to investigate the real direction of the vibrations. During the measurements, the subject was laying comfortably with the head turned right. This allowed the measurement on the neck at the left common carotid artery. A simultaneous electrocardiogram (ECG) signal (II lead) was recorded with the ECG BioAmp ® (ADInstrument) to have a time reference signal. Three measuring heads of the MPV-800 were pointed on the same spot from three independent directions $\boldsymbol{i} = \boldsymbol{i_n}(\alpha_n, \beta_n)$ defined by the angles α_n, elevation angle, and β_n, azimuth angle, where $n = 1,2,3$ (Figure 2). We selected the following measurement directions for the measurements: $\boldsymbol{i_1}$, is oriented with $\alpha_1 = 70{,}5°$ and $\beta_1 = 0°$, $\boldsymbol{i_2}$,is along the z-axis (defined by $\alpha_2 = 90°$) and $\boldsymbol{i_3}$, is defined by $\alpha_3 = 72°$ and $\beta_3 = 270°$ (Figure 2). Thus, we obtained three simultaneous velocity vectors from the directions $\boldsymbol{i_1}, \boldsymbol{i_2}$ and $\boldsymbol{i_3}$, respectively $\boldsymbol{v_1} = v_1\boldsymbol{i_1}$, $\boldsymbol{v_2} = v_2\boldsymbol{i_2}$ and $\boldsymbol{v_3} = v_3\boldsymbol{i_3}$. The selected measuring directions represent a good compromise between a good in-plane resolution and high light collection. The measurements were performed on nine subjects. The measurement point was chosen with palpation method.

2.2. 3D-vector reconstruction

The acquired velocity signals were integrated in order to reconstruct the 3D displacement vector of the heartbeat on the carotid. Noise and vibration interferences like breathing were removed from the signals.

IOP Conf. Series: Journal of Physics: Conf. Series **1149** (2018) 012025 doi:10.1088/1742-6596/1149/1/012025

Thus, only the vibrations generated by the pulse wave are considered. The three simultaneous displacement signals from the directions $\boldsymbol{i_1}, \boldsymbol{i_2}$ and $\boldsymbol{i_3}$, $\boldsymbol{s_1} = s_1\boldsymbol{i_1}$, $\boldsymbol{s_2} = s_2\boldsymbol{i_2}$ and $\boldsymbol{s_3} = s_3\boldsymbol{i_3}$ respectively, were used to reconstruct the 3D-vector of the heart-beat on the carotid artery in the defined Cartesian coordinate system $\boldsymbol{s} = (s_x, s_y, s_z)$ by the following coordinate transformation [10]

$$\boldsymbol{s_{meas}} = \begin{Bmatrix} s_1 \\ s_2 \\ s_3 \end{Bmatrix} = \begin{bmatrix} \cos\alpha_1\cos\beta_1 & \cos\alpha_1\sin\beta_1 & \sin\alpha_1 \\ \cos\alpha_2\cos\beta_2 & \cos\alpha_2\sin\beta_2 & \sin\alpha_2 \\ \cos\alpha_3\cos\beta_3 & \cos\alpha_3\sin\beta_3 & \sin\alpha_3 \end{bmatrix} \cdot \begin{Bmatrix} s_x \\ s_y \\ s_z \end{Bmatrix} = \mathbf{E}\cdot\boldsymbol{s} \Leftrightarrow \boldsymbol{s} = \mathbf{E}^{-1}\cdot\begin{Bmatrix} s_1 \\ s_2 \\ s_3 \end{Bmatrix}. \tag{1}$$

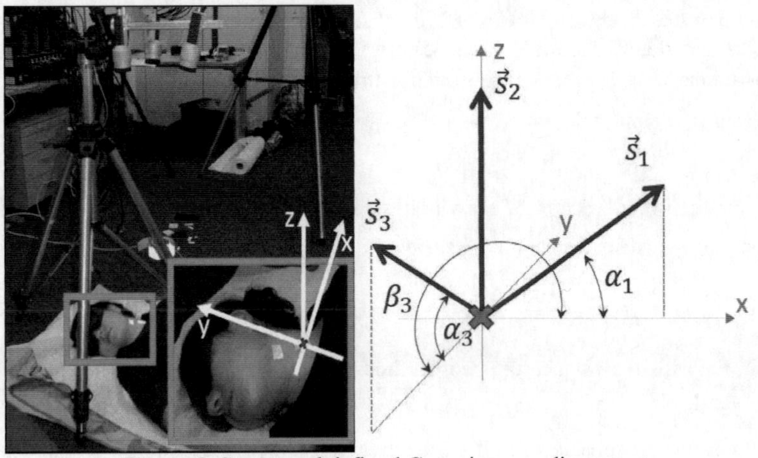

Figure 2. Set-up and defined Cartesian coordinate system

The typical LDV displacement signals acquired on the carotid present several characteristic points [5]. In our work we investigate only the maximum of the flow point (i.e. maximum of the signal in figure 1b), which is related to the maximal pressure value of the pulse wave obtained with the calibration models [7-9].

The 3D displacement vector is reconstructed in the Cartesian coordinate system for every beat. The time instant, where the selected beats present its maximum amplitude, has been chosen. The displacement vector with maximal length $\boldsymbol{s_M}$ is the displacement vector at maximum amplitude. Its length results from

$$s_M = |\boldsymbol{s_M}| = \left(s_{Mx}^2 + s_{My}^2 + s_{Mz}^2\right)^{1/2}, \tag{2}$$

where s_{Mx}, s_{My} and s_{Mz} are the amplitude components in the Cartesian coordinate system.

The direction of $\boldsymbol{s_M}$ is defined by the angles α_M (elevation) and β_M (azimuth)

$$\alpha_M = \arcsin\left(\frac{s_{Mz}}{s_M}\right), \tag{3}$$

$$\beta_M = \mathrm{atan2}\left(\frac{s_{My}}{s_M}\right). \tag{4}$$

This analysis is performed for twenty beats of all the nine subjects.

The estimation of the uncertainty has been performed as described in the following paragraph only for the maximal amplitude point and for the case of a 1D measurement in the direction perpendicular to the skin, which is the most used set-up for this application. In the defined coordinate system, the direction perpendicular to the skin corresponds to the z-axis (Figure 2).

2.3. Uncertainty determination for 1D LDV-measurement at the carotid artery

To calculate the combined uncertainty for the displacement amplitude, we made the assumption of a simple displacement model

$$s = s_0 + \Delta s_A + \Delta s_B, \tag{5}$$

where s is the measured displacement amplitude, s_0 is the real value of the displacement, Δs_{A} is the type A uncertainty deviation and Δs_{B} is the type B uncertainty deviation according with the guide to the expression of uncertainty in measurement GUM [14]. Thus, the combined uncertainty for the displacement $u_{\mathrm{C,1D}}$ is

$$u_{\mathrm{C,1D}} = \left(u_{\mathrm{A,1D}}^2 + u_{\mathrm{B,1D}}^2\right)^{1/2}, \tag{6}$$

where $u_{\mathrm{A,1D}}$ for a subject is calculated by multiplying the student-t factor for 20 values for 68% confidence bound, $t_{0.68}(20)$, by the empirical standard deviation of s_{Mz}, $\mathrm{std}(s_{Mz})$ for the selected 20 beats. For the calculation of the uncertainty of 1D-LDV displacement measurements at the carotid artery we have to consider the uncertainty due to the laser beam orientation as type B uncertainty according with GUM [14]. In particular if we are measuring from the direction $\boldsymbol{i}_p$ (Figure 3a), at the maximum of the flow point we measure the amplitude $s = s_p$

$$s_p = |\boldsymbol{s}_p| = |\boldsymbol{s}_M|\cos\gamma, \tag{7}$$

where γ is the angle between the direction of the vibration and the measuring direction. In order to calculate the uncertainty for any displacement it is helpful to define the relative amplitude error $E(\gamma)$

$$E(\gamma) = \frac{|\boldsymbol{s}_M| - |\boldsymbol{s}_p|}{|\boldsymbol{s}_M|} = 1 - \cos\gamma. \tag{8}$$

We evaluated the γ for the case of a 1D measurement on the z-axis (Figure 3b), thus $s_p = s_M$ and

$$\gamma = 90° - \alpha_M. \tag{9}$$

We calculated γ for 20 beats of 9 subjects to find $\gamma_{\max}$. The uncertainty contribution $u_{\mathrm{B,1D}}$ calculated for the maximal value of $\gamma_{\max}$ with the assumption of a rectangular distribution of γ, results in

$$u_{B,1D} = \frac{E(\gamma_{\max}) \cdot s_{Mz}}{\sqrt{3}}. \tag{10}$$

2.3.1. Uncertainty contribution of the displacement measurement to the blood pressure. The displacement amplitude acquired with LDV is related to the blood pressure inside the vessel as explained in the introduction. The displacement uncertainty has a contribution to the combined uncertainty of the pressure P. The pressure waveform $P(t)$ can be expressed with one of the calibration models described in [7-9]. The most accurate one is the exponential model [9]

$$P(t) = p_{\mathrm{d}} \exp\left[q\left(\frac{A(t)}{A_{\mathrm{d}}} - 1\right)\right], \tag{11}$$

where p_{d} is the diastolic pressure, A_{d} the diastolic arterial cross-section and q is the wall rigidity coefficient which depends on the systolic and diastolic pressure and cross-section area. The variable $A(t)$ is the section at the time instant t. For our analysis we take into consideration only the time instant t of the upstroke, $t(s_{Mz})$. According with [7-9]

$$A(t(s_{Mz})) = \pi\frac{(d_0 + 2 \cdot s_{Mz})^2}{4} \tag{12}$$

where d_0 is the diameter of the vessel at rest. Thus, the contribution of s_{Mz} to the combined uncertainty of the pressure P is $u_{\mathrm{C,P},s_{Mz}}^2$

$$u_{\mathrm{C,P},s_{Mz}}^2 = \left(\frac{\partial P}{\partial s_{Mz}} \cdot u_{\mathrm{C,1D}}\right)^2 = \left\{p_{\mathrm{d}} \exp\left[q\left(\frac{A(t(s_{Mz}))}{A_{\mathrm{d}}} - 1\right)\right] \cdot \pi q \frac{(d_0 + 2 \cdot s_{Mz})^2}{A_{\mathrm{d}}} u_{\mathrm{C,1D}}\right\}^2. \tag{13}$$

For the calculation, we took the typical values of the systolic and diastolic pressure and cross-section area from references [15].

2.4. *Uncertainty determination for 3D LDV-measurement at the carotid artery*

In a 3D-LDV measurement, the amplitude of the vibrations is measured along its direction. Thus, we neglected the uncertainties of type B.

By applying the same approach of the uncertainty of 1D case to the 3D case, we have that the combined uncertainty for the displacement u_{3D} presents only the contribution of the type A uncertainty

$$u_{3D} = u_{A,3D}, \tag{14}$$

where $u_{A,3D}$ for a subject is calculated by multiplying $t_{0.68}(20)$ by the empirical standard deviation of s_M, $\mathrm{std}(s_M)$ for the selected 20 beats. The contribution of s_M to the combined uncertainty of the pressure P is u^2_{C,P,s_M}

$$u^2_{C,P,s_M} = \left(\frac{\partial P}{\partial s_M}\cdot u_{C,1D}\right)^2 = \left\{ p_d \exp\left[q\left(\frac{A(t(s_M))}{A_d} - 1\right)\right]\cdot \pi q \frac{(d_0 + 2\cdot s_M)^2}{A_d} u_{C,1D}\right\}^2. \tag{15}$$

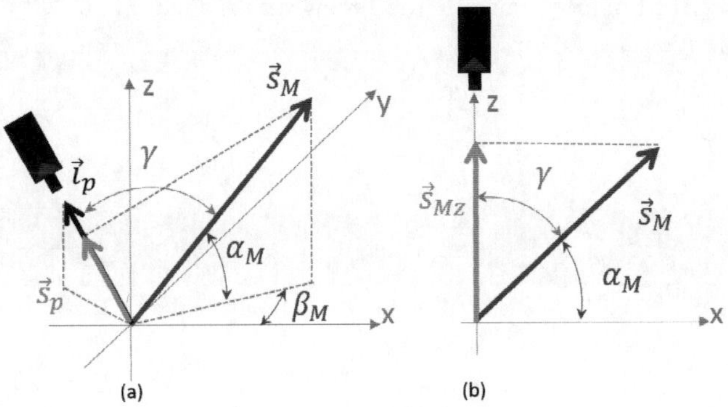

Figure 3. 1D-LDV measurement from an arbitrary direction (a), from the z-axis of the defined Cartesian coordinate system (b)

3. Results

Figure 4 shows the reconstruction of the 3D displacement vector. The amplitude of 3D-displacement vector over the time is presented in the upper part of the figure. The displacement vector in the 3D space is presented in the bottom of figure 4. The paths of the 3D-vector from the beginning till three different time instants are showed in the defined Cartesian coordinate system. The three selected time instants are: the maximum of the flow point t_{max} (right), the end of the heart beat $t_{end,beat1}$ (middle) and the end of the following beat $t_{end,beat2}$ (left). From the beginning of the beat till t_{max}, i.e. during the upstroke, the vibrations are almost along the same direction. After the upstroke (Figure 4 middle) the vector deviates and changes its direction. This behavior could depend on the expansion and contraction of the vessel due to the blood flow passing through the vessel. The pattern delivered by the second beat (Figure 4 right) seems to follow the same path of the previous beat. Thus, the pattern delivered by a beat is very reproducible for each subject.

Figure 5 shows the amplitude of the 3D vector and its projection on the z-axis. It is possible to notice that the amplitude of the projection in z-axis is smaller than the real 3D amplitude. The amplitude difference is not constant during the beat. This can be explained since the direction of the 3D vector changes (Figure 4), and therefore also its projection in the z-axis.

The calculation of α_M, and thus γ, performed for 20 beats of each subject leaded to a $\gamma_{max} = 39.97°$. The obtained value of γ_{max} is used to calculate the type B uncertainty for 1D measurement with LDV. In particular from equation (8), the relative error is $E(\gamma_{max}) = 0.23$ and thus, the uncertainty contribution $u_{B,1D}$ from equation (10) becomes

$$u_{B,1D} = 0.13 \cdot s_{Mz}. \tag{16}$$

The contribution of the laser beam orientation at the maximum point is 13% of the measured amplitude with the measuring conditions described in the paragraph 2.1.

For the calculation of the combined uncertainty, the example of one subject is reported.

In particular for a selected subject the type A uncertainty is

$$u_{\text{A,1D}} = t_{0.68}(20) \cdot \text{std}(s_{Mz}) = 0.011 \text{ mm}. \tag{17}$$

For the same subject is for example $s_{Mz} = 0.301$ mm, thus the type A uncertainty in this case is about 3.7%. Therefore the uncertainty due to the laser beam orientation is not negligible respect to the statistical determined uncertainty.

The contribution of the displacement uncertainty to the uncertainty of the pressure P can be calculate as described in equation (13)

$$u_{\text{C,P},s_{Mz}} = 3.43 \text{ mmHg}. \tag{18}$$

$u_{\text{C,P},s_{Mz}}$ results in 2.9% of the pressure at the maximal point $P(s_{Mz}) = 118.28$ mmHg calculated according with equation (11).

In the case of a 3D measurement $u_{\text{A,3D}} = 0.012$ mm and the contribution of the displacement uncertainty to the uncertainty of the pressure is described in equation (15). It results in

$$u_{\text{C,P},s_M} = 1.04 \text{ mmHg} \tag{19}$$

and it represents the 0.9% of $P(s_M) = 121.61$ mmHg, calculated according with equation (11).

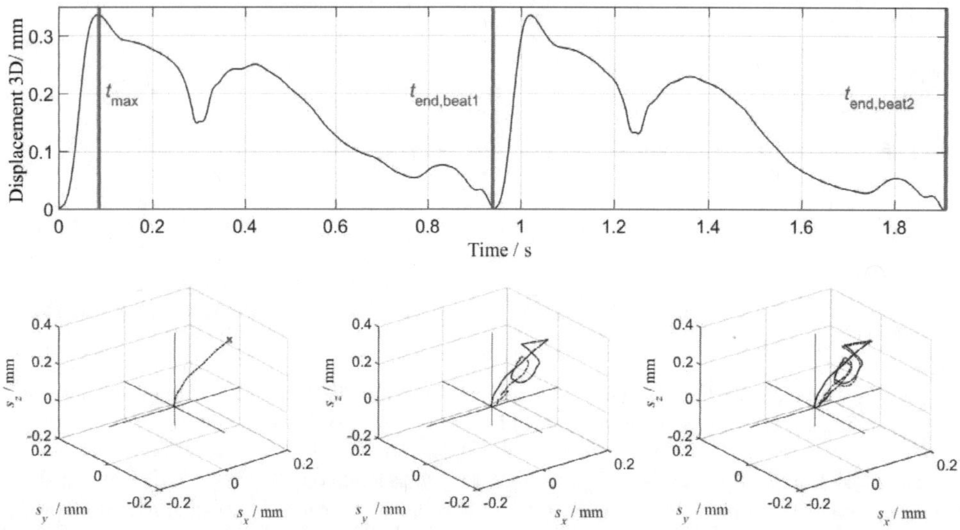

Figure 4. 3D displacement amplitude over the time of two consecutive beats (above). 3D representation of the vector at three different time instant: t_{max} is the maximum flow point (left), which is identified by the red cross, $t_{\text{end,beat1}}$ is the end of the first beat (middle) and $t_{\text{end,beat2}}$ is the end of the second beat (right). s_x, s_y and s_z indicate the displacement in x, y and z axis respectively.

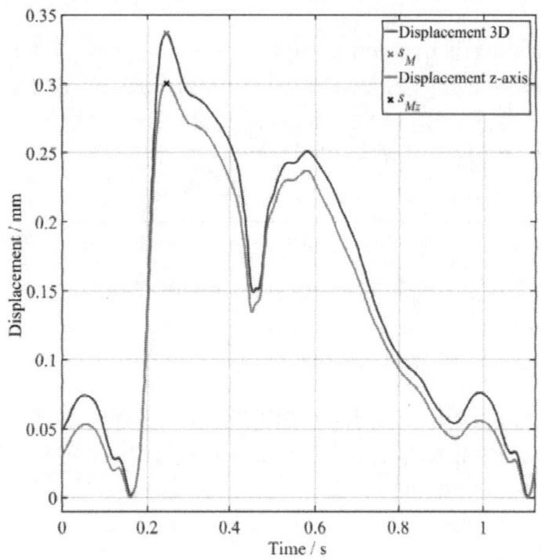

Figure 5. 3D displacement amplitude (blue) and its projection in z-axis (magenta). Their maxima s_M and s_{Mz} are marked with a red and a black cross respectively.

The uncertainty in the determination of the peak of the pressure are lower in the case of a 3D measurement. However, a normal individual that breaths spontaneously shows changes of the maximal pressure of about 5 mmHg [16], thus $u_{C,P,s_{Mz}}$ is still acceptable. If the type A uncertainty of a subject is comparable to the type B uncertainty then also the contribution to the uncertainty of the pressure increases. However, in the exponential model of the pressure P, the displacement appears on the exponent; therefore to a great variation of the displacement amplitude corresponds a small variation of the pressure P.Thus, even a large measurement error of the displacement-amplitude measurements results in a rather small error of the determined pressure amplitude.

4. Conclusion

In this work, we investigated the uncertainty contribution due to the laser beam direction on the displacement amplitude of vibration signals acquired on the carotid artery. We quantified the uncertainty due to the laser beam orientation for 1D-LDV measurements at the carotid artery and its effect on the detection of the blood pressure by measurement performed with the multipoint vibrometer. From our investigation the uncertainty contribution due to the laser beam direction is 13% of the displacement amplitude. A 3D-LDV measurement leads to a lower uncertainty in the determination of the displacement amplitude, and thus of the blood pressure. However, if the exponential model of the pressure is used, the repercussion of the uncertainty of the displacement to the value of the pressure, is not critical.

In our analysis we did not consider the influence of the set-up for the calculation of the uncertainty in the case of a 3D measurement. This analysis is addressed to future works. A further interesting analysis is the quantification of the uncertainty of the laser beam orientation for the extraction of other cardiovascular parameters.

References

[1] Scalise L and Morbiducci U 2008, *Medical Engineering & Physics* **30**(4) 490-97
[2] Mignanelli L, Rembe C, Kroschel K, Luik A, Castellini P and Scalise L 2014 *11th Int. Conf. on Vibration Measurements by Laser and Noncontact Techniques (Ancona)* ed E P Tomasini (New York: AIP Publishing) 1600 pp 313-22
[3] Campo A and Dirckx J 2014 *11th Int. Conf. on Vibration Measurements by Laser and Noncontact Techniques (Ancona)* ed E P Tomasini (New York: AIP Publishing) 1600 pp 338–41
[4] Luik A, Mignanelli L, Kroschel K, Schmitt C,Rembe C and Scalise L. 2016 *Future Cardiology 2016* 12 (3) pp 269–79.
[5] Sirevaag J, Casaccia S, Richter E a, O'Sullivan J A,Scalise L and Rohrbaugh J W 2016 *Psychophysiology* **53** (6) pp 847-67
[6] Kaplan A 2011 Information Processing for Biological Signals: Application to Laser Doppler Vibrometry All thesis and Dissertations (ETDs) Washington University in St.Louis. Paper 890
[7] Scalise L, Cosoli G, Casacanditella L, Casaccia S and Rohrbaugh J W 2017 *IEEE International Symposium on Medical Measurements and Applications (MeMeA)* pp 245-250
[8] Casacanditella L, Cosoli G, Casaccia S, Rohrbaugh J W, Scalise L and Tomasini E P 2016 *12th Int. Conf. on Vibration Measurements by Laser and Noncontact Techniques (Ancona)* ed E P Tomasini (New York: AIP Publishing)1740
[9] Casacanditella L 2017 Development of a measurement procedure for the assessment of carotid blood pressure by means of laser Doppler vibrometry, Università politecnica delle Marche
[10] Mignanelli L, Bauer G, Klarmann M, Wang H and Rembe C 2017 *Clinical and Preclinical Optical Diagnostics* ed J Brown SPIE Proc. 10411 (OSA, 2017), paper 1041105
[11] Haist T, Lingel C, Osten W, Winter M, Giesel M, Ritter F, Sandfort K, Rembe C and Bendel K 2013 *The Review of Scientific Instrument* ,**84,** 121701
[12] Cupido E,Smith D and Morel S 2002 XVI-A novel multipoint laser Doppler vibrometer *ISMA2002:Intl Conf on Noise and Vibration Engineering (Lueven)* 655-662
[13] Du Y, Guo M and Phua P B *Applied Optics* **. 50,** 1280-88
[14] ISO/IEC (2008). Guide 98-3:2008 Uncertainty of measurement—Part 3: Guide to the expression of uncertainty in measurement, Technical report, Geneva, Switzerland, pp. 1–120
[15] Krejza J, Arkuszewsky M,Kasner S E,Wigele J, Ustymowicz A,Hurst R W, Cucchiara B L Messe S R 2006 Stroke; 37:1103–5.
[16] Michard F 2005 *Anestesiology 8*,**103**, 419-28

3D deformation and strain measurement of an intact eardrum using digital image correlation

Kilian Gladiné[a], **Joris J.J. Dirckx**[a]

[a]University of Antwerp, Laboratory of Biophysics and Biomedical Physics, Groenenborgerlaan 171, 2020 Antwerp, Belgium

Kilian.Gladine@uantwerpen.be

Abstract. In the field of hearing research, strain measurements are scarce. Such measurements can however provide valuable data on several middle ear pathologies. They could also help to gain a better understanding of the properties of the eardrum and further validate middle ear finite element models. We built a digital image correlation setup in order to assess three dimensional deformation and strain measurements of an intact eardrum. Different types of speckle pattern were evaluated. We then measured the deformation and strain of an eardrum subjected to large static pressures. The results illustrate that the application of digital image correlation to measure deformation and strain in intact eardrums is feasible. In the future, the setup will be further developed and eardrum strain measurements will be performed for different pathologies.

1. Introduction

Digital image correlation (DIC) is a powerful non-contact measurement technique allowing three dimensional local deformation and strain measurements. In engineering, the popularity of DIC has increased greatly over the last few decades. In the field of hearing research, the application of DIC is however still limited. DIC has been used to determine the mechanical properties of the stapedial tendon [1], anterior mallear ligament [2] and isolated strips of the human eardrum [3]. However, measurements on an intact eardrum have not been done before. Such measurements can provide valuable data on several middle ear pathologies.

The eardrum does not lend itself as an easy subject for DIC measurements. It is very thin and semi-transparent. The transparency can lead to light reflections from tissues behind the eardrum, resulting in correlation issues. Reflections can be avoided by first coating the eardrum with white paint, however, because the eardrum is so thin this should be avoided as the drying of the paint can affect the properties of the membrane. We will therefore investigate whether reflections actually degrade the correlation results when using a toner speckle pattern. We will also evaluate if reflections can be avoided when using fluorescent powder to generate a speckle pattern. This powder can be excited using laser light. Only the speckle pattern will then emit light of a longer wave length than the excitation source. By filtering out the laser light that reaches the camera, we avoid capturing any light originating from somewhere else than the speckle pattern, thus eliminating any reflections from behind the eardrum.

We will show full field displacement and principal strain maps for an eardrum subjected to high static pressures.

2. Materials and methods

2.1. Temporal bone & pressure generation

One rabbit temporal bone was used in this study. The middle ear cavity was opened, granting visual access to the eardrum from the medial side. The cochlea was removed, eliminating all fluid pressure effects on the ossicle chain. A plastic tube was glued onto the bony part of the ear canal. This tube was connected to a pressure generator, allowing us to pressurize the middle ear. We will refer to pressure as the applied pressure difference between the ear canal and the middle ear cavity. A negative pressure then signifies a lower pressure in the ear canal in comparison with the middle ear cavity.

Measurements were performed at positive and negative pressures of 0, 1, 2, 4 kPa. One full pressure cycle was performed. Several seconds passed between subsequent measurement steps.

2.2. Camera setup & software

Two cameras (Manta G-507, Allied Vision) with 2464 x 2056 pixels resolution were used. A lens (FL-CC7528-2M, Ricoh) and a 60 mm extension tube were mounted onto the camera. The cameras were placed at a 30° angle, approximately 12 cm from the temporal bone.

Images were processed using commercial software (ISTRA 4D, Dantec dynamics). The cameras were calibrated in the same software using a 5x5mm glass calibration plate (KL501005, Dantec dynamics).

2.3. Speckle pattern & illumination

Two types of speckle pattern materials were used. We applied toner and fluorescent powder to create speckle patterns (figure 1a and 1b respectively). Depending on the type of material that was used, different illumination sources were utilized. For the toner speckle pattern, we illuminated the sample with white light (KL 2500 LCD, Schott). When using the fluorescent powder, a 532 nm laser was employed as light source instead. Additionally, 532 nm filters were placed in front of the cameras. That way, the light perceived by the cameras could only originate from the fluorescent speckle pattern on the surface of the semi-transparent eardrum and not from reflections occurring behind the eardrum. Since only one sample was used in this study, both types of materials were applied to the same eardrum and measurements were performed for each.

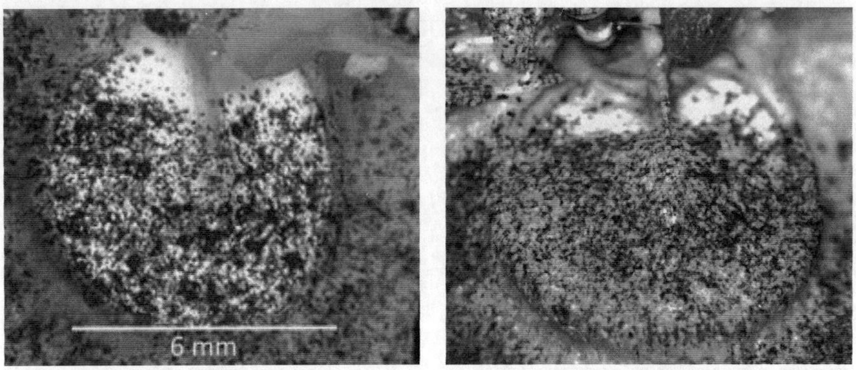

(a) (b)

Figure 1. Speckle patterns using (a) toner, (b) fluorescent powder

IOP Conf. Series: Journal of Physics: Conf. Series **1149** (2018) 012026 doi:10.1088/1742-6596/1149/1/012026

3. Results

3.1. Speckle pattern & correlation

Figure 2 shows photographs of the illuminated eardrum covered with a toner speckle pattern. The images show reflections originating from behind the eardrum. The locations of these reflections are not the same for both cameras. Reflections were not observed using the fluorescent speckle pattern.

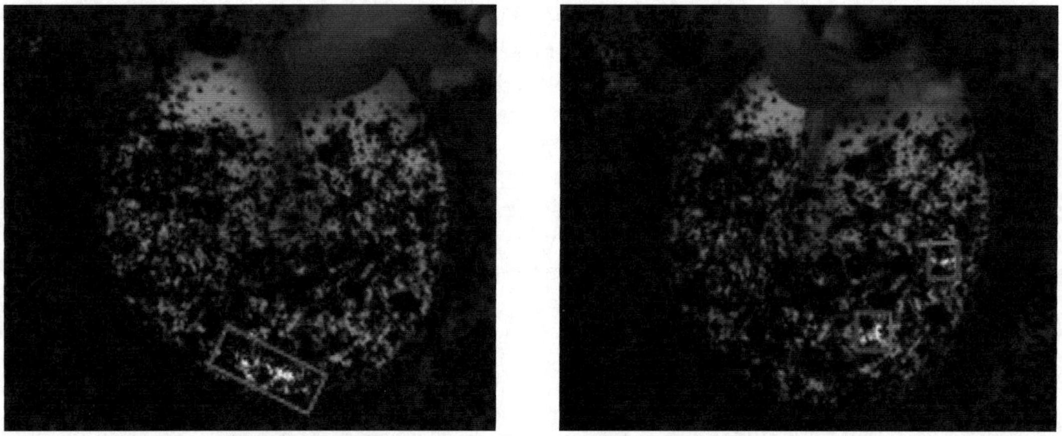

Figure 2. Different reflections are observed by each camera.

Figure 3 shows the correlation agreement between the two cameras for both speckle pattern types. The correlation failed for four large zones in the image when using the toner speckle pattern. Three of these zones overlap with the locations of the reflections seen in figure 2. The correlation result obtained when using the fluorescent speckle pattern has only one zone missing due to a lack of speckles in this zone.

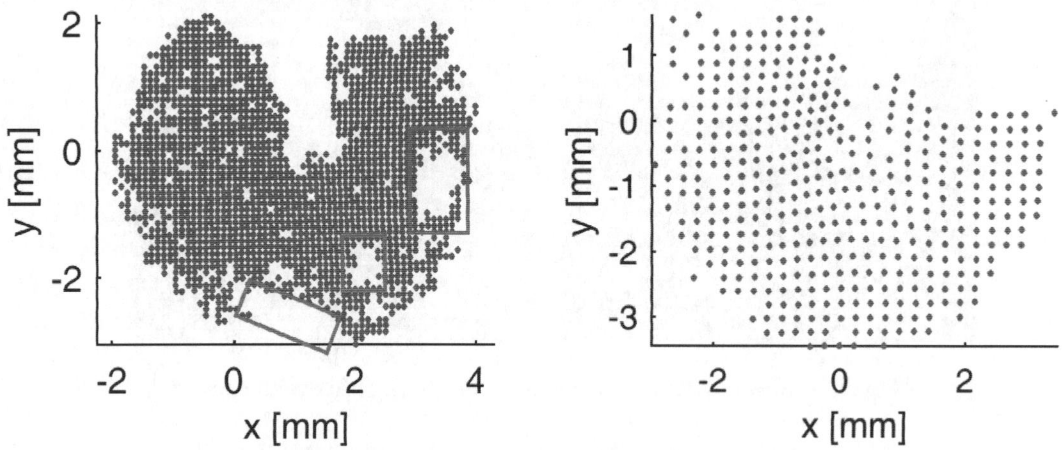

(a) (b)

Figure 3. Representation of the correlation between the images of the two different cameras using (a) toner, (b) fluorescent powder. The red zones indicate a failed correlation due to different reflections reaching each camera resulting in very different gray values.

Figure 4 shows two photographs taken at different pressure steps. We see that the reflection spots move relative to the speckle pattern.

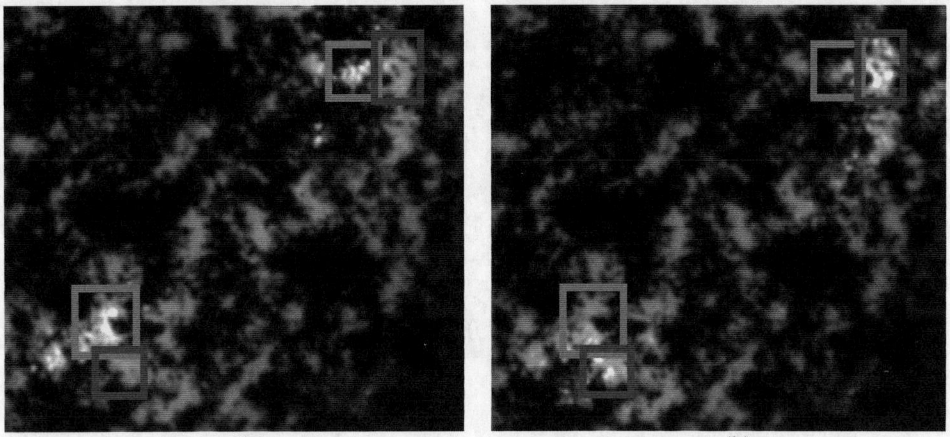

(a) (b)

Figure 4. Relative movement of reflections originating from behind the eardrum to the toner speckle pattern between pressure steps (a) +1 kPa and (b) +2 kPa (Camera 2).

3.2. Displacement

Figure 5 shows the full field Z-displacement when the eardrum is exposed to different static pressures. The fixed edge of the tympanic membrane is correlated as well, indicated by the non-moving boundary in the image.

Figure 6 shows the Z-displacement for a single point of the eardrum. We can see a hysteresis occurring when the eardrum undergoes a full cycle of the applied pressure steps. The displacement is slightly larger in the negative Z direction.

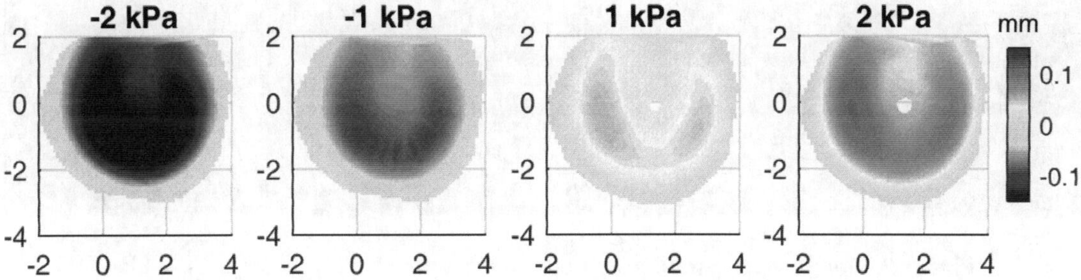

Figure 5. Interpolated full field Z-displacement shown for different applied static pressures.

IOP Conf. Series: Journal of Physics: Conf. Series **1149** (2018) 012026 doi:10.1088/1742-6596/1149/1/012026

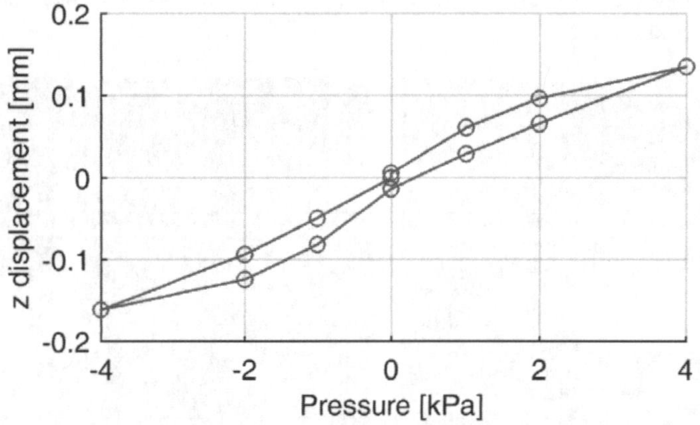

Figure 6. Single point Z-displacement.

3.3. Principal strains

Figure 7 shows the first principal strain of the eardrum for an over- and underpressure. The principal strain direction for overpressure is circumferential. The magnitude varies between 0 and 40 millistrain. The strain in the case of underpressure is almost zero.

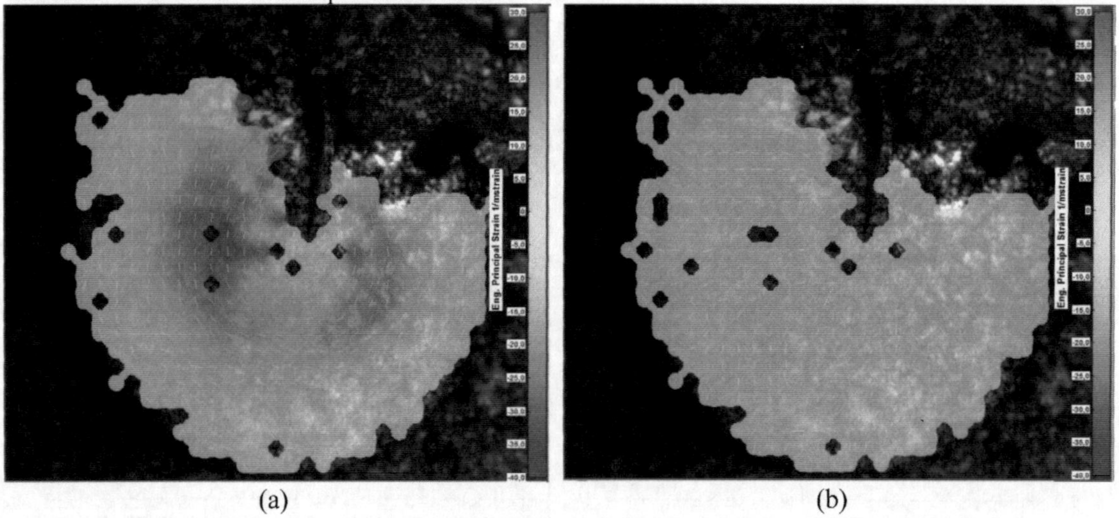

(a) (b)

Figure 7. First principal strain magnitude and direction for (a) +4 kPa, (b) -4 kPa

4. Discussion

4.1. Speckle pattern & correlation

We see that different reflections, originating from behind the semi-transparent eardrum, are observed for each camera when using the toner speckle pattern (figure 2). This will result in very different grey values in some areas of the image for each camera. This results in failing of the correlation algorithm for these areas (figure 3a). Similarly, for a single camera, the reflections move relatively to the speckle pattern when the eardrum deforms (figure 4). The correlation between images at different pressure steps from one camera will not succeed. As a consequence, the displacement cannot be calculated in these areas.

Using the fluorescent powder in combination with the laser illumination and filtering solves this problem (figure 3b). Although a different face size and spacing was used, it is clear that there are no big missing zones due to reflections as there are no reflections possible.

The eardrum could also be coated with a white layer of paint after which the toner could be applied. However, since the eardrum is so thin, we avoid using a full layer of paint, which will dry up and might stiffen the eardrum.

4.2. Displacement

We can see from the full field Z-displacement that the deformation of the eardrum is quite uniform under static loads. The areas between the malleus and the eardrum edge are slightly more mobile, as expected (figure 5).

The single point displacement illustrates the typical hysteresis loop associated with the viscoelastic properties of the middle ear system (figure 6).

The mobility is higher for negative pressures than for positive pressures. This is as expected, as the membrane stretches when being exposed to positive pressures, but 'collapses' when being subjected to large negative pressures. However, from Dirckx [4] we know that this asymmetry of the displacement curve should be larger. It is possible that air leakage occurred during the experiment. Our future experimental setup will include an extra pressure measurement in the ear canal to ensure that the desired pressure is actually applied to the eardrum.

4.3. Principal strains

As mentioned in section 4.2, the eardrum stretches when being subjected to large positive pressures but not so much when undergoing large negative pressures. This can be observed when looking at the principal strain (figure 7). A nonzero strain exists when applying an overpressure, but the strain is nearly zero on the entire membrane when applying underpressure.

We can see that the direction of the first principal strain for overpressures is circumferential. This may be due the lesser movement of the superior part of the membrane (top of the image).

5. Conclusion

A DIC setup was built capable of measuring deformation and strain in an intact eardrum. Toner and fluorescent speckle patterns were compared regarding correlation issues. It has been found that fluorescent powder works better to avoid reflections.

Future work will involve further development of the setup and measuring eardrum strain under static and dynamic loads of middle ears with different pathologies.

References

[1] Cheng T and Gan RZ 2007 *J. Biomech. Eng.* **129(6)** 913-18
[2] Cheng T and Gan RZ 2007 *Biomechanics and modeling in Mechanobiology* **7(5)** 387-94
[3] Cheng T, Dai C, Gan RZ 2007 *Ann. Biomed. Eng.* **35(2)** 305-14
[4] Dirckx JJJ, Buytaert JAN, Decraemer WF 2006 *JARO* 7 339-51

IOP Conf. Series: Journal of Physics: Conf. Series **1149** (2018) 012027 doi:10.1088/1742-6596/1149/1/012027

Noncontact Sensing of Facial Muscle Activity Using Laser Doppler Vibrometry: Time Domain Data Analysis

S. Casaccia[1, a)], E. J. Sirevaag[2], M. G. Frank[4], J. A. O'Sullivan[3], L. Scalise[1, b)] and J.W. Rohrbaugh[2, c)]

[1]*Università Politecnica delle Marche, Department of Industrial Engineering and Mathematical Sciences, v. Brecce Bianche 12, 60131 Ancona, Italy*

[2]*Department of Psychiatry Washington University School of Medicine, Saint Louis, Missouri, USA*

[3]*Preston M. Green Department of Electrical and Systems Engineering, Washington University in Saint Louis, Saint Louis, Missouri, USA*

[4]*Department of Communication, University at Buffalo, Buffalo, New York, USA*

[a)] s.casaccia@univpm.it, [b)] l.scalise@univpm.it, [c)] jwrohrba@wustl.edu

Abstract. Movements of the facial muscles offer a promising avenue for assessing stress, fatigue and emotion, and appear useful for a number of applied and clinical purposes. This paper describes a novel application of laser Doppler vibrometry (LDV) as a noncontact method for assessing facial myographic activity. The principle of the LDV method involves detection of the minute vibrations of contracting muscles, associated with the activation of individual motor units. Data were obtained from 11 participants who received 15-20 min of training before the measurement acquisitions. Participants produced several standardized facial expressions involving activation of the upper face (lowering and raising the eyebrows) and the lower face (raising the upper lip, stretching the lip corners, and clenching the jaw). The associated muscle vibratory activity was assessed using the LDV method, within context of the simultaneous electromyogram (EMG). A separate condition entailed study of the jaw muscle signals during repetitive chewing. The temporal, spatial and measurement sensitivity aspects were studied in separate tests; the present report focuses on the temporal aspects of the response, in comparison to the onsets and offsets of the simultaneous EMG and gross facial surface displacement. The LDV signals were obtained from a site overlying the principal involved muscle for the various movements. Results showed that LDV myographic signals (LDV-MMG) could be recorded from all facial muscles studied, although they were relatively small from the muscles of the upper face. LDV-MMG signals were especially prominent at times of contraction onset and offset, indicating that the method may be particularly useful for the study of dynamic activity as would be associated with brief changes in facial expression. The LDV-MMG signals generally were found to lead the onset of the EMG signal by about 100 ms, and to lag the offset of the EMG signal by about 200 ms. The LDV-MMG response associated with chewing was associated in time with the EMG and displacement signs of chewing, but were generally more polyphasic in form. The findings generally support the potential use of the LDV method as a non-contact and non-obtrusive method for assessing activity of the facial muscles.

1. Introduction

Considerable attention is given to the assessment of facial muscle activity, which is important in multiple domains. These include basic studies aimed at expression and coding of affect, applied issues such as affective computing and detection of fatigue and stress, the study of psychiatric and communication disorders, and clinical evaluation of disorders of the facial nerves and mastication. There are two primary modes of assessment: 1) coding of the visible deformations of the face, and 2) direct electromyographic (EMG) measurement of muscle activity. These assessment modes each has well-recognized strengths and weakness, in terms of application issues and measurement properties.

In the present report, we describe a non-contact method for direct measurement of muscle activity, which may offer the basis for yet another assessment opportunity. We present some preliminary data bearing on the effectiveness of the non-contact method—in terms of timing and amplitude characteristics, for multiple facial muscles. The non-contact nature of the method offers several potential advantages, insofar as there is no strict requirement for preparation of the skin, or encumbrance with electrodes or other attached transducers.

Laser Doppler vibrometry (LDV) as a physiological assessment method. The non-contact technique described here is an extension of the method of laser Doppler vibrometry (LDV). LDV is a technically mature method for non-destructive testing of mechanical vibrations, usually in industrial and engineering applications [1].

The application of LDV to the study of human physiology is based on the premise that many relevant physiological functions at the system level (including tremor, muscle contraction, and cardiorespiratory activities) are accompanied by mechanical activities which are reflected (however minutely) at the surface of the body. This uncontroversial premise underlies a number of techniques that are routinely used in clinical and laboratory investigations, including auscultatory and palpatory methods; microphone-based methods for recording muscle activity as well as cardiovascular, respiratory and gut sounds; and accelerometry as a technique for characterizing precordial, vascular and ballistocardiac movements, as well as tremor, hyperkinesias and gross body movements. A now substantial body of evidence attests to the effectiveness of the LDV method for evaluating mechanophysiological activity, particularly cardiorespiratory signals [2-7].

LDV recording of muscle activity: The mechanical myogram (MMG). The present application is based on observations that muscle contraction is accompanied by mechanical vibration activity, which can be detected at the skin surface. Following recommendations of Orizio et al. [8], we adopt here the nomenclature "mechanical myogram (MMG)" to describe these muscle vibrations. Prior evidence from our group has demonstrated the effectiveness of LDV in detecting and quantifying these vibrations, in somatic muscles including a small hand muscle [9], and bilateral muscles in the arms and legs [10-11]. The LDV-MMG signals were shown to vary systematically as a function of force under elastic tension, and to be roughly coextensive with the simultaneous EMG signals (as considered in greater detail below).

Overview of experiment. The principal goals of this experiment were to describe the basic aspects of facial myographic activity measured using the LDV method, emphasizing the temporal domain properties. There appear to have been only limited prior studies of the facial MMG signal (using conventional mechanical sensors), focusing on the muscles involved in mastication [12-17] and brow lifting [18-19]. The LDV signals were obtained here under posed facial expressions, guided by the Facial Action Coding System (FACS), which identifies 46 elemental action units (AUs) [20-21]. Although the AUs are formally described in terms of deformations of the facial surface, involvement of the underlying facial musculature is fairly well understood. Conditions were also included that supported comparison with more conventional methods based on the EMG as well as FACs coding of the video images.

Participants were tested in several sequences involving posed expressions (defined in terms of standardized AUs) involving the upper and lower face. In the conditions described here, LDV data were obtained from individual skin sites overlying the principal engaged muscles, for comparison with the simultaneous EMG as well as an LDV-obtained measure of gross surface displacement during contraction. Particular attention was paid to the transitional aspects of the LDV and EMG signals at the times of contraction onset and offset.

2. Materials and Methods

Participants. Data were obtained from 15 participants within the age range of 20-28 years (9 female). The majority of analyses described here were based on a subset of 11 participants (6 female) for whom complete sets of data were available. Comparison of data from these 11 participants with the full data set, where possible, did not show any appreciable qualitative or quantitative differences. Exclusionary criteria included body mass index > 25; facial hair, piercing or severe acne that would interfere with any of the recording procedures; history of psychiatric illness; and any illness, use of medication or cosmetic procedure that would interfere with normal neuromuscular function. Participation entailed a single session, approximately 2½ hrs in length. Following a detailed explanation of the purpose and experimental procedures, informed consent (including specific consent to reproduce images containing the face) was obtained using a protocol approved by the Washington University School of Medicine.
Facial posing. Five facial actions involving the upper and lower face were studied, selected in large part on the basis of prior reports regarding their utility signs of emotion and stress. The gestures, which were standardized and labeled according to the Facial Action Coding System [20-21], included:

- AU1, which consists of raising the inner brows, and involves contraction of the medial frontalis muscles.
- AU4, which consists of lowering and pulling together the brows, and includes involvement of the corrugator supercilii muscles.
- AU10, which consists of raising the upper lip and involves contraction of the levator labii superioris muscles.
- AU20, which consists of pulling the corners of the lips backward, and involves contraction of the risorius muscles.
- AU31, which consists of clenching the teeth and includes involvement of the masseter muscles.

Participants received 15 to 20 min of training in producing the target facial expressions. Training followed the methods and guidelines described in the Facial Action Coding System (FACS) manuals [20-21]. Training included detailed instruction and supervised production, assisted by videos and still photos drawn from the FACS manuals, and with an opportunity to practice in front of a mirror. Participants were encouraged to produce the facial actions symmetrically with an emphasis on avoiding co-contraction of other muscles—especially the common linkages cited in the FACS manuals. Participants were instructed to produce the facial actions with a force level that would be readily detectable by an observer. In general, the instructions refrained from any attempt to link the facial actions to specific emotions. In the case of AU31, the teeth were clenched against a dental cotton roll placed between the left molars, with a force level described as "3—neither weak nor strong" on a scale of 1 to 5 [22]. The instructions and video examples were repeated, with opportunity for additional practice, immediately prior to the production of individual facial actions during the course of the experiment. Review of the physiological and video records indicated that the participants were generally successful in producing the target gestures, although with a fairly high incidence of co-contraction as might be expected given the limited training received.
Procedure. The experiment involved two series of facial action production, given in the same sequence for all participants. For the single point measurements described here, two productions of

each target gesture were obtained for 5 s each. The productions were cued by pre-recorded speech signals to "tense" and "relax". The two productions were separated by a 5 s rest period, and preceded and followed by 10 s resting periods. The delivery of the speech signals was under computer control. Training emphasized the importance of making the two productions equivalent, and constant during the 5 s periods. EMG and LDV data were obtained during this series, and video data were obtained throughout. A test involving masseter activity while chewing, which focused on rapid transitional aspects of the LDV-MMG signals, was also conducted as described below.

Physiological recording methods. Physiological data were acquired using a Biopac MP150 system, with the EMG and LDV data integrated into a common file with a 1 kHz sampling rate. Data were obtained from the left side of the face.

EMG recording. EMG was recorded using shielded 4 mm Ag/AgCl electrodes affixed with double-sided tape collars, from bipolar pairs with inter-electrode spacing of 2 cm. Sites were prepared by cleaning with alcohol and abrading with a sterile lancet. Signals were amplified with a gain of 5k over the band of 0.5 to 500 Hz. The four electrodes across the forehead were originally recorded referentially to the left earlobe (A1), and subsequently recombined to produce differential signals. (A horizontal configuration was adopted, to permit evaluation of the course of the frontalis EMG signal over the nasion and eye-brows [23]). For AU1 (brow raise), the electrode pair was centered above the pupil, over the frontalis muscle 1-2 cm above the upper margin of the eyebrow. For AU4 (brow lower), the pair was horizontal and centered on the midline, 1 cm above the brow (at a site intended to overlie the conjunction of the corrugator supercilii, procerus and orbicularis oculii muscles). For AU10 (upper lip raise), the electrode pair was oriented vertically, approximately 1 cm lateral to the wing of the nose, over the course of the levator labii superioris muscle. For AU20 (lip stretch), the electrode pair was oriented horizontally over the course of the risorius muscle, with the anterior electrode located 2 cm lateral to the angle of the lips. For AU31 (teeth clench), the electrode pair was oriented vertically along the palpated superficial bulge of the masseter muscle. A ground electrode was affixed to the mid-forehead.

Laser Doppler vibrometry (LDV) recording. LDV data were obtained using a Polytec PSV-300 scanning vibrometer system (with a OFV-505 sensor head and OFV 3001 controller). This instrument utilizes a 633 nm laser beam, which at an output power < 1 mW classifies it as an "eye safe" laser (Class 2 by ANSI standards, "Standard, A. N. S. I. (2000). Z136. 1-2000, for Safe Use of Lasers. Published by the Laser."). Data were obtained with the beam direction approximately radial to the target skin site (with the laser head repositioned as necessary), from an offset of 826 mm (corresponding with a laser beam coherence node), with a sensitivity of 5 mm/s/V and an anti-aliasing filter at 500 Hz. The beam was targeted at a site midway between the EMG electrodes overlying the principally engaged muscle. (Prior to recording, the reflectivity of the skin was enhanced by treating it with a compound of 80 □m glass retroreflective beads mixed in suspension with mineral oil, with additional beads applied to the surface using a cosmetics brush. (The retroreflective bead treatment was applied to decrease the incidence of speckle-related artifact, while avoiding any appreciable surface loading or areal summation of the signal that might be associated with use of retroreflective tape. Subsequent improvements in equipment and signal processing methods indicate that this skin treatment is not necessary.)

Preprocessing of LDV signal. Prior to specific analyses described below, the LDV signal was processed to suppress speckle dropout artifacts. Speckle effects are inherent with highly coherent and monochromatic light sources as are used by the LDV method [24], particularly when recording from non-cooperative surfaces such as the skin that have high micro-roughness (with associated areas of destructive interference in the reflected signal), as well as a relatively low reflectivity and lack of translational stationarity. The speckle artifact is manifest in the LDV signal in the form of aperiodic large amplitude high frequency events in the waveform. The artifact was suppressed using a non-linear method that was designed to minimize the suppression of genuine higher frequency aspects of the signal. Speckle-affected data points were defined in terms of the abrupt (and aphysiological) transition points in the waveform, as identified in the thresholded second derivative of the original velocity signal. These points were then removed from the waveform, and the missing points were resampled using linear interpolation. In general, the speckle artifact affected less than 1% of all points.

3. Results

Basic character of response. Figure 1 illustrates a typical LDV response obtained during a single-point measurement condition—in this case from the skin overlying the masseter during AU31. Shown in the bottom trace (e.) is the accompanying EMG. The intent of this figure is to depict the overall frequency composition of a typical signal and the rationale for selecting a narrow band-pass filtered signal for analyses. The original velocity signal in the top trace (a.) contains activity related to multiple sources, including gross surface deformation during contraction, physiological tremor, a readily visible ballistocardiac signal, and nonspecific, adventitious movement of the body and face. As described below, the ratio of activity generated intramuscularly to these other signals is generally improved at the higher frequencies, even though the absolute amplitude there is small. In the second trace (b.) the velocity signal has been converted by differentiation to an acceleration signal, to facilitate comparison with other MMG findings in the literature which are generally based on accelerometry recordings. The non-contraction related activities are evident in the acceleration signal as well. The third panel (c.) depicts a time-frequency spectrogram for the acceleration signal, produced by the Welch technique using 512 points, with 50% overlap.

The results reported here are based principally on activity within a narrow band of 30-50 Hz (or 32-48 Hz in some analyses). A band-bass filter of finite impulse design (999 taps) was used. In addition to providing a favorable signal to noise ratio, this band supported a high temporal resolution with respect to identifying signal onset and offset during transitional periods. It should be noted in this context that many of the findings described below are correspondingly restricted to this band. This is a potentially important caveat in view of indications that the MMG (at least as recorded from muscles of the limbs and hand) appears to incorporate activities in at least 3 functionally distinct frequency ranges [27]. The 32-48 Hz band encompasses the range that is often referred to as the "Piper rhythm" [28]

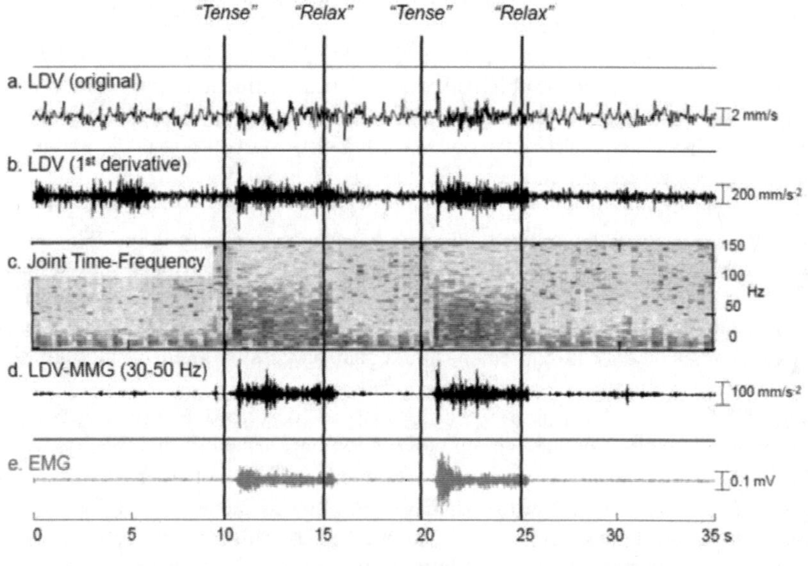

Figure 1. Illustration of an original LDV signal during two repetitions of AU31 (a.), acceleration signal derived by differentiation of the native velocity signal (b.), joint time-frequency depiction of the acceleration signal (c.), plus the narrow band filtered LDV-MMG signal (middle) and the accompanying EMG (bottom).

It should also be noted that the absolute magnitude of the LDV-MMG at higher frequencies is small. If converted to a displacement signal, the high frequency MMG shown in Figure 1 (d.) is on the order of 6 μm peak to peak, indicating that a high level of measurement sensitivity is required. This small vibratory component compares with the much larger gross surface displacements (associated with gross surface deformation including lateral expansion of the underlying muscle masseter bundle during contraction) in the mm range (see below).

The amplitude characteristics of the LDV and EMG signals associated with the five facial action units were investigated in detail from the records obtained during the single point conditions. Four segments, each 4 s in duration, were selected from each condition for more detailed analysis. Data associated with active expression were extracted during 2.048 s segments beginning 2 s after the commands to begin contraction. Two resting segments (also of 2.048 s) were also extracted, one from the first 9 s of the baseline period, and one from the final 9 s of the post-contraction period. The precise resting segments were selected as those periods in which the root-mean square (RMS) of the LDV-MMG signal was lowest (attesting to quiescence). Power spectral density functions were computed for each of the four segments (Hanning windowed), and individual spectra were normalized with respect to the total power of a mean spectrum computed across all four segments.

In Figure 2 (left), the normalized LDV-MMG power specifically within the 30-50 Hz band for each AU is depicted, for both resting and contraction conditions.

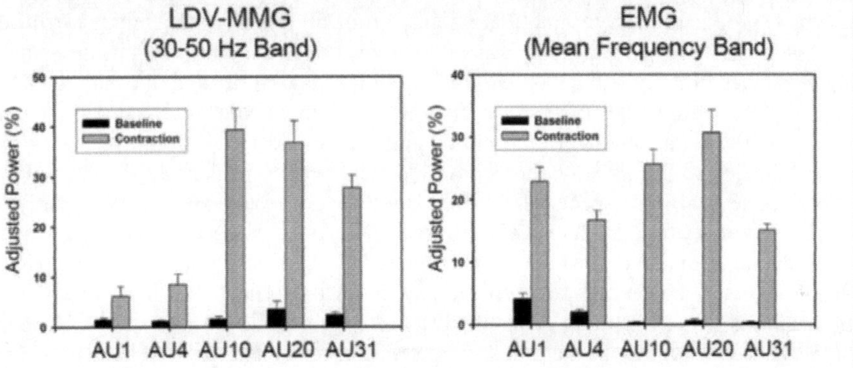

Figure 2. Adjusted LDV-MMG spectral power within the 30-50 Hz band, under baseline and contraction conditions for each of the five AUs (left), and corresponding EMG values (right).

The increase in contraction values (in comparison to resting values) there is evident and was observed without exception—for all participants, for all AUs, and for all replications. The greatest LDV-MMG activation was seen in the lower face AUs. Shown at the right of Figure 2 are the corresponding EMG values, for a 20 Hz band centered on the mean EMG frequency (which was generally in the range of 21 Hz (AU1) to 34 Hz (AU31). The EMG signal amplitudes were more uniformly distributed among the various AUs than were the LDV-MMG effects, although the resting-activation differences were readily apparent for both measurement methods, for all AUs.

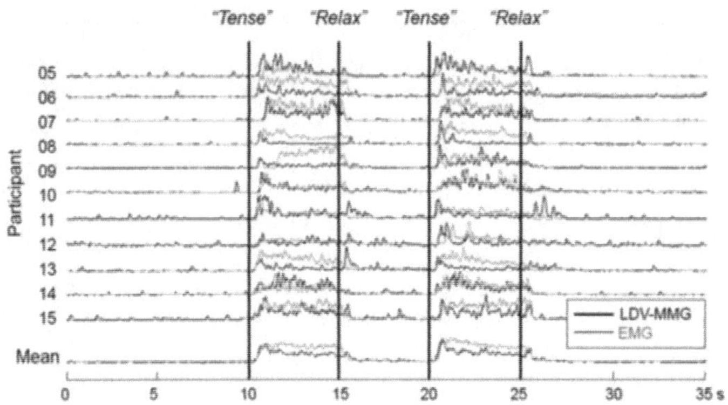

Figure 3. Overlaid integrated LDV-MMG and EMG activity for eleven participants (05 through 15), for two 5 s expressions of AU10. Vertical scaling is arbitrary.

Time domain characteristics. Temporal aspects of the LDV-MMG signals were analyzed in detail, with particular emphasis on the onsets and offsets. The following presentation focuses on the AU10 movement, which produced a large amplitude signal from all participants. Figure 3 illustrates the overlaid processed LDV-MMG and EMG signals obtained during the AU10 posing sequence. The LDV-MMG signal was filtered with a pass band of 32-48 Hz, as described above, and the EMG signal was high pass filtered with a filter of IIR design at 10 Hz. The signals were full wave rectified, and integrated using a 100 point moving window. For purposes of this illustration, the waveforms have been individually scaled so that all waveforms have identical peak to peak scale values.
Figure 3 illustrates several key aspects of the LDV and EMG signals. Both were reliably produced during the AU10 productions. Both types of signal showed substantial variability, both between consecutive productions, and over the course of individual contractions. For some participants (05, 07, 11, 12, 14) the temporal coherence between the two types of signals was high, whereas for other participants there was less agreement. A major difference appears to be that the LDV-MMG signal contained large bursts of activity at the times of contraction onset and, to lesser extent, offset, which were not reliably seen in the EMG signal.

The onset and offset segments of the integrated LDV-MMG and EMG signals associated with AU10 are illustrated in greater detail in Figure 4. The record of surface position (dotted trace) is the integrated representation of the original (unfiltered) velocity signal, and can be interpreted in terms of the gross in plane displacement toward and away from the laser sensor head—reflecting lateral expansion of the skin overlying the levator labii superioris muscle during contraction and subsequent relaxation. The position waveform has been formed by averaging across the two repetitions for each of the eleven participants (total number of records = 22). Each of the individual records was synchronized before averaging to the moment of peak expansion velocity (for the onset), and separately for peak retraction velocity (for the offset). These peaks were clearly evident in the unprocessed velocity waveforms. The total displacement amplitude for this averaged record is approximately 1.75 mm.

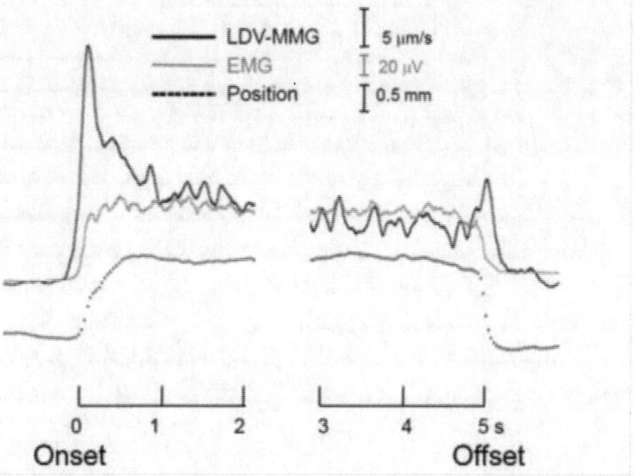

Figure 4. Averaged displacement, and integrated LDV-MMG and EMG signals associated with AU10 onset and offset. These records are averaged over the 11 participants (two replications for each participant).

The associated integrated LDV-MMG (black trace) and EMG (grey trace) signals were similarly time locked to the peak velocities in the displacement records and averaged over repetitions and participants. These records disclose two major differences between the EMG and LDV records. One is that the LDV velocity record shows prominent peaks at the times of the gross dimensional changes associated with the onset and, to lesser extent, offset of muscle contraction. This property would suggest that the LDV-MMG would be well suited for identifying rapidly transient events, such as facial micro expressions. The second property is that there are appreciable differences in timing of the LDV-MMG and EMG signals, with the LDV signal leading the onset of the EMG signal by about 100 ms, and lagging the offset of the EMG signal by about 200 ms. Although the integration process would tend to shift the peaks of these waveforms toward later values and thereby affect the apparent relationship with the position signal, both the LDV-MMG and EMG signals were processed using the same procedures so the relationships between them would be unaffected. The physiological interpretation of these latency differences remains unclear.

Time domain characteristics: Masseter muscle signals during chewing. Temporal relationships among LDV-MMG signals, EMG and surface movement associated with muscle contraction were examined in detail from the masseter muscle while chewing. Participants chewed a single standard 3g piece of gum of their desired flavor. Participants were given 3 min preliminary chewing, then a 30 s standard chewing period during which they were instructed to chew between their left molars at an approximate rate of 1 chew per s (approximating a natural rate observed for gum chewing [29]) and to avoid swallowing to the extent possible. The actual rate varied from 0.7 chews per s (participant 14) to 1.5 chews per s (participant 10) (mean = 0.9 chews per s). The chewing period was preceded and followed by 5 s recording periods of rest.

The raw EMG signals and band pass filtered LDV-MMG signals (32-48 Hz) from 11 participants are illustrated in Figure 5. Consistent with prior observations [30] there were substantial individual differences in the kinematic and myographic signals. The EMG signal generally consisted of a monophasic burst associated with each chew, whereas the LDV myographic signal was less consistent across chews and also more complex, with indications of additional bursts within the chewing cycle.

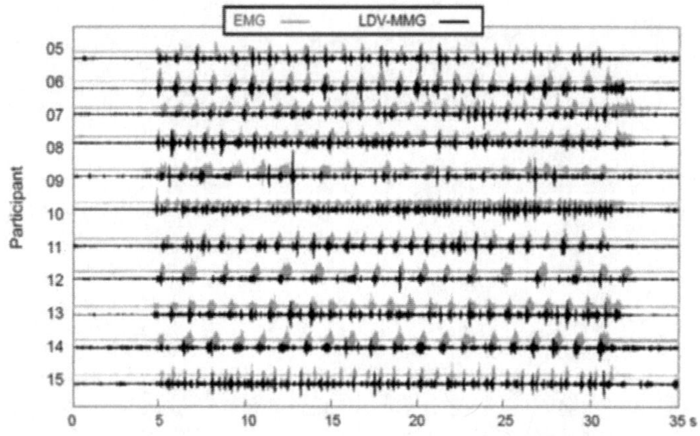

Figure 5. EMG and LDV-MMG signals from the masseter while chewing, for 11 participants. Vertical scaling is arbitrary.

Processed signals are illustrated in Figure 6. The signals were transformed by full wave rectification, followed by smoothing using a 100 point symmetric average. Symmetric smoothing was used, rather than integration, to preserve the absolute timing of peaks in the waveform with respect to gross dimensional changes. The dimensional changes were analyzed in terms of the original (unfiltered) LDV waveform, which included low frequency surface movements related to lateral expansion of the masseter during contraction. A position (displacement) waveform was generated by integrating the original velocity signal. The peak velocity of the expansion movement was used as a synchronization point, for purposes of averaging the waveforms across all available chews. The overlaid records in Figure 6 include the position, smoothed EMG, and smoothed LDV-MMG within the 32-48 Hz band. Vertical scaling is arbitrary.

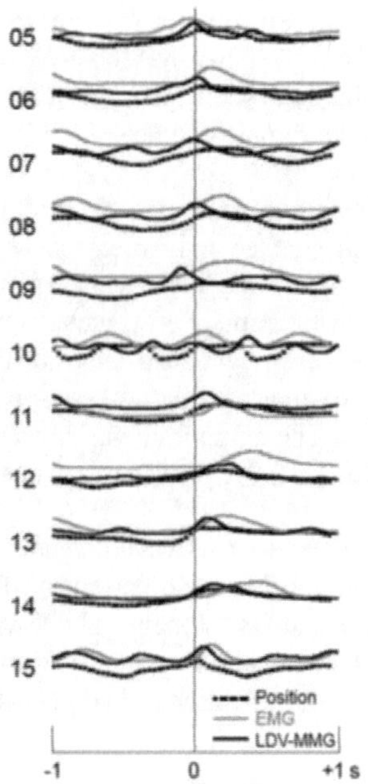

Figure 6. Masseter muscle surface position (displacement), smoothed EMG and smoothed LDV-MMG signals during chewing for 11 participants. Records are synchronized to a point corresponding to the peak velocity of the displacement signal (not illustrated).

Consistent with the known complexity of mastication movements [31], the gross displacement signals, which reflect the visible superficial bulge of the masseter muscle during chewing, were typically polyphasic with multiple inflection points during the opening and closing phases. The peak to peak amplitude of the gross displacement movements averaged 1.89 mm (range 0.44 to 3.26 mm), and the displacement peaked on average 182 ms following the velocity peak (see Table 1). The smoothed EMG waveforms were generally simpler, with a single monomorphic inflection that peaked in most participants later than the peak velocity (mean = 196 ms) and nearly coincided with the time of peak displacement. (The exception was participant 05, whose EMG waveform showed a peak that slightly preceded the velocity peak, and a second, smaller peak about 350 ms later; see Figure 6 top.)

Table 1. Peak latencies (in ms) of key features in chewing-related EMG and LDV signals. Times are measured with respect to the peak of the gross LDV velocity peak.

	Mean	Range
Latency of displacement peak	182	45 to 512
Latency of EMG peak	196	-19 to 440
Latency of LDV-MMG1 (peak 1)	46	-97 to 174
Latency of LDV-MMG2 (peak 2)	523	262 to 807
Interpeak latency: LDV-MMG1 to EMG	150	-15 to 318
Interpeak latency: LDV-MMG1 to LDV-MMG2	477	178 to 706

In contrast, the smoothed LDV-MMG in all participants contained two peaks. The first peak was close to the time of the gross velocity peak (following it on average by 46 ms). The first LDV-MMG peak was significantly earlier than the EMG peak (mean difference = 150 ms, $t[10] = 5.022$, $p < .001$). The latencies of the first LDV-MMG and EMG peaks were significantly but modestly correlated ($r = 0.64$, $p < .05$). The second LDV-MMG peak followed the first by a mean of 477 ms. In general, none of the latency variables was significantly correlated to chew rate ($p > .05$).

The functional significance of the second LDV-MMG peak is not clear. Although it is difficult to appreciate from the records in Figure 6, which have been smeared temporally by the smoothing process, there was no consistent overlap between this peak and the EMG burst. The EMG burst peaked on average 327 prior to the second LDV-MMG peak, and inspection of the records indicated that in many records the EMG burst was terminated prior to the onset of the second LDV burst. This would suggest that the second LDV-MMG component was not a simple product of masseter muscle relaxation during the jaw opening phase of the chewing cycle.

Considerable latency differences between the LDV-MMG and EMG signals were thus evident in the chewing condition, in which the two were separated in time but not to a consistent degree across individuals. Neither the LDV-MMG onset nor offset was consistently related to the timing of the EMG onset and offset. The LDV-MMG signal was again biphasic, with a distinct chewing-related component appearing mid-cycle. The physiological origins of this are, again, uncertain, since it did not appear to be reliably associated in time with either the offset of the EMG burst or the mechanical signs of masseter relaxation as evidenced in the retraction of the gross surface displacement. Chewing is a complex movement involving many muscles in addition to the masseter (including the anterior temporalis, lateral pterygoid, digastric and sternocleidomastoid muscles [32-33], several of which are active during the jaw opening phase of mastication. It is not clear, however, that the LDV recording site used would pick up the associated mid-cycle activities, unless they were transmitted through bone or were referred from distant sites.

4. Discussion and Conclusions

The results described here indicate that the LDV method can be used to provide a sensitive measure of facial muscle activity:

1) The LDV-MMG signal was reliably obtained from all participants, from all muscle groups studied. The variations in amplitude from different muscle groups corresponded with similar variations in the amplitude of the simultaneous EMG signals.

2) The LDV signals corresponded in time with the durations of muscle contraction, but were especially prominent at the times of contraction onset and offset indicating that they would be particularly well suited for the study of dynamic aspects of facial expression including micro-expressions.

Temporal aspects of the facial LDV-MMG. The LDV-MMG signal generally agreed well with that of the simultaneous EMG. The principal difference was that the LDV signal showed bursts of activity coinciding with contraction onset and, to lesser extent, offset. Careful examination of the temporal relationships for the AU10 signals showed that there were also appreciable latency differences in comparison to the EMG signal, with the LDV-MMG onset preceding the EMG onset by about 100 ms, and the LDV-MMG offset lagging the EMG offset by about 200 ms. The physiological origins of these effects are uncertain. The large LDV myographic burst at contraction onset could be ascribed to the additional effort required at onset to produce the gross dimensional muscle change [34-35], although the terminal burst is more difficult to explain on this basis. It may be that it is associated with some active eccentric process during cessation of the posed expression, although the latency differences in comparison to the EMG remain anomalous. Collectively, the differences between the LDV-MMG and EMG signals are consistent with suggestions that the two types of measures are to some extent non-redundant, and that the LDV-MMG signal may harbor unique information regarding muscle function.

General considerations. Several limitations, and possible directions for future research and development, can be identified. A major limitation in the procedures used here likely lies in the need to maintain the gesture at a constant level over the course of the measurement period lasting several seconds. Facial gestures are normally fluid in nature, and the requirement here to sustain a static pose likely placed unrealistic and artificial demands—ones not properly served by the limited training regimen.

A limitation of the technology used here lies in the use of a visible laser beam (at 633 nm) in the Polytec PSV-300 instrument. Although the instrument is rated “eye safe” (Class 2), concerns for eye safety nevertheless prompted us generally to avoid potentially informative periorbital facial sites including the belly of the corrugator supercilii and the orbicularis oculii muscles. Fortunately, this is no longer an issue in future studies, given the recent availability of commercial Class 1 vibrometers (i.e., no concerns with eye or skin exposure) based on infrared lasers [36]. It also is worth noting that there are options for increasing the simultaneous spatial sampling among multiple facial sites, including rapid dithering or continuous scan of the beam among selected sites, multi-point systems, and the technical feasibility of full-field methods. Computer vision-based dynamic tracking capabilities are now available, to accommodate the natural range of movement during interviews and other assessment situations.

The non-contact basis of the LDV method is an inherent strength that presents unique assessment opportunities. A major disadvantage of the EMG method in this context, as noted by Ekman et al. [20] is that applied surface electrodes and the accompanying leads are inherently obtrusive. The constant awareness of the recording sensors almost certainly interferes with the natural production of facial expressions. An additional problem is that the surface EMG signals are rather poorly focalized, because of the conduction by volume of bioelectric signals. High resolution methods utilizing dense sensor arrays have been described [37-39] but the sensor apparatus is cumbersome and would be incompatible with the study of naturally produced facial displays. Further research will be needed to establish the spatial discreteness of the LDV-MMG signal.

References

[1] Castellini, P., Revel, G. M., & Tomasini, E. P. (2009). Laser doppler vibrometry. *An Introduction to Optoelectronic Sensors, 7*, 216-229.

[2] Casaccia, S., Sirevaag, E. J., Richter, E. J., O'Sullivan, J. A., Scalise, L., & Rohrbaugh, J. W. (2014). *Decoding carotid waveforms recorded by laser Doppler vibrometry: Effects of rebreathing*. Paper presented at the 11th International Conference on Vibration Measurements by Laser and Noncontact Techniques - AIVELA 2014: Advances and Applications, Ancona, Italy.

[3] Kaplan, A. D., O'Sullivan, J. A., Sirevaag, E. J., Lai, P.-H., & Rohrbaugh, J. W. (2012). Hidden state models for noncontact measurements of the carotid pulse using a laser Doppler vibrometer. *IEEE Transactions on Biomedical Engineering, 59*(3), 744-753.

[4] Marchionni, P., Scalise, L., Ercoli, I., & Tomasini, E. P. (2013). An optical measurement method for the simultaneous assessment of respiration and heart rates in preterm infants. *Review of Scientific Instruments, 84*, 121705-121701 - 121705-121709.

[5] Scalise, L., & Morbiducci, U. (2008). Non-contact cardiac monitoring from carotid artery using optical vibrocardiography. *Medical Engineering & Physics, 30*(4), 490-497.

[6] Tabatabai, H., Oliver, D. E., Rohrbaugh, J. W., & Papadopoulos, C. (2013). Novel applicaitions of laser Doppler vibration measurements to medical imaging. *Sensing and Imaging, 14*, 13-28.

[7] Sirevaag, E. J., Casaccia, S., Richter, E. A., O'Sullivan, J. A., Scalise, L., & Rohrbaugh, J. W. (2016). Cardiorespiratory interactions: Noncontact assessment using laser Doppler vibrometry. *Psychophysiology, 53*(6), 847-867.

[8] Orizio, C., Gobbo, M., Diemont, B., Esposito, F., & Veicsteinas, A. (2003). The surface mechanomyogram as a tool to describe the infuence of fatigue on biceps brachii motor unit activation strategy. Historical basis and novel evidence. *European Journal of Applied Physiology, 90*, 326-336.

[9] Rohrbaugh, J. W., Sirevaag, E. J., & Richter, E. J. (2013). Laser Doppler vibrometry measurement of the mechanical myogram. *Review of Scientific Instruments, 84*, 121706-121701 - 121706-121709.

[10] Scalise, L., Casaccia, S., Marchionni, P., Ercoli, I., & Tomasini, E. P. (2013). Laser Doppler myography (LDMi): A novel non-contact measurement method for the muscle activity. *Laser Therapy, 22*(4), 261-268. doi: 10.5978/islsm.13-OR-21.

[11] Casaccia, S., Scalise, L., Casacanditella, L., Tomasini, E. P., & Rohrbaugh, J. W. (2015, May). Non-contact assessment of muscle contraction: Laser Doppler Myography. In *Medical Measurements and Applications (MeMeA), 2015 IEEE International Symposium on* (pp. 610-615). IEEE.

[12] Ioi, H., Kawakatsu, M., Nakasima, A., & Counts, A. L. (2008). Mechanomyogram and electromyogram analyses during isometric contraction in human masseter muscle. *Australian Orthodontic Journal, 24*, 116-120.

[13] Ioi, H., Kawakatsu, M., Nakata, S., Nakasima, A., & Counts, A. L. (2006). Mechanomyogram and electromyogram analyses for investigating human masseter muscle fatigue. *Orthodontic Waves, 65*, 15-20.

[14] Kawakami, S., Kodama, N., Maeda, N., Sakamoto, S., Oki, K., Yanagi, Y., . . . Minagi, S. (2012). Mechanomyographic activity in the huan lateral pterygoid muscle during mandibular movement. *Journal of Neuroscience Methods, 203*, 157-162.
[15] L'Estrange, P. R., Rowell, J., & Stokes, M. J. (1993). Acoustic myography in the assessment of human masseter muscle. *Journal of Oral Rehabilitation, 20*, 353-362.
[16] Stiles, R., & Pham, D. (1991). *Acoustic- and surface electro-myography of human jaw elevator muscles*. Paper presented at the Annual International Conference of the IEEE Engineering in Medicine and Biology Society.
[17] Tortopidis, D., Lyons, M. F., & Baxendale, R. H. (1998). Acoustic myography, electromyography and bite force in the masseter muscle. *Journal of Oral Rehabilitation, 25*, 940-945.
[18] Alves, N., & Chau, T. (2010). The design and testing of a novel mechanomyogram-driven switch controlled by small eyebrow movements. *Journal of Neuroengineering and Rehabilitation, 7*(22), 1-10.
[19] Alves, N., Falk, T. H., & Chau, T. (2010). A novel integrated mechanomyogram-vocalization access solution. *Medical Engineering & Physics, 32*, 940-944.
[20] Ekman, P., Friesen, W. V., & Hager, J. C. (2002a). *Facial Action Coding System. Investigator's Guide* (2nd ed.). Salt Lake City, UT: Research Nexus.
[21] Ekman, P., Friesen, W. V., & Hager, J. C. (2002b). *Facial Action Coding System. The Manual*. Salt Lake City, UT: Research Nexus.
[22] Hagberg, C. (1987). Assessments of bite force: A review. *Journal of Craniomadibular Disorders: Facial & Oral Pain, 1*(3), 162-169.
[23] Goncharova, I. I., McFarland, D. J., Vaughan, T. M., & Wolpaw, J. R. (2003). EMG contamination of EEG: Spectral and topographical characteristics. *Clinical Neurophysiology, 114*, 1580-1593.
[24] Goodman, J. S. (1976). Some fundamental properties of speckle. *Journal of the Optical Society of America, 66*(11), 1145-1150.
[25] Sowman, P. F., & Turker, K. S. (2005). Methods of time and frequency ddomain examination of physiological tremor in the human jaw. *Human Movement Science, 24*, 657-666.
[26] Coulson, S. E., Croxson, G. R., & Gilleard, W. L. (1999). Three-dimensional quantification of "still" points during normal facial movement. *Annals of Otology, Rhinology and Laryngology, 108*, 265-268.
[27] McAuley, J. H., Rothwell, J. C., & Marsden, C. D. (1997). Frequency peaks of tremor, muscle vibration and electromyographic activity at 10 Hz, 20 Hz and 40 Hz during human finger muscle contraction may reflect rhythmicities of central neural firing. *Experimental Brain Research, 114*, 525-541.
[28] Hagbarth, K.-E., Jessop, J., Eklund, G., & Wallin, E. U. (1983). The Piper rhythm-a phenomon related to muscle resonance characteristics? *Acta Physiologia Scandinavica, 117*, 263-271.
[29] Christensen, L. V., Tran, K. T., & Mohamed, S. E. (1996). Gum chewing and jaw muscle fatigue and pains. Journal of Oral Rehabilitation, 23, 424-437.
[30] Ahlgren, J. (1966). Mechanism of mastication: A quantitative cinematographic and electromyographic study of masticatory movements in children, with special reference to occlusion of the teeth. Acta Odontologica Scandinavica, 24, 5-109.
[31] Shimazaki, K., Matsubara, N., Hisano, M., & Soma, K. (2006). Functional relationships between the masseter and sternocleidomastoid muscle activities during gum chewing: The effect of experimental muscle fatigue. Angle Orthodontist, 76(3), 452-458.
[32] Mooller, E. (1966). The chewing apparatus. Acta Physiologia Scandinavica, 69, Supplement 280, 1-232.
[33] Saifuddin, M., Miyamoto, K., Ueda, H. M., Shikata, N., & Tanne, K. (2001). A quantitative electromyographic analysis of masticatory muscle activity in usual daily life. Oral Diseases, 7, 94-100.

[34] Celichowski, J., Grottel, K., & Bichler, E. (1998). Relationship between mechanomyogram signals and changes in force of human forefinger flexor muscles during voluntary contraction. *European Journal of Applied Physiology, 78*, 283-288.
[35] Mananas, M. A., Jane, R., Fiz, J. A., Morera, J., & Caminal, P. (2002). Influence of estimators of spectral density on the analysis of electromyographic and vibromyographic signals. *Medical & Biomedical Engineering &Computing, 40*, 90-98.
[36] Mignanelli, L., Rembe, C., Kroschel, K., Luik, A., Castellini, P., & Scalise, L. (2014). *Medical diagnosis of the cardiovascular system on the carotid artery with IR laser Doppler vibrometer*. Paper presented at the 11th International Conference on Vibration Measurements by Laser and Noncontact Techniques - AIVELA 2014: Advances and Applications, Ancona, Italy.
[37] Lapatki, B. G., Oostenveld, R., Van Dijk, J. P., Jonas, I. E., Zwarts, M. J., & Stegeman, D. F. (2006). Topographical characteristics of motor units of the lower facial musculature revealed by means of high-density surface EMG. *Journal of Neurophysiology, 95*, 342-354.
[38] Lapatki, B. G., Stegeman, D. F., & Jonas, I. E. (2003). A surface EMG electrode for the simultaneous observation of multiple facial muscles. *Journal of Neuroscience Methods, 123*, 117-128.
[39] Lapatki, B. G., van Dijk, J. P., Jonas, I. E., Zwarts, M. J., & Stegeman, D. F. (2004). A thin, flexible multielectrode grid for high-density surface EMG. *Journal of Applied Physiology, 96*, 327-336.

IOP Conf. Series: Journal of Physics: Conf. Series **1149** (2018) 012028 doi:10.1088/1742-6596/1149/1/012028

Feasibility study of the employment of laser Doppler Vibrometry for photoacoustic imaging

L Mignanelli[1] and C Rembe

Clausthal University of Technology,Leibnizstraße 28, Clausthal-Zellerfeld, Germany

[1]mignanelli@iei.tu-clausthal.de

Abstract. Photoacoustic imaging (PAI) is a biomedical imaging method which seems very promising for the detection of tumors in breast or in brain. Currently the detection of these signals is performed with piezoelectric transducers, which are very sensitive but have several limitations. Laser Doppler Vibrometry (LDV) represents a good alternative because it is a noncontact technique. The purpose of this paper is to investigate the potentiality of LDV for PAI in comparison with current transducers, and understand, in which measuring condition it is reasonable to adopt it. First, we make an introduction about the theory of PAI, we present a model of PAI for LDV and then, we investigate the theoretical limits for PAI with LDV. We derive a minimal detectable tumour radius of 390 μm for a commercial LDV with He-Ne laser.

1. Introduction

Photoacoustic imaging (PAI) is an imaging technique based on the photoacoustic effect. The main applications of PAI are in imaging of molecules, microvasculature, tumors, brain, and small animals [1-3]. This technique provides the use of a short-pulse light source for the irradiation of the tissue, this leads to the generation of a broadband photoacoustic (PA) wave. The PA wave propagates through the tissue and it is detected with ultrasound (US) transducers, and an image is computed.

The ultrasound transducers used for PAI present a high sensitivity but they have limitations due to the narrow frequency bandwidth, the finite aperture size, the contact nature of the device and the need of a coupling medium [4]. Optical detection of ultrasound could represent a valid alternative, thus several research groups investigated different interferometric techniques for PA applications [4-7]. In particular reference [4] presents a study of an all-optical non-contact PAI by using a commercial laser Doppler vibrometer (LDV) [8]. In reference [4] the authors present experiments on a pig brain with an artificial tumor as absorbing object, showing that LDV offers a promising method to achieve a non–contact PAI.

In this work, we investigated for which conditions it is reasonable to use LDV for PAI. We investigated which are the limits of detection for LDV depending on the metrological characteristics of the device and on the geometry of the investigated object. We introduce at first the typical signals generated from a spherical absorbing object and we examined the suitability of the LDV in comparison with ultrasonic transducers.

2. Photoacoustic imaging theory

Photoacoustic imaging is based on the photoacoustic effect. A short laser pulse irradiates the tissue, the light is locally absorbed, and converted into heat. The temperature rise of the irradiated object induces a pressure rise through thermoelastic expansion. This process is widely described in [1-3,9-14]. In particular the fractional volume expansion of the heated region is

$$\frac{\mathrm{d}V}{V} = -\kappa p + \beta T, \qquad (1)$$

IOP Conf. Series: Journal of Physics: Conf. Series **1149** (2018) 012028 doi:10.1088/1742-6596/1149/1/012028

κ is the isothermal compressibility, β is the coefficient of volume expansion, and T is the change in temperature. Neglecting volume expansions, $dV/V=0$ during the fast heating, i.e. assuming that the laser pulse is shorter than the thermal and stress relaxation times, [1-3] the initial pressure can be calculated as

$$p_0=\frac{\beta T}{\kappa}=\frac{\beta A_e}{\kappa\rho C_V}=\Gamma A_e. \tag{2}$$

The temperature increase T can be expressed as the specific optical energy deposition A_e divided by the density of the material ρ and the heat capacity at constant volume C_V. $\Gamma=\beta/\kappa\rho C_V$ is the Grüneisen parameter. If all absorbed energy is converted into heat [3], A_e can be expressed as the optical fluence F of the laser beam multiplied by the optical absorption coefficient μ_a

$$p_0=\Gamma\mu_a F. \tag{3}$$

The originated pressure propagates as an ultrasound wave in the tissue according with the wave equation

$$\left(\nabla^2-\frac{1}{v_s}\frac{\partial^2}{\partial t^2}\right)p(\boldsymbol{r},t)=-\frac{\beta}{\kappa v_s^2}\frac{\partial^2 T(\boldsymbol{r},t)}{\partial t^2}, \tag{4}$$

where v_s is the sound velocity in the material and $p(\boldsymbol{r},t)$ is the pressure at the location $\boldsymbol{r}$ and time instant t. Equation (4) can be solved with the Green's function approach, as the response to a temporal and spatial impulse [1-3] leading to

$$p(\boldsymbol{r},t)=\frac{1}{4\pi v_s^2}\frac{\partial}{\partial t}\left[\frac{1}{v_s t}\int d\boldsymbol{r}' p_0(\boldsymbol{r}')\delta\left(t-\frac{\boldsymbol{r}-\boldsymbol{r}'}{v_s}\right)\right], \tag{5}$$

where δ is the delta function and $\boldsymbol{r}'$ is the position of the source location. This equation is expressed in spherical coordinates and it is valid to calculate the photoacoustic pressure generated by an arbitrary heterogeneous optically absorbing object heated by a short laser pulse in stress and thermal confinement conditions. A detailed description of the mathematics can be found in [1-3].

2.1. Typical photoacoustic signal generated from a spherical absorbing object

Equation (6) describes the pressure propagation of a homogeneously heated sphere which could represent an early stage tumor in breast tissue [9]. For the case of a small sphere, the pressure p at the time instant t and at the distance r from the center of the sphere (Figure 1) [1], can be derived from equation (5) by

$$p(r,t)=p_0\left[U(R_S-v_s t-r)+\frac{r-v_s t}{2r}U\left(r-\left|R_S-v_s t\right|\right)U(R_S+v_s t-r)\right], \tag{6}$$

where U is the Heaviside function and R_S is the radius of the small sphere [1-3]. Therefore, the signal generated by a small sphere with higher absorption has a bipolar shape. The amplitude is inversely proportional to the distance r from the sphere/tumor and directly proportional to the initial pressure p_0. In particular, the duration τ of the bipolar signal depends on R_S

$$\tau=\frac{2R_S}{v_s}. \tag{7}$$

Figure 1 shows the set-up for measurements performed in transmission, i.e. the transducer is positioned in the opposite side of the excitation at the distance r_d from the sphere. Figure 2 shows the signal generated by a small heated and expanded sphere with $R_S=0.5$ mm acquired for different detector distances $(r=r_d)$ according with equation (6). Acoustic attenuation was not considered for this simulation. The duration of the signal τ remains the same since R_S and v_s are fixed, but the amplitude is damped by a factor $1/2r$.

The amplitude of the positive peak of the bipolar signal when the photoacoustic wave reaches the detector position r_d is [9-11]

$$p_{peak}(r_d)=\frac{1}{2r_d}\Gamma\mu_a F(z_S)R_S, \tag{8}$$

where $F(z_S)$ is the laser fluence at the position of the object (z_S), F_0 is the fluence on the surface and μ_{eff} is the effective optical attenuation in the tissue

$$F(z_S) = F_0 e^{(-\mu_{\text{eff}} z_S)}. \tag{9}$$

The acoustic attenuation of the wave travelling through a medium plays an important role in ultrasound imaging. Soft tissue has a typical acoustic attenuation coefficient [2, 10]

$$\mu_{\text{ac}}(f) = 0.1\left[\frac{\text{dB}}{\text{cm MHz}}\right] f_{(\text{MHz})}, \tag{10}$$

where $f_{(\text{MHz})}$ is the frequency of the signal in MHz. Equation

$$p_{peak}(r_d) = \frac{1}{2r_d}\Gamma\mu_a F(z_S) R_S \cdot \exp(-\mu_{\text{ac}}(f) r_d) \tag{11}$$

includes also the acoustic attenuation through the tissue by multiplying equation (8) with the damping. The frequency spectrum of the bipolar signal presents a characteristic sequence of oscillations which decrease in amplitude. In particular the central frequency f_u depends on the sound velocity and the dimension of the object [10-12]

$$f_u = 0.33\frac{v_s}{R_S}. \tag{12}$$

The major portion of acoustic energy resides in the half-power bandwidth (-3dB), i.e. between $f_{\text{low,-3dB}} = 0.16 v_s / R_S$ and $f_{\text{upper,-3dB}} = 0.51 v_s / R_S$ [2,10-12]. If we take for example, $R_S = 1\ \text{cm}$, $r_d = 2\ \text{cm}$ and $v_s = 1545\ \text{m/s}$, the attenuation factor $\exp(-\mu_{\text{ac}}(f)\cdot r_d)$ at $f = f_u$ results in $\exp(-\mu_{\text{ac}}(f_u)\cdot r_d) = 0.9899$, at $f = f_{\text{low,-3dB}}$ and $f = f_{\text{upper,-3dB}}$ is $\exp(-\mu_{\text{ac}}(f_{\text{low,-3dB}})\cdot r_d) = 0.9951$ and $\exp(-\mu_{\text{ac}}(f_{\text{upper,-3dB}})\cdot r_d) = 0.9844$ respectively. The error of the attenuation factor by considering f_u instead of the frequencies inside the half power bandwidth is : 1% in this example. In this first essay we assume the attenuation factor $\exp(-\mu_{\text{ac}}(f_u)\cdot r_d)$ not only for the central frequency but for the whole spectrum of the pulse. Thus, the frequency dependence is only considered for the center frequency of the signal in our model.

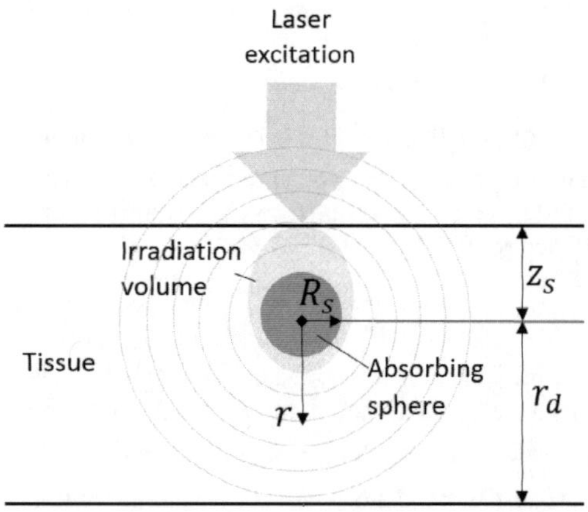

Figure 1. Setup for photoacoustic measurement performed in transmission. The acquisition sensor is positioned at r_d [9]

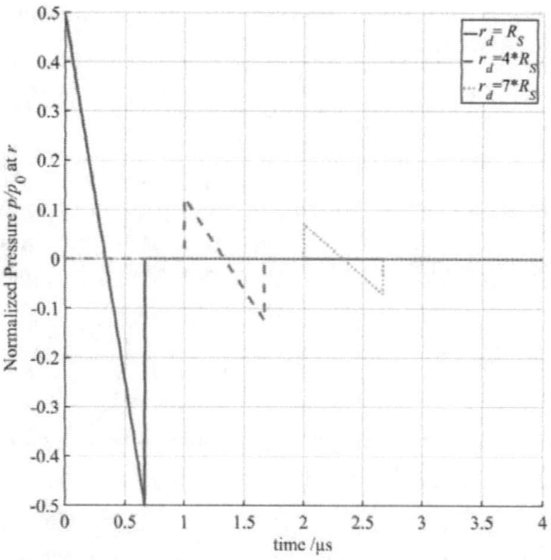

Figure 2. Simulation of the signal generate by a small sphere with $R_S = 0.5\ \text{mm}$ [1-3] at different distances r_d

3. Photoacoustic signal detection with LDV

The LDV is a noncontact interferometric technique that employs a heterodyne detection scheme [8]. It allow the detection of real time velocity and displacement signals with a resolution down to picometer range. An LDV sensor measures the Doppler shift of the back-scattered measurement beam which impinges the moving object. Reference [8] provides a detailed description of the measurement principle. LDV sensors are widely used for noncontact surface vibration measurements in mechanical, civil and industrial engineering. Due to its metrological properties and its non-contact nature, this technique finds also several application in the biomedical field like the identification of the cardiovascular parameter, studies of the ear, and dentistry. Recently LDV was employed for PA detection [4].

3.1. Measurement condition for the detection of PAI signals with LDV

Photoacoustic detectors are usually broadband ultrasound sensor, here we present the measurement condition for the alternative solution with a LDV proposed in reference [4]. While ultrasound sensors measure the pressure, LDV sensors measure the velocity or the displacement. In order to be able to detect the PA signal without degradation, two main conditions need to be fulfilled:

a) The maximal detectable frequency of the sensor has to be greater than the characteristic frequencies of the photoacoustic signals, which are strictly related to the minimal dimension of the absorbing object. According with [2], to acquire PA signals properly a bandwidth of 150% in respect to the central frequency f_u is necessary

$$\%B = \frac{f_H - f_L}{f_u}. \tag{13}$$

This represent a good compromise between the detection of high frequency components and the noise level. The typical bandwidth of LDV has a lower frequency f_L=0 Hz and an upper frequency of f_H=2 MHz [8]. The minimal detectable absorbing object is calculated by combining equation (12) and (13).

b) The pressure, the velocity or the displacement at the boundary needs to be larger than the resolution of the sensor. In the case of the LDV the minimal detectable velocity or the minimal detectable displacement has to be greater than the velocity/displacement at the boundary.

The minimal detectable displacement of a heterodyne interferometer is related to the signal to noise ratio referred to 1 Hz bandwidth (SNR'). In particular the noise equivalent mean square displacement $d'_{\min}$ after demodulation [8] is

$$d'_{\min} = \frac{\lambda}{4\pi}(SNR')^{-1/2}, \tag{14}$$

λ is the wavelength of the laser source. In theoretic calculations $d'_{\min}$ can reach $4\,\text{fm}\cdot\text{Hz}^{-1/2}$ [8]. However, commercially available class II LDVs are in combination with digital decoding that yields to a resolution of about $d'_{\min} = 50\,\text{fm}\cdot\text{Hz}^{-1/2}$ [8]. Thus we made the assumption that the displacement resolution of LDV at a given bandwidth B, $d_{\min@B}$, can be expressed in terms of power spectral density (PSD) of a white noise

$$d_{\min@B} = \left(\int_{\nu_1}^{\nu_2=\nu_1+B} PSD(\nu)d\nu\right)^{1/2} = d'_{\min}\cdot(B)^{1/2}, \tag{15}$$

where ν_1 and ν_2 are respectively the lower and the upper frequencies of the bandwidth. As mentioned before, the typical bandwidth of commercial LDV is 0-2 MHz, thus

$$d_{\min@B=2\text{MHz}} = 50\frac{\text{fm}}{\text{Hz}^{1/2}}\cdot(2\text{ MHz})^{1/2} = 70.7\text{ pm}. \tag{16}$$

We investigate for which combination of parameters (R_S, r_d, z_S) and metrological characteristics $(B, d_{\min@B})$ the detection with LDV is feasible.

4. Discussion

4.1. Resolution comparison between US sensors and LDV

An important aspect is to understand when it is reasonable to apply the LDV for PA detection in comparison with US sensors. While the typical resolution for commercial LDV is $d'_{\min} = 50\,\text{fm}\cdot\text{Hz}^{-1/2}$, the typical resolution for broadband ultrasound sensors is in the range of 0.2-0.6 $\mathrm{mPa}\cdot\mathrm{Hz}^{-1/2}$ [12]. The pressure resolution p'_{min} can be expressed in displacement resolution $d_{\mathrm{min,p}}$ in $\mathrm{m}\cdot\mathrm{Hz}^{-1/2}$ and compared with the LDV one, by considering the resolution for a sinusoid with frequency f at the boundary skin/air [13]

$$p'_{\mathrm{min}} = \frac{Z}{2} 2\pi f d'_{\mathrm{min,p}} \Rightarrow d'_{\mathrm{min,p}} = \frac{p'_{\mathrm{min}}}{Z\pi f}. \tag{17}$$

where Z is the acoustic impedance of the tissue. The factor 2 depends on the impedance mismatch between skin and air which results in an asymmetric displacement in direction of the skin-air interface [14].

In figure 3 the resolution of the different sensors are compared. In particular the resolution of a broadband ultrasound sensor (0.6 $\mathrm{mPa}\cdot\mathrm{Hz}^{-1/2}$) and of a commercial class II He-Ne LDV are reported. In addition, since the newest LDVs have an infrared (IR) laser(1550 nm) and this technology seem very promising, we show in figure 4 also the theoretical resolution limit for a 1550 nm heterodyne interferometer $d'_{\mathrm{min,IR}}$ =0.88 $\mathrm{fm}\cdot\mathrm{Hz}^{-1/2}$ [4].

Figure 3 reveals that for frequencies lower than ~1 kHz the commercial LDV has better resolution than broadband US transducers. This limit can go theoretically up to almost 100 kHz with IR technology. Since typically the ultrasonic frequency spectrum of PA signals has components between 100 kHz-10 MHz [12], US transducers have generally better performances than LDV for the detection of PA signals. However the detection of PA signals with LDV is possible according to the conditions described in paragraph 3.1.

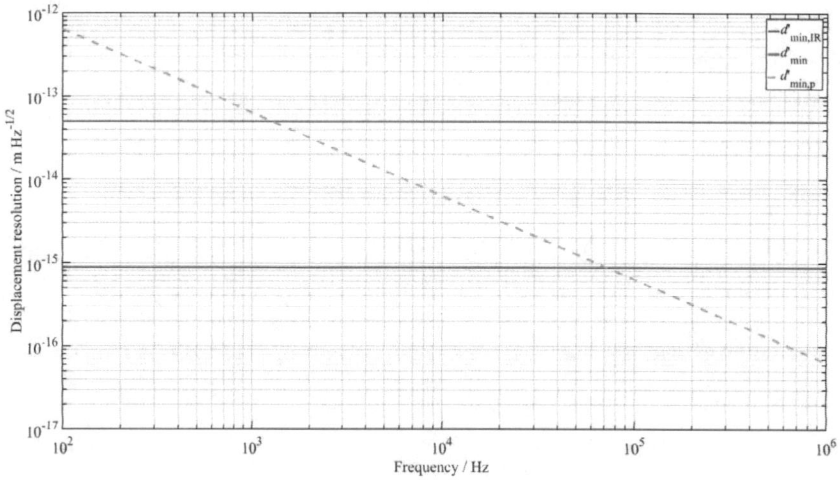

Figure 3. Comparison between the resolution of LDV and broad band US transducers

4.2. Limits of LDV for the detection of PA signals

According to the condition a) in paragraph 3.1, for a proper reconstruction of a PA signal with LDV, its central frequency has to be lower as $f_{\mathrm{u,max}}$

$$f_{\mathrm{u,max}} = \frac{100}{150} \cdot (f_{\mathrm{H}} - f_{\mathrm{L}}) = 1.3 \text{ MHz}. \tag{18}$$

From equation (12) it is possible to relate $f_{\mathrm{u,max}}$ to the corresponding dimension of the object $R_{S,\mathrm{min}}$ as

$$R_{S,\mathrm{min}} = 0.33 \frac{v_s}{f_{\mathrm{u,max}}}, \tag{19}$$

with the assumption of $v_s = 1545$ m/s (typical sound velocity for soft tissue). The minimal detectable absorbing object with LDV is $R_{S,\mathrm{min}} = 0.39$ mm.

Conform to the second detection condition, $d_{\min@B}$ has to be lower than the maximal displacement amplitude at the boundary. At the boundary between skin and air at $r = r_d$ the relation, between pressure p and velocity v [13] is

$$v(r_d,t) = \frac{2 \cdot p(r_d,t)}{Z}. \tag{20}$$

The equation (20) can be expressed in terms of displacement d

$$d(r_d,t) = \int \frac{2 \cdot p(r_d,t)}{Z} dt . \tag{21}$$

The velocity signal is directly proportional to the pressure and thus, except for an amplitude factor, it has the same characteristics in time and in frequency domain of the pressure signal. The displacement signal is presented in figure 4. Its maximum amplitude is $d_{\max}$

$$d_{\max} = \frac{1}{2}\left(\frac{2p_{peak}(r_d)}{Z} \cdot \frac{\tau}{2} \right). \tag{22}$$

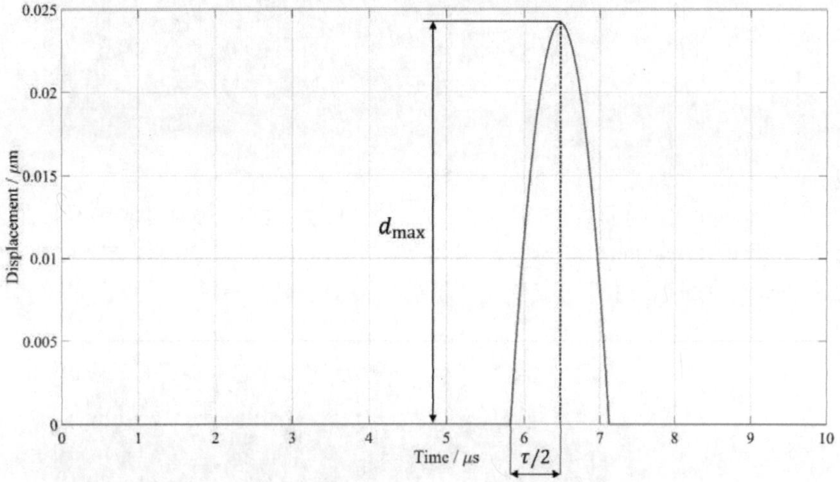

Figure 4. Example of a displacement signal $d(r_d,t)$ generated by $R_S = 1$ mm, $r_d = 1$ cm and $z_S = 1$ cm. The value of the other parameters are the same as in figure 5.

We calculate the value of $d_{\max}$ in dependence on all relevant parameters (R_S, r_d, z_S and μ_a) of the set up and the excitation. From (11) and (22) we have

$$d_{\max} = \frac{R_S}{Z \cdot v_s} \frac{1}{2 \cdot r_d} \Gamma \mu_a F(z_S) R_S \cdot \exp[-\mu_{ac}(f) r_d]. \tag{23}$$

In figure 5 we presented a map of $d_{\max}$ generated by an absorbing object with dimension $R_{S,\min} = 0.39$ mm in function of the depth z_S and the distance between the tumor and the acquisition system r_d. In particular we assumed $v_s = 1545$ m/s, $Z = 1.5$ MRayl, $\Gamma = 0.8$, $\mu_a = 0.6\ \text{cm}^{-1}$, $\mu_{\text{eff}} = 1.2\ \text{cm}^{-1}$ and $\mu_{ac}(f_u)$ [9-11,14]. We chose a pulsed laser with 1064 nm wavelength as laser excitation. The maximal allowed laser fluence for application in human body according with the laser safety norm [15] is $F_0 = 100\ \text{mJ cm}^{-2}$.

By comparing the value of $d_{\min@B=2\text{MHz}}$ with the values of $d_{\max}$ in figure 5 it is possible to notice that a sphere with a 0.39 mm radius and $\mu_a = 0.6\ \text{cm}^{-1}$ can be detected for the combination of z_S and r_d in the yellow range of the figure $\left(d_{\max} > d_{\min@B=2\text{MHz}} = 70.7\ \text{pm}\right)$. The interval is delimited by the distance to the detector $r_d < 16.5$ mm and the tumour-depth $z_S < 33.4$ mm.

We analysed also the maximal depth z_S where the simulated spheres with $R_{S,\min}$ are still detectable depending on the absorption coefficient μ_a for $z_S = r_d$. In figure 6 the results are displayed. With an absorption coefficient of $\mu_a = 1\ \text{cm}^{-1}$ (typical value for a tumor with 12% blood [14]) the maximal depth z_S where the signal generated by a sphere with $R_{S,\min}$ is still detectable (yellow range) is ~9.7 mm.

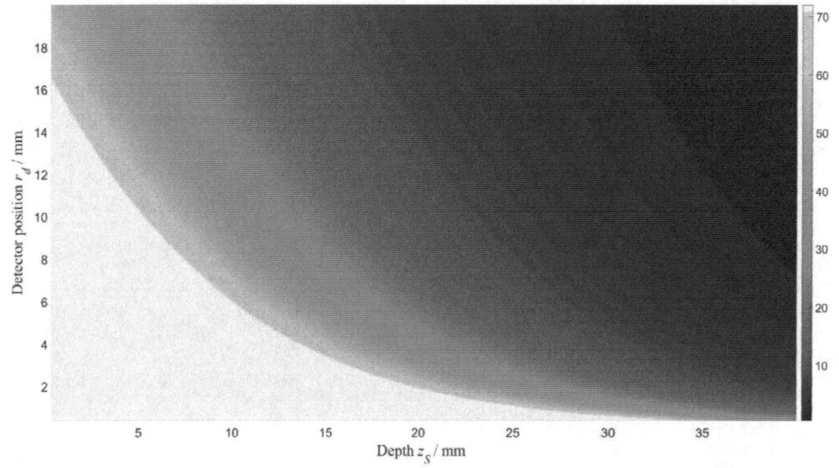

Figure 5. The colourmap presents the value of $d_{\max}$ in pm generated by a sphere with $R_{S,\min}$ in function of the depth z_S and the distance from the acquisition system r_d. The yellow range indicates where the sphere is detectable $\left(d_{\max} > d_{\min@B=2\text{MHz}} = 70.7\ \text{pm}\right)$.

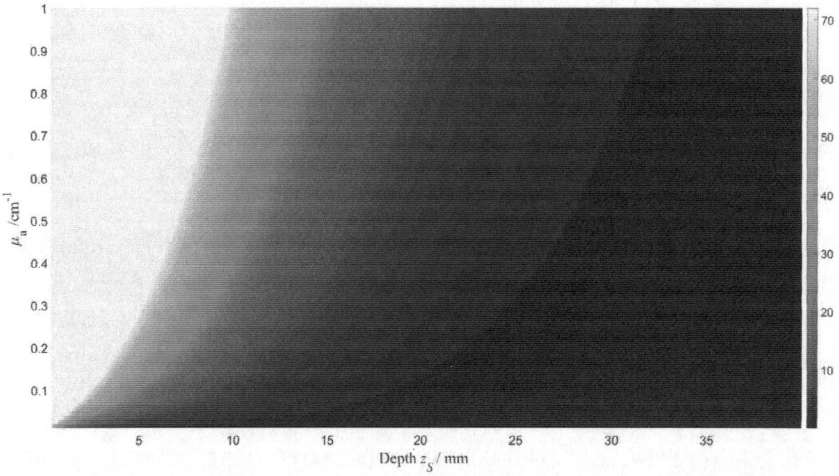

Figure 6. The colormap presents the value of $d_{\max}$ in pm generated by a sphere with $R_{S,\min}$ in function of the absorption coefficient μ_a and the depth z_S. The distance of the detector for this simulation is $r_d = z_S$. The yellow range indicates where the sphere is detectable.

Figure 7 shows a comparison between the minimal detectable displacement with the LDV and the maximal displacement $d_{\max}$ generated by a heated sphere. For the LDV minimal displacement we present three cases:

a) Commercial He-Ne LDV with $B = 2$ MHz, thus $d_{\min@B=2\text{MHz}}$.

b) According with equation (13) for objects bigger than $R_{S,\min}$, the necessary detection bandwidth is smaller than 2 MHz. With a smaller bandwidth the noise level is lower, and thus, if possible, it is convenient to measure with a bandwidth $B(R_S)$

$$B(R_S) = \frac{150}{100} \cdot f_u = \frac{150}{100} \cdot 0.33 \cdot \frac{v_s}{R_S}. \quad (24)$$

Therefore, we obtain a minimal displacement $d_{\min@B(R_S)}$ that varies with the necessary bandwidth

$$d_{\min@B(R_S)} = d'_{\min} \cdot \left(B(R_S)\right)^{1/2}. \quad (25)$$

c) The theoretical resolution limit of a 1550 nm heterodyne interferometer with the bandwidth varying with the radius of the sphere $B(R_S)$

The curves of case a), b) and c) are graphically, compared with the curves of $d_{\max}$ over the radius R_S at different depths z_S, $d_{\max}(z_S)$. In particular we chose $z_S = 1$ cm, $z_S = 2$ cm and $z_S = 3$ cm. The detector distance is fixed at $r_d = 1$ cm and the other parameters have the same values as the simulation performed in figure 5. Figure 7 allows to determine when a measurement with a commercial He-Ne LDV is feasible depending on the size of the absorbing object R_S. A measure is feasible if the value of the displacement originated from the heated sphere is, for a given R_S, above the displacement resolution limit of the LDV. In case b), due to the reduction of the bandwidth depending on R_S, the resolution increases allowing the detection of the same object at greater z_S respect to the case a). Case c) has always the highest resolution. For example, for a sphere with $R_S = 0.8$ mm placed at $z_S = 1$ cm, $d_{\max}(z_S\!=\!1\text{ cm})$ is detectable in all three cases. If the depth increases, $z_S = 2$ cm, only case b) and c) can detect $d_{\max}(z_S\!=\!2\text{ cm})$. The sphere at $d_{\max}(z_S\!=\!3\text{ cm})$ is detectable only in case c). Case c) presents the best measuring conditions and it can be considered the theoretical limited for the acquisition of PA signals with LDV.

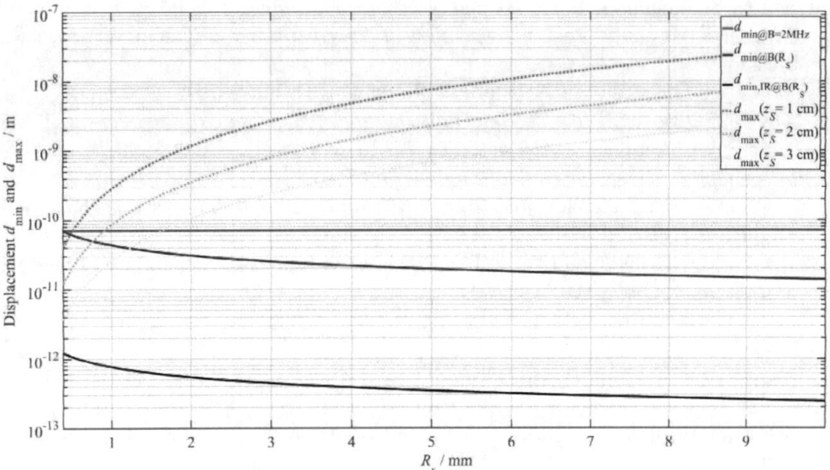

Figure 7. Comparison of the minimal detectable displacement with the maximal displacement originated by an absorbing sphere at different values of z_S. The three continuous lines represent the LDV minimal displacement calculated as described in case a), b) and c). The upper one is $d_{\min@B=2\text{MHz}}$, the middle one is $d_{\min@B(R_S)}$ and the lower one is $d_{\min,IR@B(R_S)}$. The three dotted lines represent the displacement $d_{\max}$ over the radius R_S at different values of z_S. Starting from the top we have respectively $d_{\max}(z_S\!=\!1\text{ cm})$, $d_{\max}(z_S\!=\!2\text{ cm})$ and $d_{\max}(z_S\!=\!3\text{ cm})$.

5. Conclusion

In this paper we present a model of PAI for LDV and the potential of LDV for PAI in comparison with current transducers. We explore the measuring conditions for a reasonable adoption of the LDV. In addition, we derived the theoretical limits for PAI with LDV. The analysis indicates that the commercial LDV has better resolution than broadband US transducer for frequencies lower then 1 kHz. This limit can go theoretically up to almost 100 kHz with IR technology. Since typically the ultrasonic frequency spectrum of PA signals are about between 100 kHz -10 MHz [2], US broadband transducers have generally better resolution than LDV for the detection of PA signals. However, since LDV presents advantages due to the non-contact nature of the technology and it has sufficient resolution to detect PA signals, it is still reasonable to use LDV for PAI acquisition. LDV allows the detection of absorbing spherical objects with a radius down to 0.39 mm, depending on its depth in the tissue, and the absorption coefficient at the excitation wavelength.

Future works are oriented on experimental tests to verify the theoretical limits described in this work. We would like also to include in our model the frequency dependent attenuation and evaluate its effects on the limits of LDV for the detection of PA signals.

References

[1] Wang L V and Wu H 2007 *Biomedical optics: principles and imaging* (New York: Wiley-Interscience
[2] Wang, L. V. (Ed.), Thompson, B. 2009 Photoacoustic Imaging and Spectroscopy. (Boca Raton: CRC Press)
[3] Wang L V 2008 *IEEE J. Sel. Top. Quantum Electron* **14** 171
[4] Tian C, Feng T, Wang C, Liu S,Cheng Q, Oliver D E,Wang X and Xu G *2016 IEEE Sensors Journal* **16** (23) 8381-88
[5] Payne B P, Venugopalan V, Mikić B B and Nishioka N S2003 *Journal of Biomedical Optics* **8** (2) 273-80
[6] Nuster R, Holotta M, Kremser K, Grossauer H, Burgholzer P and Paltauf 2010 *Journal of Biomedical Optics* **15** (2) 021307
[7] Speirs R W and Bishop A I 2013 *Applied Phisics Letters* **103** 053501
[8] Rembe C, Siegmund G, Steger H and Wörtge 2006 Measuring MEMS in motion by Laser Doppler Vibrometry *Optical Inspection of Microsystem* , ed Osten W (Boca Raton: CRC Press)
[9] Esenaliev R O, Karabutov A A,Tittel F K, Fornage B D, Thomsen S L, Stelling C and Oraevsky A A 1997 *Proc. of SPIE - The International Society for Optical Engineering* **2979** 71-82
[10] Oraevsky A A, Andreev V A, Karabutov A A and Esenaliev R O 1999 *Proc. of SPIE - The International Society for Optical Engineering* **3601** 256-67
[11] Oraevsky A A and Karabutov A A 2000 *Proc. of SPIE - The International Society for Optical Engineering* **3916** 228-39
[12] Winkler A M, Maslov K and Wang L V 2013 Noise Equivalent sensitivity of photoacoustics *J. of Biomedical Optics* **18** (9) 097003
[13] Rousseau G, Blouin A and Monchalin J P 2012 *Biomed Optics Express* **3** (1) 16-25
[14] Oraevsky AA and Karabutov A A 2003 Optoacoustic *Tomography Biomedical Photonics Handbook* ed Vo-Dinh T (Boca Raton: CRC Press) Chapter 34, pp. 34/1-34/34.
[15] American National Standard for Safe use of Lasers. Laser Institue of America, Amer. Nat. Standards Institute, Inc; 2007. ANSI Standards Institute, Inc; 2007. ANSI Standard Z136.1-2007

IOP Conf. Series: Journal of Physics: Conf. Series **1149** (2018) 012029 doi:10.1088/1742-6596/1149/1/012029

Experimental and Numerical dynamic characterization of a human tibia

Lorenzo Verdenelli, Robert Rossetti, Paolo Chiariotti, Milena Martarelli and Lorenzo Scalise

Università Politecnica delle Marche, via Brecce Bianche, 60131, Ancona, Italy

E-mail: l.scalise@univpm.it, p.chiariotti@univpm.it

Abstract. Many studies form the early 70s have determined the possibility, through the use of vibrations, to analyze the response of a human tibia in order to study its dynamic behavior under different excitation conditions and for different purposes. Comparative studies between contact and non-contact techniques have been carried out over the years, but in the case of non-uniform structures in form and materials, such as a human tibia, , there is a lack of information. The aim of this paper is to dynamically characterize a human tibia replica,with the same mechanical and morphological proprieties, with and without an external fixation system highlighting limits and advantages of contact and non-contact approaches.

This work proposes a comparison, in terms of FRFs and modal parameters, between the tibia alone and in the presence of an external fixation system, moreover the experimental data are compared also with a numerical model of such structures. Tests have been carried out using a mono-axial accelerometer for the contact measurement approaches and a laser Doppler vibrometer (632nm wavelength) for the non-contact one. The tibia has been placed on a foam support in order to simulate free-free conditions. The input was supplied both by a shaker and by a micro-hammer, keeping the same excitation direction. Results show that shaker-based excitation and micro-hammer excitation method can cause issues on the response measured with lasers. Modal analysis results tend to smooth these issues, even though laser-based data tend to senses a vertical in-plane in correspondence of the third horizontal mode because multiple components are sensed simultaneously during the scan.

1. Introduction

The fractures of the lower limbs, femur and / or tibia, are fractures which are among the most common in the event of road accidents, osteoporosis, sport or among injuries reported in war. In particular, tibial fractures are very important since, more than femur, they are more subject to support a greater amount of body weight. Weight is supported not only statically, when the person is at rest, but also and above all dynamically, when the person is walking or doing physical activity. For these reasons tibia fractures can be very impactful on every day life, especially those diaphyseal fractures that are displaced and/or exposed and for which the need of an external fixation system is required in order to provide a correct stabilization and healing process.

This research activity is part of a bigger work which final aim is to validate a non-invasive and non-ionizing method able to detect through the vibrations transmitted via the pins of the external fixation system, the status of the healing process in a tibia fracture, starting from past studies, where the excitation was given directly on the patient leg and the response measured

manually pressing the accelerometer on the leg [1], [2], [3], [4], until reach the more recent ones where the attention has been shifted from the patient's leg to the external fixation system. As Mattei et al.[5] and other [6] in literature have been demonstrated, using the external fixation system as a propagating medium to transmit the vibrations results in a more clean and usable output signal, since there will be less artifacts due to muscles and soft tissues and so less damping effect too and in the end using the fixation system it is possible to be least invasive on the patient. But before being able to proceed with the transmissibility tests through the fixation system, it is first necessary to understand how the tibia, a structure with a complex geometry and composite in the materials, behaves when subject to vibrations.
The aim of this research was to dynamically characterize the tibia with and without the external fixation. This means initially carrying out a characterization of the structure, experimentally and numerically. Characterization that will allow us not only to optimize the measurement system, choosing between different input/output techniques, but also to understand how the tibia, with and without the external fixation system, behaves when subject to vibrations and numerically validate the 3D models that could be then used to simulate different fracture typologies and/or different fixation configurations.

2. Materials and Methods

Different excitation and response sensing strategy used for the experimental tests are presented and compared in terms of their efficacy. The numerical simulations set-up will be also presented and carefully explained.

2.1. Experimental characterization of the tibia model

The tibia used for the experimental test was a 4th generation Sawbones large left tibia (#3402) with a mass of about 350±1 g, 405±1 mm in length (longitudinal axis; proximal to distal axis; x-axis) with an inner canal of 10 mm in diameter. The tibia is suitable for biomechanical tests since it has been proved to have a comparable mechanical behavior to cadaveric tibias. The specimen is made of composite materials in order to reproduce as much as possible the real bone structure, i.e. short fiber filled epoxy (Young's modulus between 10-16 GPa) replicating the cortical bone, and rigid polyurethane foam (Young's modulus between 137-155 MPa) replicating the cancellous bone.
The complex geometry of the tibia and its lightweight-nature make the experimental study task quite tough. For this reason, different combinations, in terms of input/output methods have been investigated, as reported in Tab. 1.

Table 1. Input/Output devices.

INPUT	OUTPUT
Instrumented Micro-Hammer	Laser Doppler Vibrometer
Shaker	Single-Axis Accelerometer

2.1.1. Input excitation

Impact excitation The micro-hammer used to excite the tibia is a Modally Tuned ICP Impulse Hammer by PCB Piezotronics (Model No: 086D80) with a sensitivity varying according to the tip used and a resonant frequency ≥100 kHz . For this specific test the tip used was a vinyl tip giving an hammer sensitivity of 15.92 (mV/N). A vinyl tip was used because the steel one would have not given a correct output due to the specific of the structure under test. The impulse was given at the lateral condyle in an oblique way in order to excite both the horizontal and vertical modes. The hammer works in conjunction with a PCB ICP Power Unit (Model No:484B) that features an AC signal decoupling mode for standard operation with ICPsystems as well as a DC mode for calibration or ultra low frequency operation.

Random excitation The shaker used to excite the tibia is a Tira GmbH (type: TV 50009) vibration exciters. These shakers are characterized by high lateral and axial stiffness. The shaker is powered by a power amplifier BAA 60 and has been connected to the tibia through a stinger, of 7 cm in length, and a PCB dynamic force sensor (208 A03). The force sensor was used to record the force generated by the shaker.
The shaker too has been mounted in correspondence of the lateral condyle with a specific angle in order to excite both the horizontal and vertical modes of the tibia.

2.1.2. Output response

Laser Doppler Vibrometer (LDV) The Laser Doppler Vibrometer used for the measurements is a Polytec LDV HeNe (wavelength 632nm). The LDV works in conjunction with a vibrometer controller that permits the operator to easily select measurement ranges, filters and controlling remote focus. Furthermore a signal strength indication is shown on the LCD panel in order to optimize the focusing of the optics head. The Polytec LDV sensor OFV-055 and the vibrometer controller OFV-3001 S that we used in the experiment can be configured to detect vibrations under different velocity ranges: 1, 5,10, 25, 125 and 1000 mm/s. To analyze the tibia the velocity range has been set on 10 mm/s.

Single axis accelerometer The accelerometer used for the experiment is a mono-axial PCB ICP Accelerometer (model No: 352C23), powered by simple, inexpensive, constant current conditioners and characterized by fixed voltage sensitivity, low-impedance output signal, low noise and intrinsic self-test feature. The sensitivity of the accelerometer is 5.11 mV/g with a resonant frequency 70 kHz and an optimal frequency range (+/- 5%) ranging from 2.0 to 10,000 Hz. In order to use the accelerometer on the tibia, a very little quantity of wax has been used so as not to dampen the signal excessively.

2.1.3. Boundary conditions The specimen has been placed on an acoustic foam bed. Then two pieces of an other acoustic foam have been placed near the lower and upper epiphyses in order to reproduce as much as possible the free-free conditions. In this way only the extremities of the specimen lean on the supports.

2.1.4. Signal Analysis The tests have been performed using the LMS Test.Lab by Siemens [7], which is a solution for experimental vibrational analysis that includes tools for data acquisition (LMS SCADAS), test execution and data processing. The LMS SCADAS have been used to acquire the signals from the load cell, the accelerometer and the LDV and to control the amplifier connected to the shaker and to the micro-hammer, through a digital-analog converter output. The Impact Testing and the Spectral Testing modules enables all test parameters to be set for impact(hammer) and shaker (random) tests. The modal Analysis module offers a solution with full functionality for experimental modal analysis. The solution allow to identify the dynamic properties of a structure with tools to easily create frequency response function (FRF) sets, carry out modal parameter evaluations, validate the experimental modal analysis modes and compare the original FRF data with summarized results.

2.1.5. External fixation system The external fixation used for the experimental tests is an Hoffmann II Tibia Shaft Frame. The fixation system total mass is about 677.04±1 g. The configuration used is the standard configuration for treat the tibia shaft fracture. The fixation is composed of different components. The only ones that interact with the bone are the pins that are screwed into the bone through the cortical and cancellous bone coming out the opposite side of about 3 mm. The materials of the single components are carbon fibers for the connecting rods, austenitic steel for the pins, alluminum for the clamps, posts and delta couplings.

2.2. Numerical model of the tibia

To get the same size and a more than good fidelity in details the first step was to make a scanning of the model. The 3D scanning of the tibia model has been made using a Non-Contact 3D Digitizer - RANGE 7 by KONICA MINOLTA [8]. The scanning system present a semiconductor laser (λ=660nm), CMOS sensor of 1,31 Mega Pixel(1280x1024) , with an accuracy of $\pm 40\mu$m (standard measures with globular objects) and a precison of $\pm 4\mu$m (standard measures with a KONICA MINOLTA reference plain chart).

The mesh of a scanned model is typically rough and not clean and for this reasons it's not suitable for being directly used in a numerical simulation suite. The refinement process on the tibia geometry has been done using a specific software for mesh optimization called Meshmixer. Once the 3D model of the tibia has been obtained and its mesh optimized for the simulation, it has been imported into COMSOL Multiphysics [9]. For our purpose the simulation was done using the Structural Mechanics Module. A frequency domain study was performed to obtain the mode shapes [10] of the tibia. In the Structural Mechanics Module it was set a point load with an harmonic perturbation, simulating the input given by the hammer and shaker in the same direction and position of the experimental tests.

3. Tibia dynamic characterization results

The following results show a comparison in terms of FRFs and modal parameters [11] obtained using the different measurement set-up presented previously. The experimental tests with the instrumented micro-hammer and with the shaker have been done considering a frequency range between 0-2000 Hz. Particularly important is the range between 200-2000 Hz because it represents our range of interest, within which are located the bending modes of the tibia.

3.1. Impact excitation

The impact test has been done with the following set-up:

- **EXCITATION INPUT**: Instrumented Micro-Hammer;

and using these response output techniques:

- **RESPONSE OUTPUT 1**: Laser Doppler Vibrometer (LDV)
- **RESPONSE OUTPUT 2**: Accelerometer

As it is possible to see from the image in Fig 1 a comparison, given in terms of FRFs amplitude and phase data, between the structure FRFs obtained measuring with the LDV and accelerometer of a single point, out of a total of 102, that corresponds to the Tibial Tuberosity is shown. The accelerometer signal has been integrated in order to obtain the velocity and so being able in this way to compare it with the LDV signal. Focusing on the range of interest (200-2000 Hz),Fig 2, is possible to see how even if at lower frequencies the two signals seem to be well correlated, moving towards the higher ones the signal, specially that of the LDV, shows

IOP Conf. Series: Journal of Physics: Conf. Series **1149** (2018) 012029 doi:10.1088/1742-6596/1149/1/012029

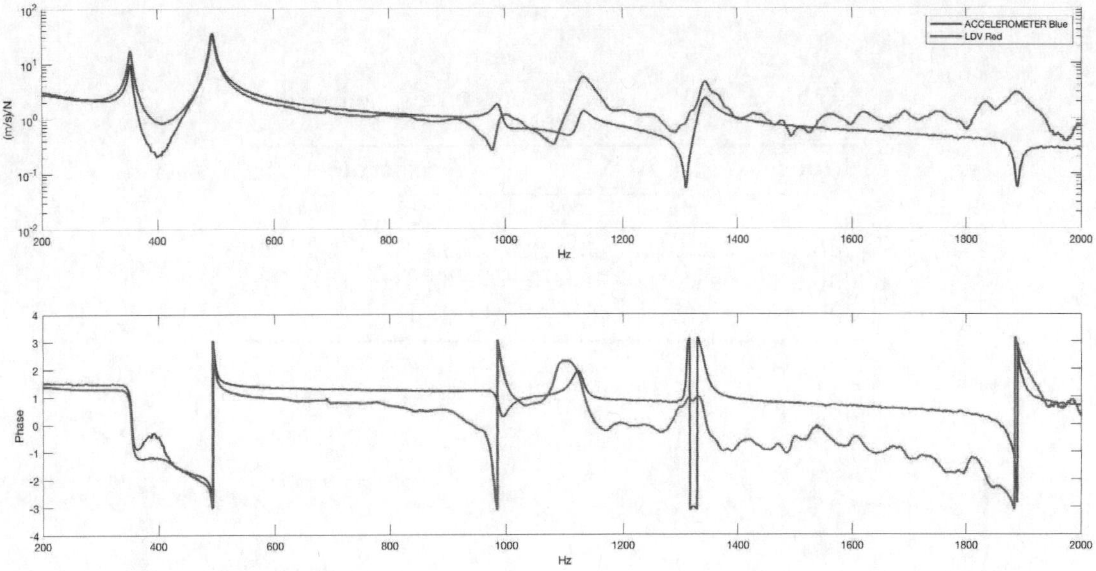

Figure 1. Micro-hammer excitation frequency response functions for accelerometer and LDV (amplitude and phase charts)

a noisy behavior. Also the signal provided by the accelerometer shows a bad behavior with a decrease in amplitude at higher frequencies. Highlighted in fig.2 by circles, it is possible to

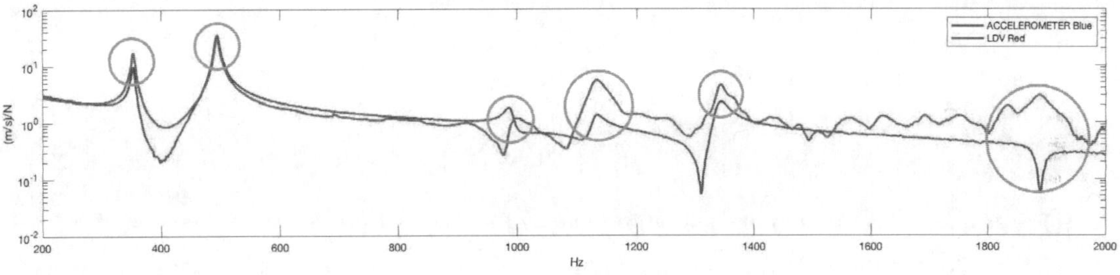

Figure 2. Frequency response functions and resonant peaks with micro-hammer excitation for accelerometer and LDV (amplitude)

see the resonant peaks related to the vibrating modes of the tibia. For frequencies higher than 1kHz, the peaks are no more clearly identifiable, especially those around the 1800 Hz, where the third horizontal mode of the tibia is present. The reason for which the laser is not able to provide a sufficient good signal at higher frequencies can be because the micro-hammer is not able to provide a good amount of energy on such a complex geometry and also because the impact of the micro-hammer causes the tibia to move excessively to ensure correct reading by the laser, generating the ring effect at higher frequencies. However, also the impossibility to guaranteed a constant repeatability for each point under investigation is the cause of this bad output signal. This lack of energy and repeatability is also present in the accelerometer signal that shows a decrease in amplitude both at lower and higher frequencies. In Tab. 2 it is possible to observe the resonant frequencies related to the vibrating modes of the tibia excited with the micro-hammer.

Table 2. LDV and Accelerometer resonant peaks

Modes	LDV		Accelerometer	
	XY	**XZ**	**XY**	**XZ**
Mode 1	353.1 Hz	493 Hz	353.1 Hz	493 Hz
Mode 2	996.9 Hz	1344 Hz	986.7 Hz	1344 Hz
Mode 3	Ø	Ø	1877 Hz	Ø

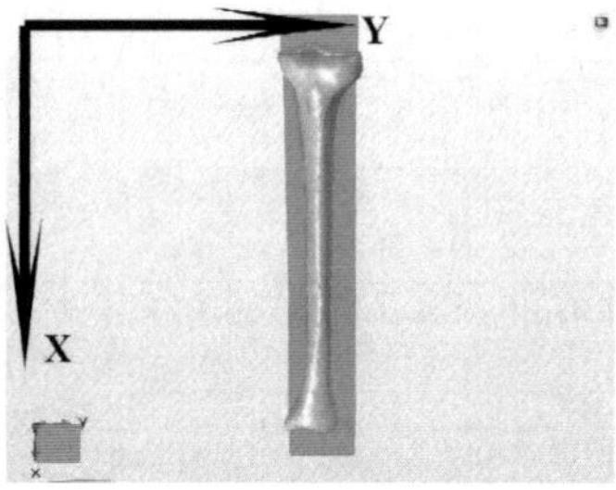

Figure 3. Tibia projection plane XY

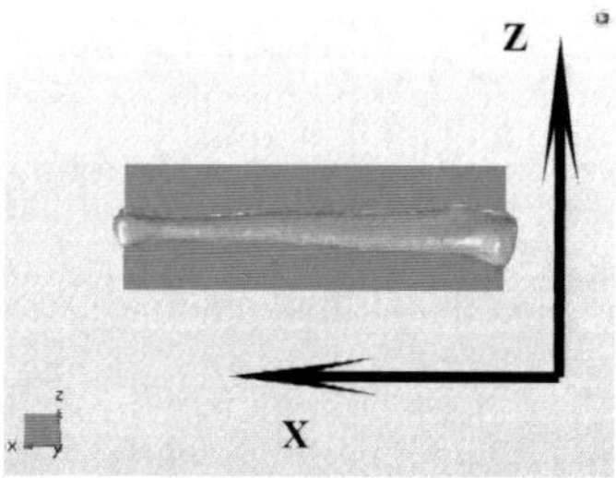

Figure 4. Tibia projection plane XZ

3.2. Random excitation
For the random test, the following set-up has been used:

- **EXCITATION INPUT**: Shaker excitation;

while for the output response:

- **RESPONSE OUTPUT 1**: Laser Doppler Vibrometer (LDV)
- **RESPONSE OUTPUT 2**: Accelerometer

In Fig 5 the comparison, in terms of FRFs and Phase, of the signals obtained exciting the tibia with a shaker driven with random noise. The shaker was attached to the structure in the same point where the impact excitation was given. As it is immediately observable the two signals appear to be much more better respect to those provided with the micro-hammer excitation, specially at higher frequencies. Restricting the frequency range to that of interest (Fig 6) The signal provided by the accelerometer too, appear to be much cleaner than that measured during the impact test. In particular if we move towards the higher frequencies, around 1827 Hz, we can see how, while before with the micro-hammer the accelerometer was not able to provide a clear signal and highlight the third mode on the horizontal plane, now, instead, through the shaker excitation it can highlight it. These better output signals were due to the fact that even if the shaker has a greater mass respect to the micro-hammer and that the shaker is linked to the tibia through a stinger, it is able to guarantee a constant repeatability in the excitation signal, for each point under investigation, and a greater amount of energy.
Using the random excitation, the FRFs of the two signals are highly correlated; in fact there are no shifts in frequency as observable in Tab. 3.

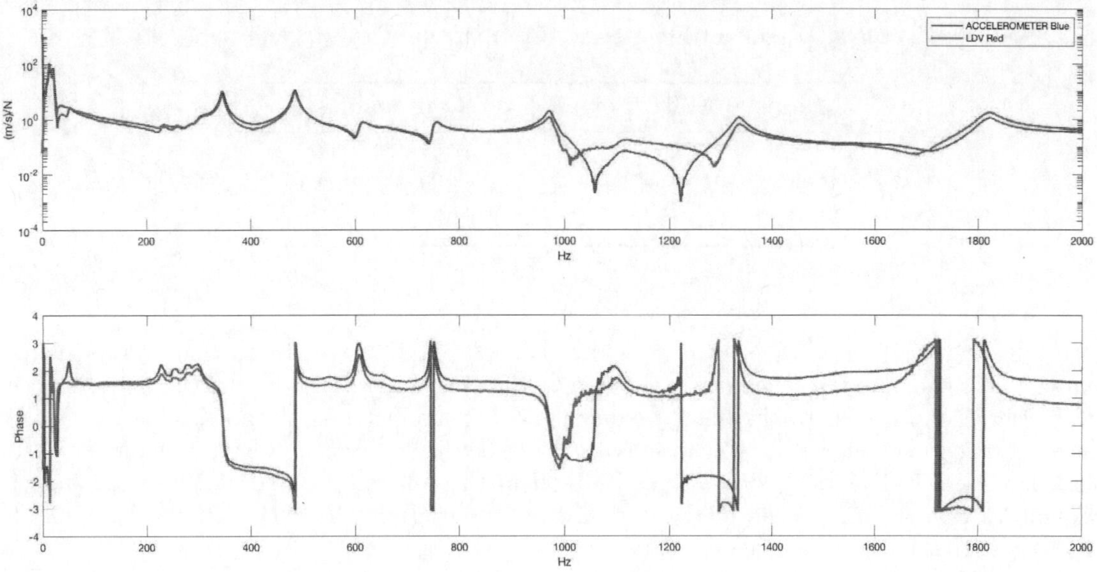

Figure 5. Shaker excitation frequency response functions for accelerometer and LDV (amplitude and phase charts)

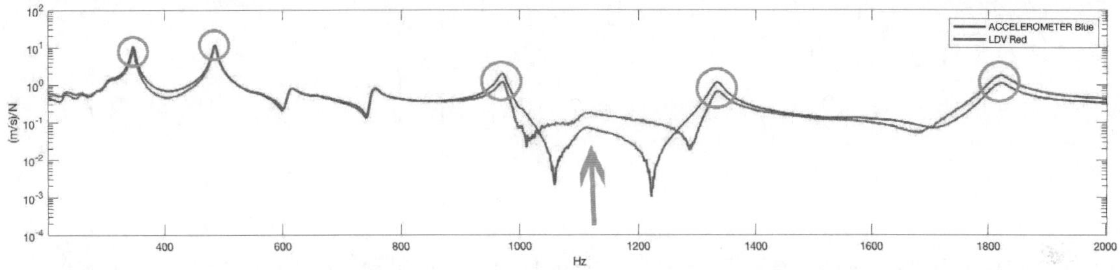

Figure 6. Frequency response functions and resonant peaks with random excitation

Table 3. LDV and Accelerometer resonant peaks

Modes	**LDV**		**Accelerometer**	
	XY	**XZ**	**XY**	**XZ**
Mode 1	345.3 Hz	485.2 Hz	345.3 Hz	485.2 Hz
Mode 2	971.9 Hz	1335 Hz	971.9 Hz	1334 Hz
Mode 3	1822 Hz	Ø	1822 Hz	Ø

The only mode that is not highlighted, because the point considered for the measurement does not make it possible to highlight it, is the torsional one that should be around the 1116 Hz (Orange arrow). The results obtained using the shaker and the accelerometer gave us the possibility to confirm also the goodness of our experimental test, and in particular the accuracy of our measurement set-up, comparing our data with those proposed by Mattei et al.(Tab. 4). They focused only on the frequency range 0-1kHz, so only the firsts modes were observable.

Table 4. Resonant Frequencies Comparison with Mattei et al.

Modes	Mattei et al.		Our results	
	XY	**XZ**	**XY**	**XZ**
Mode 1	353 Hz	496 Hz	345.3 Hz	485.2 Hz
Mode 2	965 Hz	Ø	971.9 Hz	1334 Hz

An easy way to observe the goodness of the shaker excitation respect to that of the micro-hammer is to compare the coherence of the signals((Fig 7)). Indeed, if we compare the coherence of the signal obtained with the micro-hammer and accelerometer and that obtained with the shaker and accelerometer is possible to see why the shaker is better than the micro-hammer. The blue signal (micro-hammer), shows some drop-off in the coherence signal at lower and higher frequencies that are the causes of the decrease in amplitude in the FRF and the inability to highlight correctly the resonant peaks at higher frequencies..

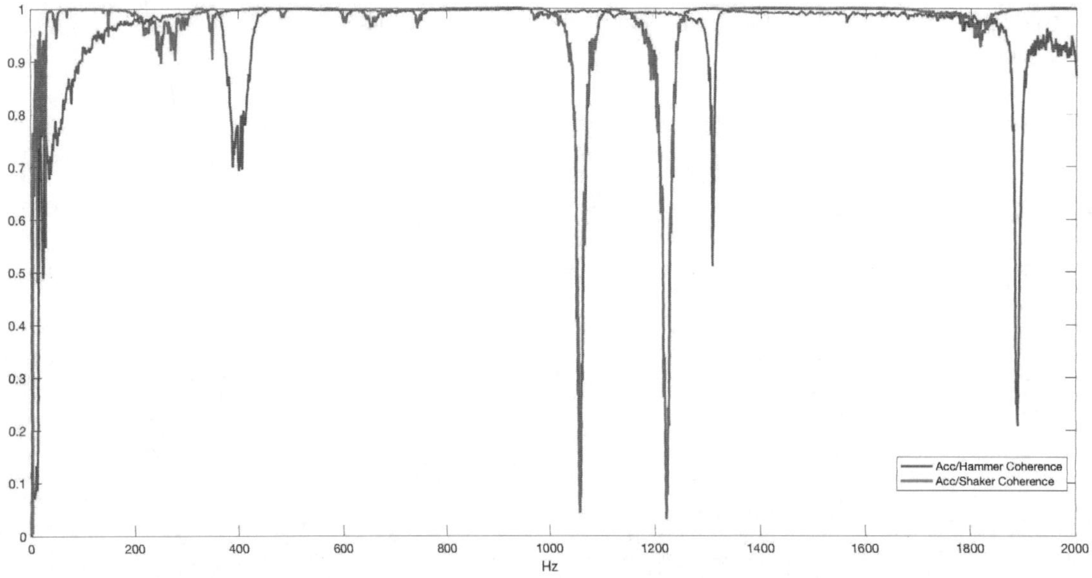

Figure 7. Coherence functions for micro-hammer and shaker excitations

3.3. Impact and random test Experimental modal analysis (EMA) comparison

Beacuse the LDV has been show a bad output signal exciting the tibia with the micro-hammer, the experimental modal analysis comparison has been done considering only the accelerometer (Fig 8) signal obtained with the micro-hammer and the two signals (Fig 14) obtained exciting with the shaker.

The main difference in the EMA of the signal excited by micro-hammer, can be found at higher frequencies, in particular for the second vertical mode and the third horizontal one that are not well stabilized as that given by the shaker. Instead, in Fig 10 the vibrating modes of the tibia obtained measuring with the LDV and exciting with the shaker are shown. Even if the LDV shows a perfect coherence in the stabilization of the modes with the accelerometer ones,

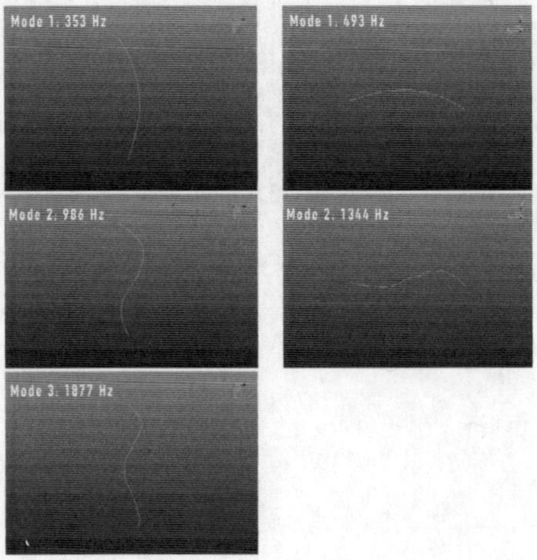

Figure 8. Micro-Hammer with Accelerometer Experimental Modal Analysis

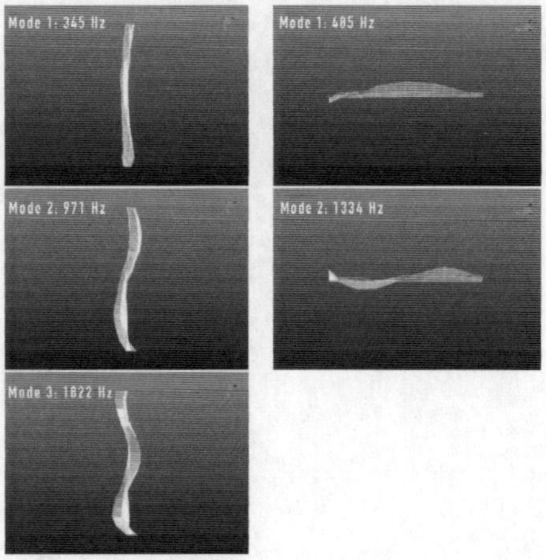

Figure 9. Shaker with Accelerometer Experimental Modal Analysis

at 1822 Hz the LDV highlights not only the third horizontal mode but it senses also a vertical in-plane due to the excessive movement of the tibia excited which causes the third vertical mode to be shown. This is an error because the third vertical mode of the tibia is observable only above the 2200 Hz. The analysis of the structure mode shapes, together with the point FRFs

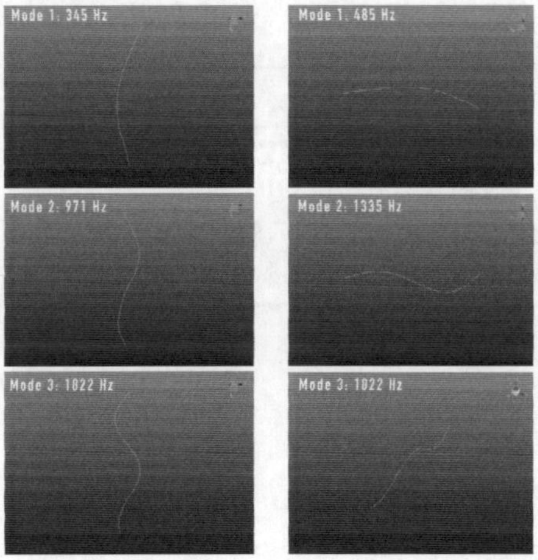

Figure 10. Shaker with LDV Experimental Modal Analysis

made it possible to identify in the shaker-accelerometer the best input-output combination to successfully characterize the tibia dynamics.

3.4. Tibia Experimental modal analysis (EMA) and Finite element analysis (FEA) comparison
The last step was to compare the EMA of the tibia with the mode shapes coming from the FEA done in COMSOL As shown in Fig 11 and in (Tab. 5) the horizontal modes of the FEA and those of the EMA match quite well. The main differences can be identified by observing the first and second vertical modes of the FEA where indeed there are some not indifferent variations in the frequencies when compared with those of the EMA. These frequency shifts for the vertical

IOP Conf. Series: Journal of Physics: Conf. Series **1149** (2018) 012029 doi:10.1088/1742-6596/1149/1/012029

modes are mainly due to the fact that in the numerical simulation the mass of the shaker is not included nor the constraint given by the stinger. Despite the frequency shift in the plane XZ

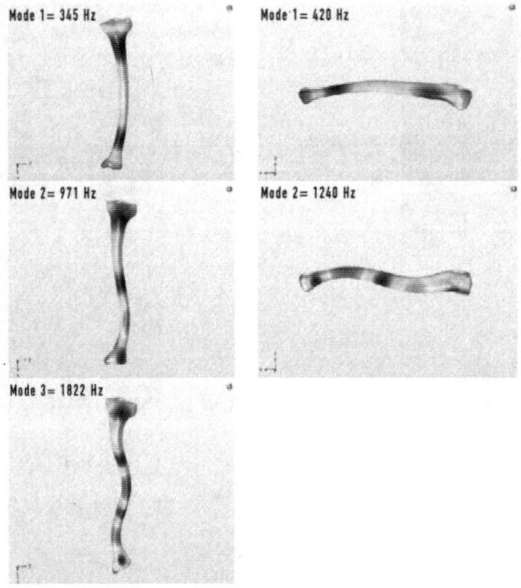

Figure 11. Numerical Mode Shapes

Table 5. Experimental modal analysis (EMA) and finite element analysis (FEA) modes

Modes	EMA		FEA	
	XY	**XZ**	**XY**	**XZ**
Mode 1	345.3 Hz	485.2 Hz	345.3 Hz	420 Hz
Mode 2	971.9 Hz	1335 Hz	971.9 Hz	1240 Hz
Mode 3	1822 Hz	Ø	1822 Hz	Ø

easily correctable, the numerical model can be considered valid.

4. Tibia with fixation system applied dynamic characterization results

The optimization of the measurement set-up allowed us to proceed with the analysis first on the fixation and then on the complete system tibia with external fixation mounted.

4.1. Tibia and tibia with external fixation system FRFs comparison

The measurements of the external fixation have been carried out experimentally using only the micro-hammer as excitation input and the mono-axial accelerometer as the response output. A total of 72 points were measured.

Once the measures on the fixation were made, it was mounted on the tibia and the experimental measurements were carried out. As expected the presence of the external fixation produces a significant effect on the tibia, going to modify the FRF in a really important way (Fig 12). The most important and relevant effects of the external fixation system on the tibia FRF are observable at lower frequencies, below the 200 Hz where many complex modes arise and at higher frequencies where the second and third horizontal modes were shifted downward of about

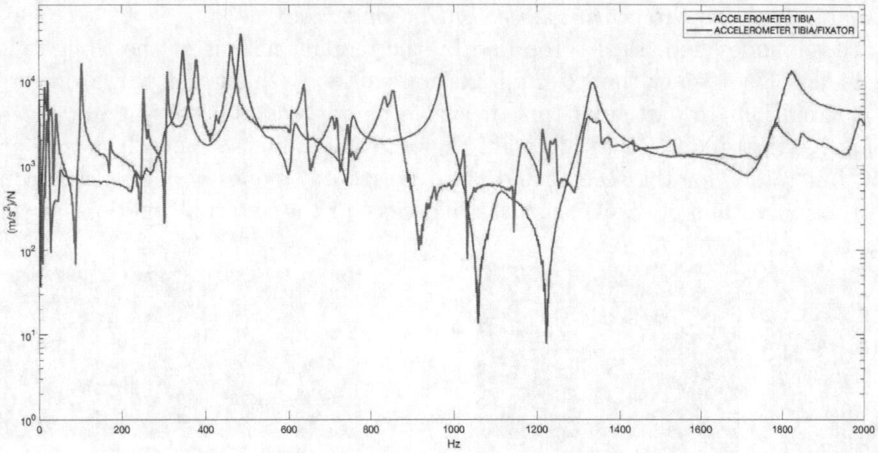

Figure 12. Frequency response functions for the tibia and the tibia with external fixation

100 Hz and 300 Hz respectively, and where the second vertical mode is completely blocked by the fixation, as can be seen in (Fig 13) and in (Tab. 6)

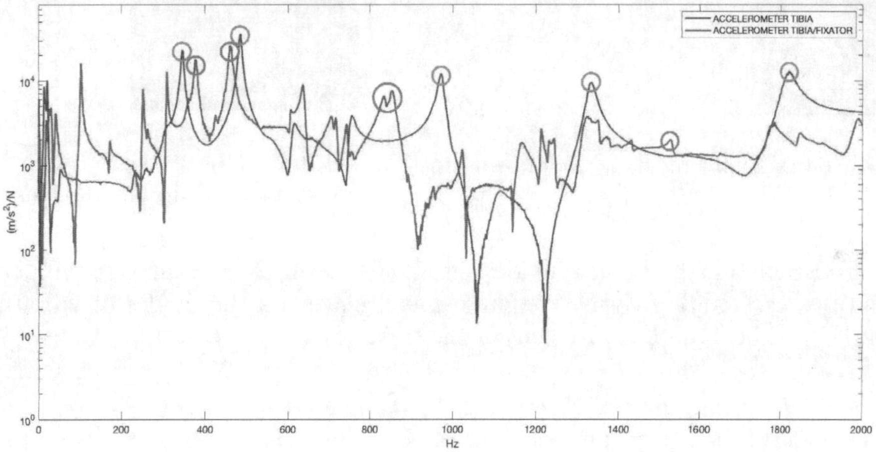

Figure 13. Comparison of the modes for the tibia and the tibia with an external fixation

Table 6. Tibia and tibia with external fixation Modes

Modes	Tibia		Tibia with external fixation	
	XY	**XZ**	**XY**	**XZ**
Mode 1	345.3 Hz	485.2 Hz	377 Hz	460 Hz
Mode 2	971.9 Hz	1335 Hz	844 Hz	Ø
Mode 3	1822 Hz	Ø	1530 Hz	Ø

4.2. Tibia and Tibia with External Fixation EMA Comparison

Comparing those modes with those proposed by the modal analysis of the tibia, it is possible to see (Fig 14 and Fig 16)how the external fixation works on shifting downwards and blocking the modes of vibration. In particular this effect can be appreciate observing not much the first horizontal and vertical modes, that after all, they did not suffer so much from the presence of the fixation, but rather for the second and third horizontal modes which show a considerable decrease in the percentage of displacement. The effect of the external fixation, as said before,

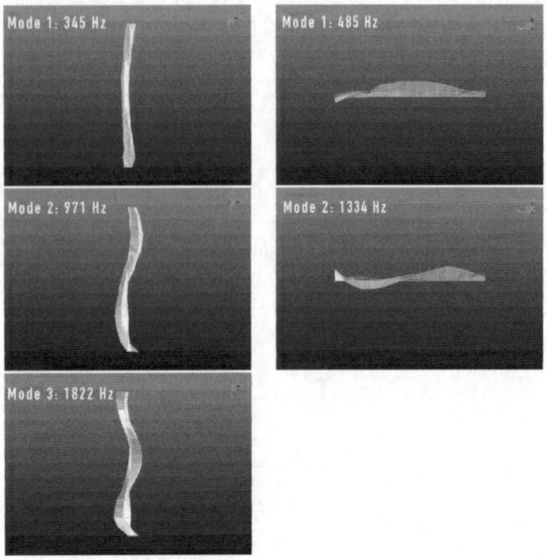

Figure 14. Tibia Experimental Modal Analysis (EMA)

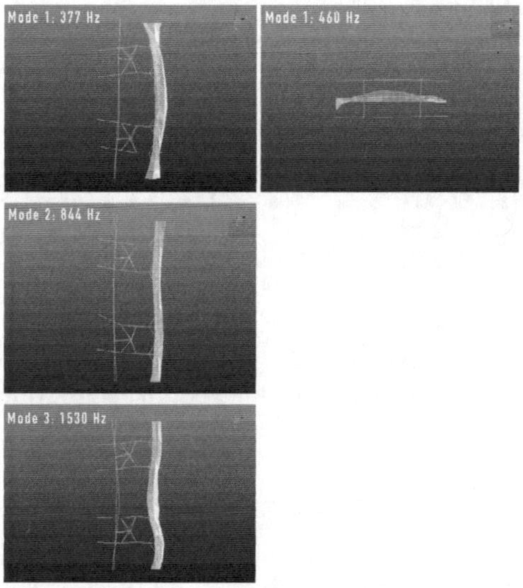

Figure 15. Tibia with External Fixation Experimental Modal Analysis

also acts considerably on the vertical modes, in particular on the second vertical mode that is not detected by the modal analysis since at higher frequencies the modes of vibrating of the fixation overhang, blocking, the movement of the tibia.

4.3. Tibia and Tibia with External Fixation EMA and FEA Comparison

With the 3D model of the external fixation validated it proceeded to the numerical simulation of the tibia with the fixation applied. As previously introduced, the fixation has been positioned in the same exact position of the real one and the numerical simulation has been done in free-free conditions due to the difficulty in recreating the perfect boundary constraints of the experimental one and without a point load applied, running in this way an eigenfrequencies study. But, even though the FEA has been done using this numerical set-up, the results obtained were very good. As shown in (Fig 17) the horizontal modes frequencies values are really close to that of the experimental data.

The shifts in frequency (Tab. 7) are easily referable to the specified numerical boundary conditions and to the fact that, as already said for the FEA of the tibia alone, in the numerical simulation the mass of the shaker was not considered.

IOP Conf. Series: Journal of Physics: Conf. Series **1149** (2018) 012029 doi:10.1088/1742-6596/1149/1/012029

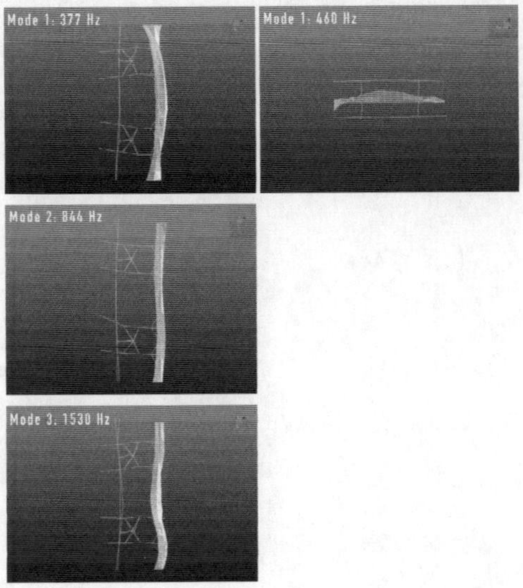

Figure 16. Tibia with external fixation Experimental Modal Analysis

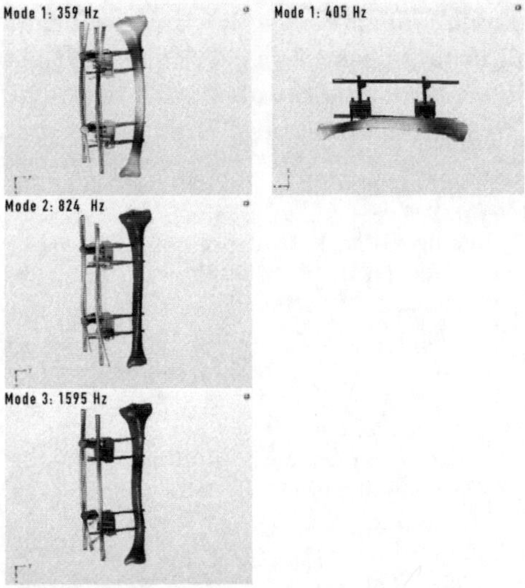

Figure 17. Tibia with External Fixation Mode Shapes

Table 7. Tibia and Tibia with Fixation Modes

Modes	**Tibia**		**Tibia with external fixation**	
Mode 1	359 Hz	405 Hz	377 Hz	460 Hz
Mode 2	824 Hz	Ø	844 Hz	Ø
Mode 3	1595 Hz	Ø	1530 Hz	Ø

5. Discussions and conclusions

The characterization on the tibia permit to highlight the advantages and disadvantages of all the input and output devices used. In particular, it highlighted that for this type of investigation, where a complex structure in geometry and materials has to be investigated, the best input/output measurement set-up is given by the shaker, due to the repeatability and the greater energy able to offer in the input signal compared to the micro-hammer and the accelerometer, since it turned out to be a more robust device, for the specific application, with respect to the Laser Doppler Vibrometer.
The results obtained exciting with the shaker and measuring with the accelerometer gave us the opportunity to validate also the goodness of our experimental test comparing it with the results of Mattei et al.[5]. In fact respect to them our results show a percentage decrease of about -2.18% for the first horizontal and of about -2.17% for the first vertical mode and a percentage increase of about 0.71% for the second horizontal one. These differences in percentage are related especially to the fact that respect to them our shaker was linked to the tibia through a stinger while they manually press the shaker on the specimen and because we excite on the lateral condyle while they excite at medial malleolus. The presence of the fixation has been identified instead to act very concretely on the vibrating modes of the tibia, observable both at the level of FRF, EMA and FEA, blocks the second vertical mode of the tibia and shifts the resonant frequencies of the second and third horizontal mode downwards; the second of about 100 Hz while the third of about 300 Hz.

Having validated the 3D tibia model, with and without the external fixation, was extremely important because, in particular the tibia model with fixation, allowed us to simulate the fracture and see how the tibia behaved, before proceeding experimentally, and what we will expect from the transmissibility test through the external fixation screws.

References

[1] Folman et al 1993 *Arch Orthop Trauma Surg* **112** 193–197
[2] Lippmann R K 1932 *J. Bone Jt Surg* **14** 118–126
[3] McGaw N H 1942 *Arch. Surg* **45** 195–205
[4] Seikiguchi T and Hirayama T 1979 *Acta orthopedics* **50** 391–398
[5] Lorenza Mattei et al 2017 *Annals of Biomedical Engineering* **45** 1111–1121
[6] Ong et al W H 2016 *Structural Control and Health Monitoring*
[7] Lms modal analysis module URL `https://www.plm.automation.siemens.com/it/products/lms/testing/test-lab/structures/modal-analysis.shtml`
[8] Konica minolta range 7 URL `https://www.konicaminolta.eu`
[9] Comsol multiphysics URL `https://www.comsol.it`
[10] Logan D L 2011 *A FIRST COURSE IN THE FINITE ELEMENT METHOD* (Cengage Learning)
[11] Peter Avitabile *Experimental Modal Analysis, Sound and Vibration Magazine*

IOP Conf. Series: Journal of Physics: Conf. Series **1149** (2018) 012030 doi:10.1088/1742-6596/1149/1/012030

Absolute interferometric angle measurement

J Tkaczyk, G Budzyn and J Rzepka

Faculty of Electronics, Wroclaw University of Technology,
Janiszewskiego 7, 50-370 Wroclaw
Tel. (4871) 3202309, Fax. (4871) 3202334

jakub.tkaczyk@pwr.wroc.pl

Abstract. A new method of absolute angle measurement based on analysis of interference fringe patterns is proposed. This approach is a bridge between interferometry, offering high accuracy, and other methods that are using position sensors or amplitude effects, which can give information about absolute value of angle. Utilisation of quadrature signals makes it possible to achieve better stability, with comparison to some other methods, thanks to reduced influence of thermal expandability. Usage of special signal processing algorithms facilitates wide measurement range and high resolution. Simulations and experiments show the possibility of achieving the range of 1 deg with arc-sec accuracy. Using commercially available optical and electronic components makes it possible to create lightweight, battery operated device that could be applied in CNC machine alignment process.

1. Introduction

In many scientific and industrial applications, there is need for a method of assessing the angular position with respect to the laser beam. The most commonly used techniques are based on laser beam position measurement or total internal reflection effect. Both are burdened with some limitations that reduce the attainable accuracy. Devices that are using position sensitive detectors, calculate the value of angle according to the lateral displacement of the beam on a known distance [1,2]. It is possible to achieve very good resolution with this method, but the accuracy is limited due to linearity of PSD and influence of thermal expandability [3,4]. It is especially evident, when a long reference arm is used. Total internal reflection does not react on lateral movement and allows for a smaller detector circuit. It can also be a lot simpler, as it requires only two photodiodes. Unfortunately, this effect is highly nonlinear and can be affected by difference in sensitivity of used sensors. Those factors effectively limit the accuracy to about 2% [5-8].

In this paper we present a novel interferometric method, that is based on the analysis of fringe pattern. This technique features lower sensitivity on standard sources of error and allows for higher range/accuracy ratio.

2. Measurement method

Two plane waves traveling at some angle to each other will interfere and generate a fringe pattern. Described situation is presented in the figure 1. From the analysis of wave functions it can be concluded that for a monochromatic waves of the same wavelength and polarization, created intensity pattern will have a sinusoidal form with a period that can be described with equation (1)[9].

IOP Conf. Series: Journal of Physics: Conf. Series **1149** (2018) 012030 doi:10.1088/1742-6596/1149/1/012030

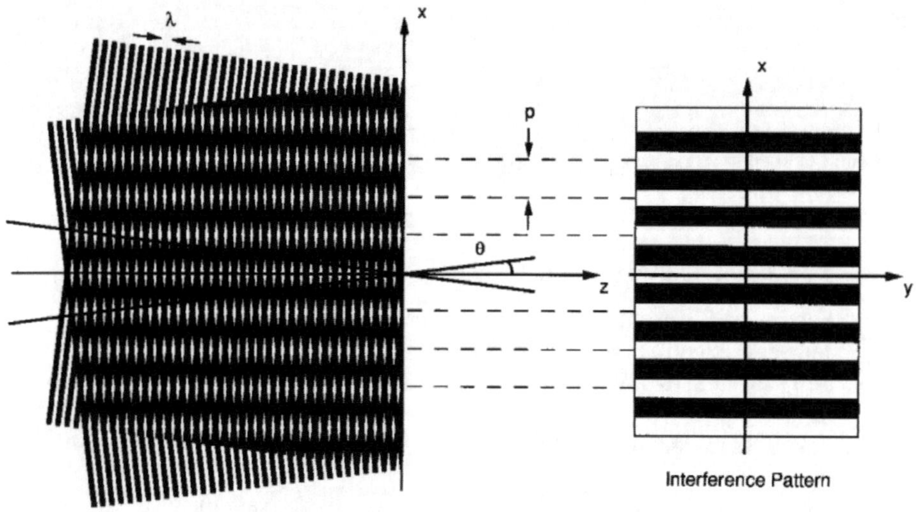

Figure 1. Interference of two plane waves and generated fringe pattern [9].

$$p = \frac{\lambda}{2\sin\theta} \quad (1)$$

We propose to perform the analysis of interference pattern in order to find the distance between fringes and calculate the angle θ. In a real situation it is necessary to use a laser beam instead of the planar wave. This implies some physical limitations to the beam shape and the wavefront flatness, that have to be considered. Those aspects are included in our further investigation.

In order to generate the fringe pattern, it is necessary to use special optical setup (two beam interferometer) that will split incoming laser beam in two and that will travel at some angle connected with angle of incidence. Some setups capable of this are showed in the figure 2. We decided to work with the setup composed of a polarization beam splitter PBS, a mirror prism and a retroreflector, as all of those components are easily available.

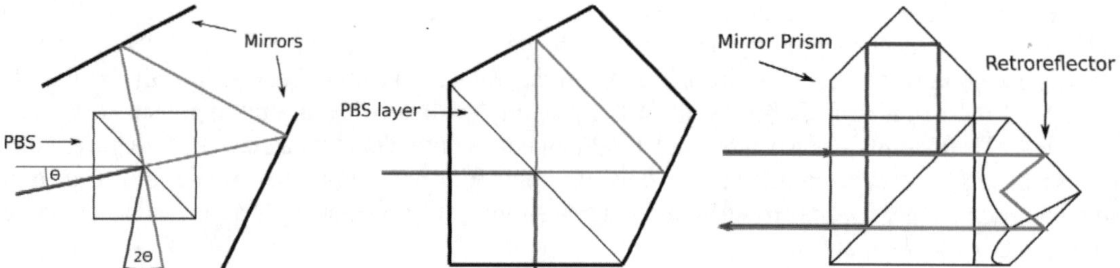

Figure 2. Optical setups for fringe pattern generation based on PBS and right-angle etalon (left), pentaprism (centre) and PBS, mirror prism and retroreflector (right).

At the output of presented optical setups, there are two beams with orthogonal polarizations. The quarter waveplate should be placed in the beam paths to change the linear polarization into the circular one. In order to generate and detect the fringe pattern there is needed a special detector presented in the figure 3.

IOP Conf. Series: Journal of Physics: Conf. Series **1149** (2018) 012030 doi:10.1088/1742-6596/1149/1/012030

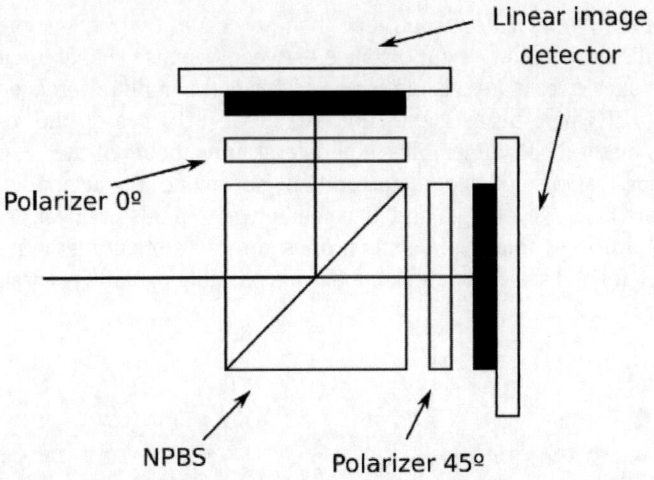

Figure 3. Schematic of the fringe pattern detector unit.

Proposed method requires usage of two linear image detectors with polarizers set so that two generated fringe patters will be shifted in phase by 90 degree, creating sinusoidal and cosinusoidal signals. Detected images may look similar to the one presented in the figure 4. It can be seen that image is composed of beam shapes and interference pattern. The only relevant information is the distance between fringes. Having two sets of such data, with identical beam shapes and orthogonal fringe patterns, it is possible to apply signal processing algorithms, in order to extract only the information about the interference. One of the possible algorithms for this purpose is Principal Component Analysis (PCA). Its usage will also decrease the influence of any beam distortions that can be seen in real systems, as they will appear identical in both images and therefore efficiently separated from signal of interest [10,11].

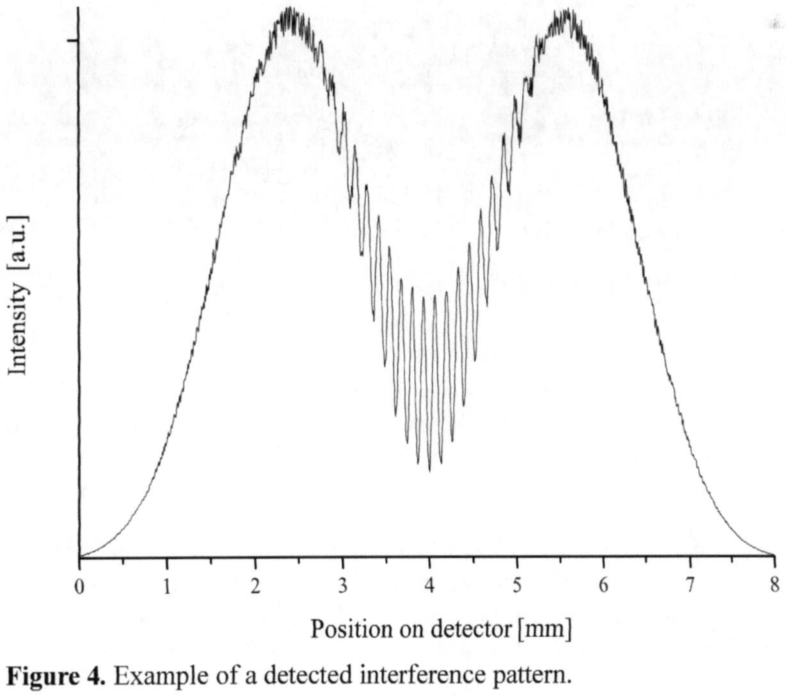

Figure 4. Example of a detected interference pattern.

Extracted image of the fringe pattern should be then processed by frequency estimation algorithms that effectively gives the information about distance between fringes. This approach utilizes the whole fringe pattern and averages over many visible fringes. This is the main difference in comparison to the method presented in [12] where beam shape can have great influence on the results. Presented data processing makes this method not vulnerable to the lateral movements of the beam or the sensor. Also the change in amplitude response of photodetectors will not impact the accuracy. Base values used in calculations of angle are laser wavelength and distance between pixels in linear detector.

We have created a prototype that implements proposed measurement method. It is presented in the figure 5 with a calliper for the scale. Small size makes it possible to build a wireless device suited for the industrial use.

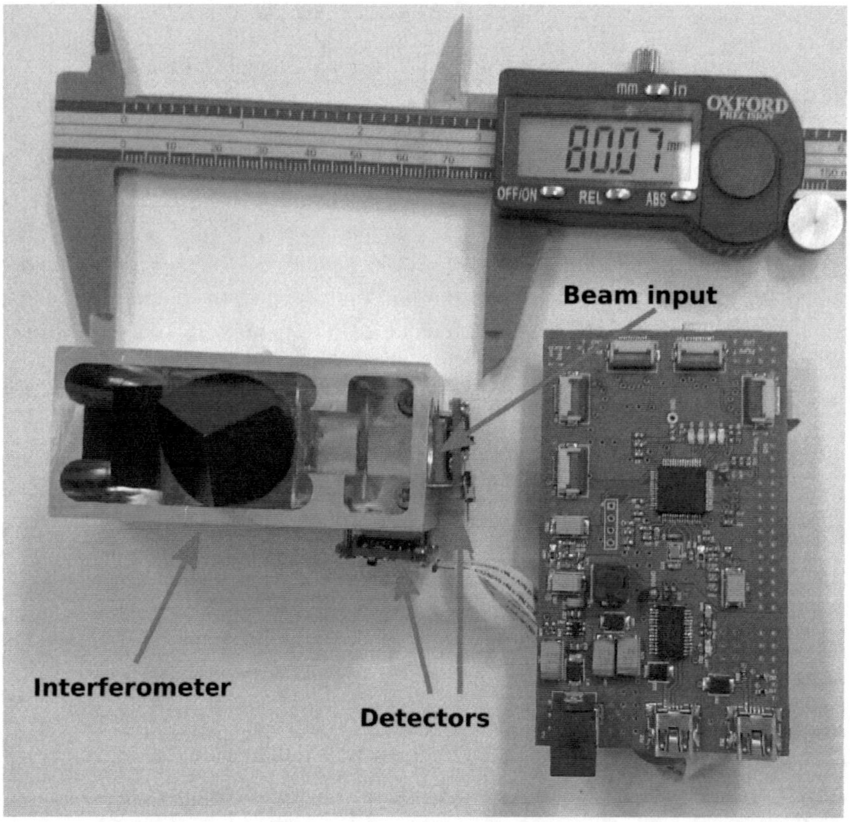

Figure 5. Created prototype of a measurement device.

3. Simulation results

We have performed many simulations to select the best data processing algorithms and asses the attainable accuracy of the presented method. Aspects of signal noise, beam shape and lateral beam shift were taken into consideration. We tried to create a model closest to the reality, therefore signal resolution, pixel count and pitch were set according to the specification of a linear photodetector that was chosen for a prototype.

For a comparison, there were selected algorithms of frequency estimation based on IFFT, SinFit and ESPRIT. Signal noise level was varied from 0 to 10% and the beam shifts of few millimeters were simulated. In the figure 6 there are presented results for a realistic situation with 5% additive noise and the beam shift.

IOP Conf. Series: Journal of Physics: Conf. Series **1149** (2018) 012030 doi:10.1088/1742-6596/1149/1/012030

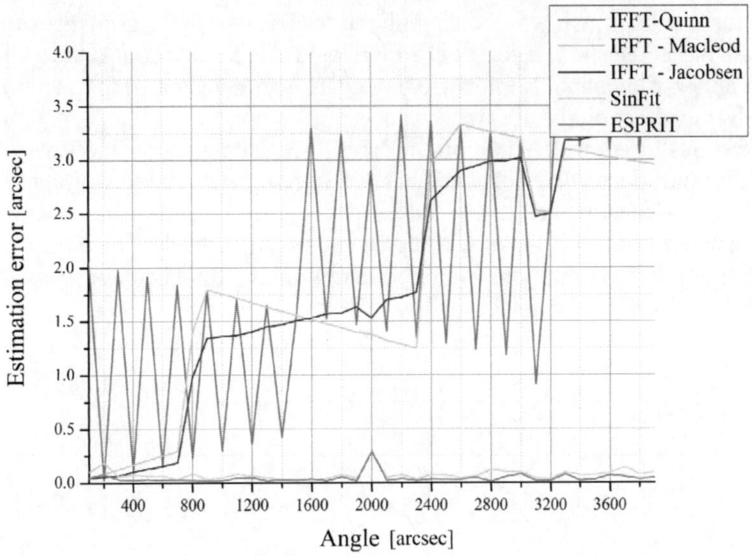

Figure 6. Angle estimation error with different algorithms in realistic simulation conditions (5% additive noise, lateral beam shift).

Algorithm ESPRIT proves to work best in this situation and was chosen for implementation on a prototype platform. Estimation error was lower than 0.5 arcsec in the measurement range of 1°. With increase of angle above the presented values, frequency in the fringe pattern is getting bigger and soon the spatial aliasing will occur.

4. Potential sources of error

Proposed method is based on interferometric effect and requires using linear detector for a reception of fringe pattern. This implies some possible sources of errors like wavelength instability, influence of thermal effects on detector, precision of mechanical mounting and wavefront flatness.

Figure 7 presents the error caused by change of laser wavelength. It can be seen that 1 nm change will introduce more than 5 arcsec of error. Therefore, this aspect have to be considered when semiconductor lasers are used but with stabilised HeNe lasers, like the one used in our tests, can be completely omitted.

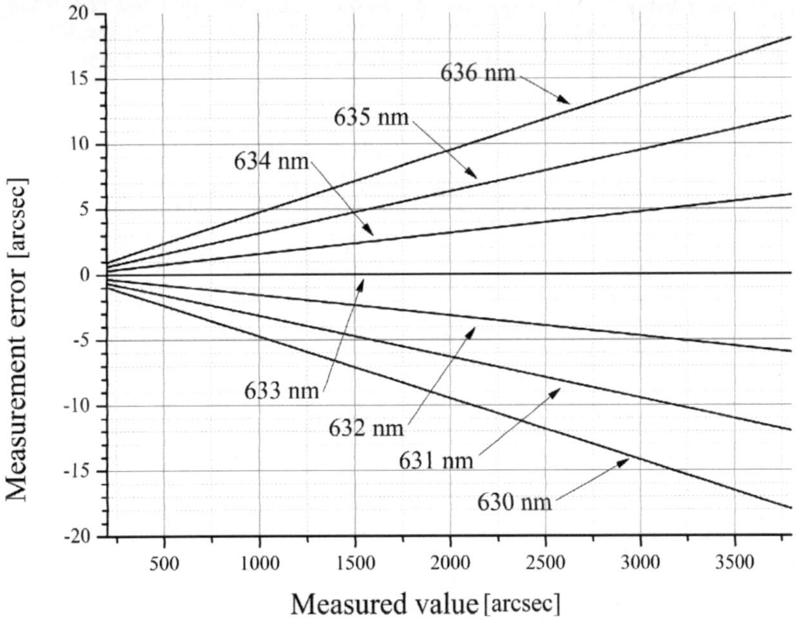

Figure 7. Angle estimation error with change of laser wavelength.

IOP Conf. Series: Journal of Physics: Conf. Series **1149** (2018) 012030 doi:10.1088/1742-6596/1149/1/012030

Thermal effects can influence the results in two ways, changing the response of detector pixels and the distance between them. As all the pixels in a detector are fabricated on one substrate, they all should be affected in the same manner. This should not cause any change in detected fringe pattern and therefore will not have any adverse effect on the accuracy.

Pixel pitch is one of the constants used in calculations of angle. Its change due to thermal expandability of a detector itself will introduce some scale mismatch and error. However its influence was found to be <0.01arcsec on every Celsius degree.

Effective pixel pitch can be changed also with tilt of a detector caused by non-ideal mechanical mountings. This is showed in the figure 8. This problem can be solved with one time calibration as it will not change in time.

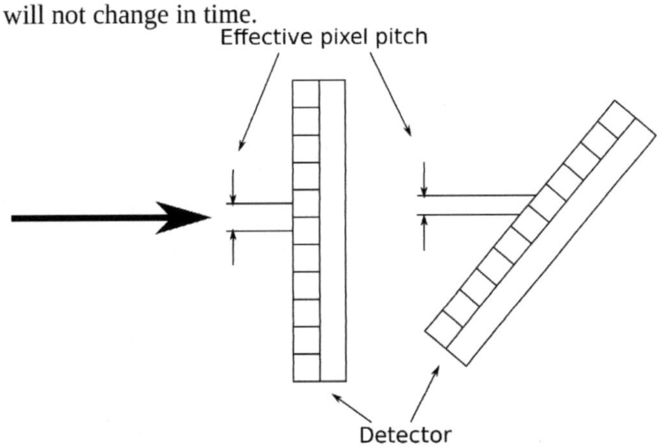

Figure 8. Change in effective pixel pitch with a detector tilt.

Equations used for calculation of the angle are derived from the interference of two plane waves. If the waves do not have planar wavefront, created fringe pattern will be disrupted. It is hard to estimate its influence but we assumed that devices based on the proposed method should be built with commercially available optical components which have surface flatness $\lambda/10$. Aside from optical components, wavefront distortions may come from a laser beam itself. Due to physical limitations of gaussian beam, wavefront have always some finite radius of curvature ROC. This is showed in the figure 9. To limit the error caused by wavefront flatness, laser beam have to have ROC high enough to introduce smaller phase disturbance than optical components. Therefore, we concluded that laser beam should have width greater than 5mm.

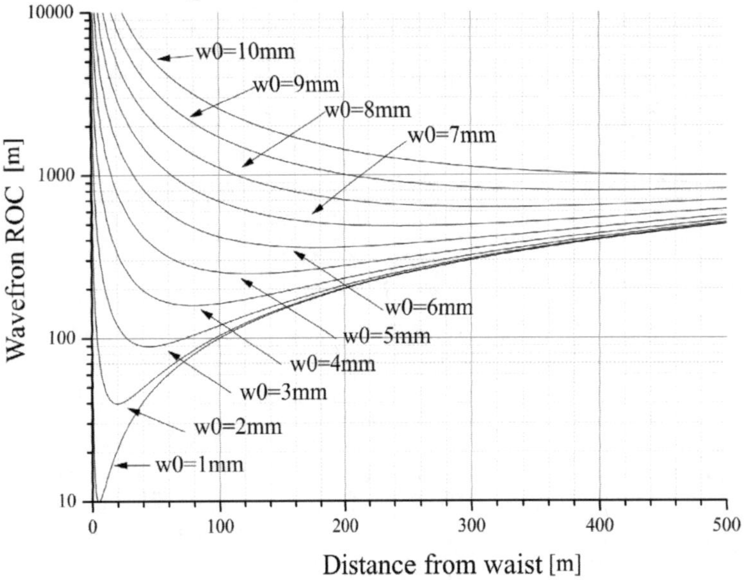

Figure 9. Wavefront ROC with different beam widths

IOP Conf. Series: Journal of Physics: Conf. Series **1149** (2018) 012030 doi:10.1088/1742-6596/1149/1/012030

5. Conclusions

In the paper we propose a new method of interferometric angle measurement. It has some significant advantages over other, well known techniques. It is less vulnerable to thermal effects and can possibly offer very good resolution and accuracy. This method can be implemented in a miniature measurement device that could be used in the out of laboratory environment. Greatest errors may be caused by a laser wavelength instability and wavefront flatness in the beam. Those two problems can be solved with a stabilised laser and a high quality optical components that are widely used in other interferometric devices.

References

[1] Fitzsimons E D et al. 2013 *Precision absolute positional measurement of laser beams* Applied Optics
[2] Cary A 2012 *Application Note: Measuring Laser Position & Pointing Stability* Ophir Photonics
[3] Ivan I A, Areleanu M, Laurent G J, Tan N and Clevy C 2012 *The metrology and applications of PSD (position sensitive detector) sensors for microrobotics* IEEE
[4] Makynen A, Ruotsalainen T and Kostamovaara J 1997 *High accuracy CMOS position-sensitive photodetector (PSD)* Electronics Letters
[5] Wang S F et al 2013 I*mproved Small-Angle Sensor Based Total-Internal Reflection and Surface Plasmon Resonance in Heterodyne Interferometry* Sensors and Materials Vol. 25
[6] Chiu M H and Su D C 1997 *Angle measurement using total-internal reflection heterodyne interferometry* Society of Photo-Optical Instrumentation Engineers
[7] Zhang W Y, Zhang J and Wu L Y 2006 *Small-Angle Measurement of Laser Beam Steering Based on Total Internal-Reflection Effect* Journal of Physics: Conference Series 48
[8] Huang P and Kiyono S 1993 *Method and apparatus for angle measurement based on the internal reflection effect* US Patent 5,220,397
[9] Bass M 2010 *Handbook of optics* Optical Society of America
[10] Kluwer M 2003 *A Practical Approach to Microarray Data Analysis: Singular value decomposition and principal component analysis* LANL LA-UR-02-4001
[11] Abdi H and Williams L J 2010 *Principal component analysis* John Wiley & Sons
[12] Dobosz M and Iwasinska-Kowalska O 2014 *Interference method for ultra-precision measurementand compensation of laser beam angular deflection* Applied Optics

IOP Conf. Series: Journal of Physics: Conf. Series **1149** (2018) 012031 doi:10.1088/1742-6596/1149/1/012031

Improvement of intermodal frequency stability of two-mode HeNe laser

G Budzyn[1], J Tkaczyk[1] and J Rzepka[1]

[1] Faculty of Electronics, Wroclaw University of Science and Technology, Janiszewskiego 7, 50-370 Wrocław

E-mail: grzegorz.budzyn@pwr.edu.pl

Abstract. A method of improving intermodal frequency stability of two-mode helium-neon laser has been developed. The approach is based on using in the stabilization circuit a signal processor for real-time analysis of data from an LF amplitude and an HF frequency detector. We propose to use the signal from the LF detector in the initial phase of the stabilization process and the signal from the HF detector afterwards. Experiments proved that with the suggested design it is possible to obtain frequency stability of the intermodal frequency better than 100 Hz or 2*10-10 for 1000s averaging. In the paper we also show the influence the intermodal frequency stability on the absolute stability of the laser frequency.

1. Introduction

Despite constant advances in semiconductor, fiber and solid state laser technologies, the gaseous helium-neon lasers are still the preferable choice in the industrial metrology. Currently the majority of commercially available high precision measurement systems are based on laser sources of this kind. Their narrow line width, excellent beam quality, good natural frequency and amplitude stability and relatively low cost make them perfect in all demanding metrology applications. Main applications of HeNe laser like shift or vibration measurements require good stability of the absolute frequency [1-8]. In other applications like optical vortices or Fizeau interferometers either amplitude or combined amplitude/frequency stability is required [9-11]. There have been reported He-Ne lasers with stabilized intermodal beat frequency [12-13] for use for example in laser distance meters [14]. The stability of the intermodal beat frequency was reported worse than 5 kHz. In [15] we presented the laser with intermodal frequency stability better than 2kHz. In this paper we show a method of construction of the two-mode He-Ne laser with the intermodal beat frequency stability value better than 100 Hz. Using light sources with such high modulation frequency stability allow omitting the influence of the source on the overall measurement accuracy when using such lasers as the source in a precision laser range finders.

2. Stabilization of the two-mode HeNe laser

The stability of a resonant cavity is determined by the change in its optical length. This change is due essentially to three causes:

1) thermal drift,
2) elastic vibrations of the structure,
3) fluctuations in the refractive index of the medium.

The cavity resonance frequency (for TEM_{00} modes) is given by

$$\nu_c = q\frac{c}{2nL} \quad (1)$$

where c is the speed of light, n is the refractive index along L, q is the number of half wavelengths in L and L is the distance between the resonator mirrors.

For the internal mirror laser the largest influence on the overall frequency stability of the laser has the thermal drift of the resonator. Far less important, but not negligible, is the influence of the other two effects. In [5] we showed the comparison of the influence of those factors on the frequency stability of two-mode HeNe laser.

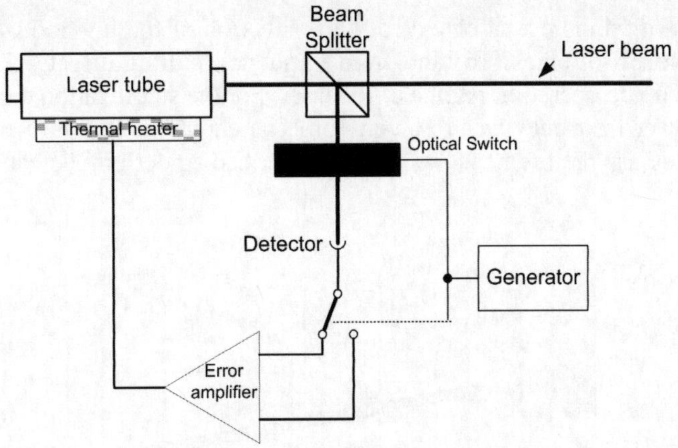

Figure 1. Block diagram of a two-mode HeNe laser stabilized with the modes' balance method

In the case of the most commonly used 1mW tubes, the usual method of obtaining the laser stability is keeping the balance between the power of two output modes of the laser [5, 7] – figure 1. In this method the modes intensity ratio is taken as the error signal for the stabilization loop. The desired situation (with error signal equal to zero) is when mode intensities become equal.

3. Experimental setup

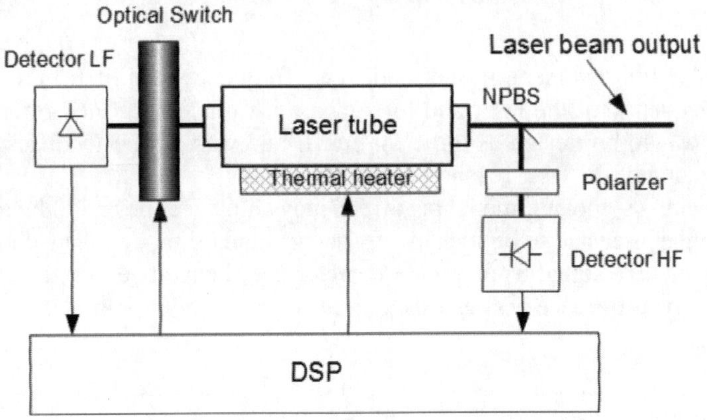

Figure 2. Schematic diagram of the experimental setup. DSP, digital signal processor; NPBS, non-polarizing beam splitter. For clarity electronic blocks interfacing DSP with the optoelectronic elements of the laser tube are omitted.

A schematic diagram of the experimental setup is shown in the figure 2. The operation of the internal-mirror 1mW He-Ne laser (Lasos model LGR3655S) was controlled by a Digital Signal Processor DSP. Around the tube a thin thermal heater was wrapped. The parasite beam of the tube was used for the mode-intensity stabilization of the laser [15].

A part of the main output beam, split by the Non Polarizing Beam Splitter NPBS was directed to the High-Frequency Detector. The Detector HF was used for measuring the intermodal frequency of the laser, while the Detector LF was used for monitoring the modes' intensity. Information from both detectors were gathered and analyzed by the DSP. Depending on the measured values the power of the thermal heater was altered.

During the laser warmup there was observed the fluctuation of the intermodal frequency resulting from intra-cavity phase anisotropies [16] and from frequency pulling effect [17,18]. Discontinuities occurred during single mode operation regime of the laser. For the stabilization either the mode spacing with higher or with lower frequency was to be chosen. The choice had no influence on the following analysis. The only necessary requirement was that the laser had smooth mode change (i.e. no so called "mode flip").

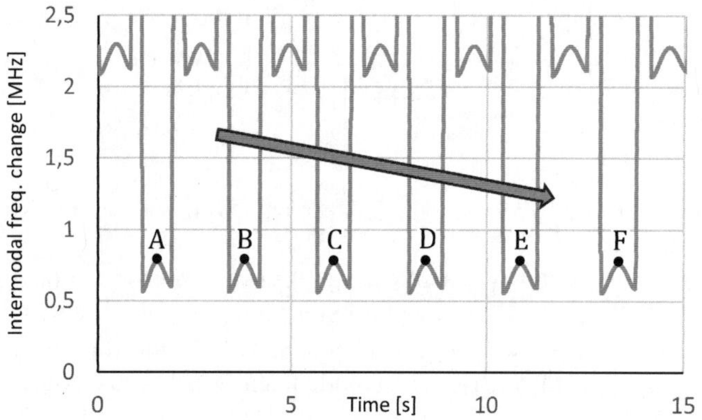

Figure 3. Intermodal frequency change of the laser during heating up of the laser. The frequency measured in points A to F is decreasing.

The idea of further improvement of intermodal beat frequency stability of the laser was to measure directly the beat frequency of the laser and to compare it to the predefined reference frequency. The comparison result was to be treated as the error signal and was, in the feedback loop, to be used for altering the temperature of the laser resonator.

As it was shown in [18] the intermodal beat frequency value declines with the resonator temperature (Fig. 3). Therefore the laser had to be locked first using standard mode balance method and only after the laser became stable the signal from the HF detector could have been used. The control loop of the laser was driven by the difference between the current value of the intermodal frequency and the value set as the reference.

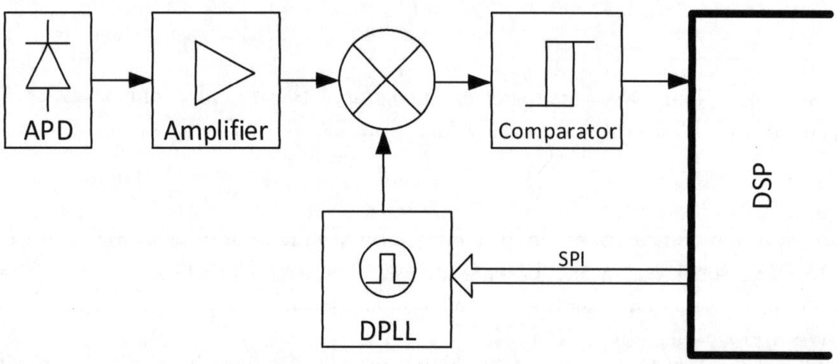

Figure 4. The functional diagram of the circuitry used for beat frequency measurement

IOP Conf. Series: Journal of Physics: Conf. Series **1149** (2018) 012031 doi:10.1088/1742-6596/1149/1/012031

In the figure 4 there is shown the functional diagram of the circuitry used for beat frequency measurement. In order to achieve higher accuracy and better measurement resolution the signal detected in the Detector HF was downconverted in the Mixer with the LO signal generated in the digital PLL. The output frequency of the DPLL was controlled by the DSP processor through a simple Serial Peripheral Interface SPI. Because of the high frequency of the detected signal and the required measurement precision special precautions were made to deliver to the mixing circuit high quality, low jitter clock from the PLL circuitry.

4. Measurements results of laser stability

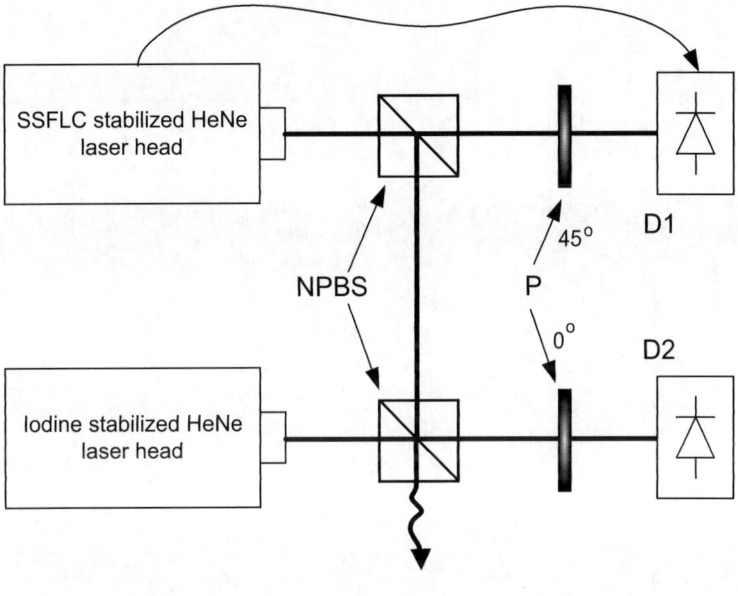

Figure 5. Measurement setup for estimation of SSFLC laser beam frequency stability and intermodal beat frequency stability. NPBS – Nonpolarizing Beam Splitter, P – Polarizer, D1, D2 – Detectors

The laser stability measurements were conducted in the circuit shown in the Figure 5. Simultaneous measurement of the intermodal and the absolute frequency of the laser were performed. The intermodal frequency was read from the detector D1 and the absolute frequency of the vertically polarized mode from the detector D2. The detectors were integrated with counters (Lasertex, model AVM-02) clocked from the same clock source as the DSP processor - the SSFLC stabilized laser and the D1 detector were clocked from the same 10 MHz clock. With this configuration of the clock path the stability of the clock source is irrelevant.

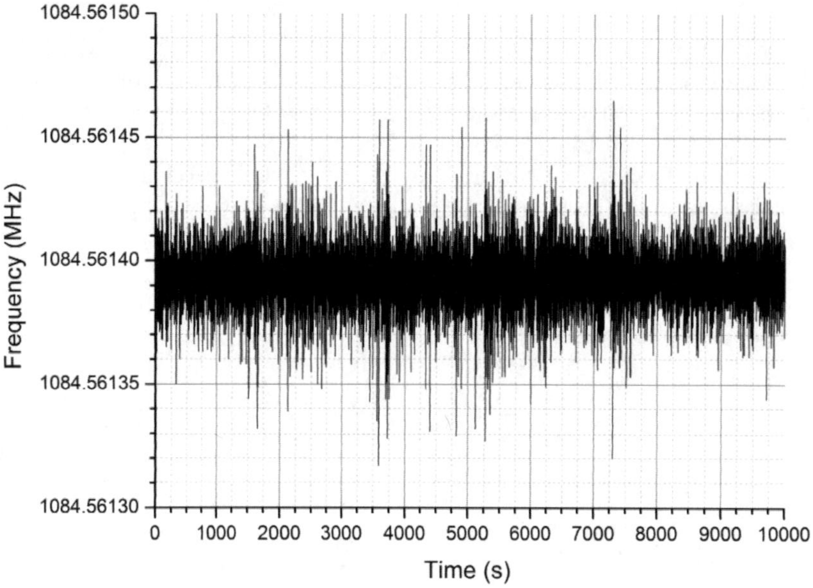

Figure 6. Intermodal frequency of the constructed laser.

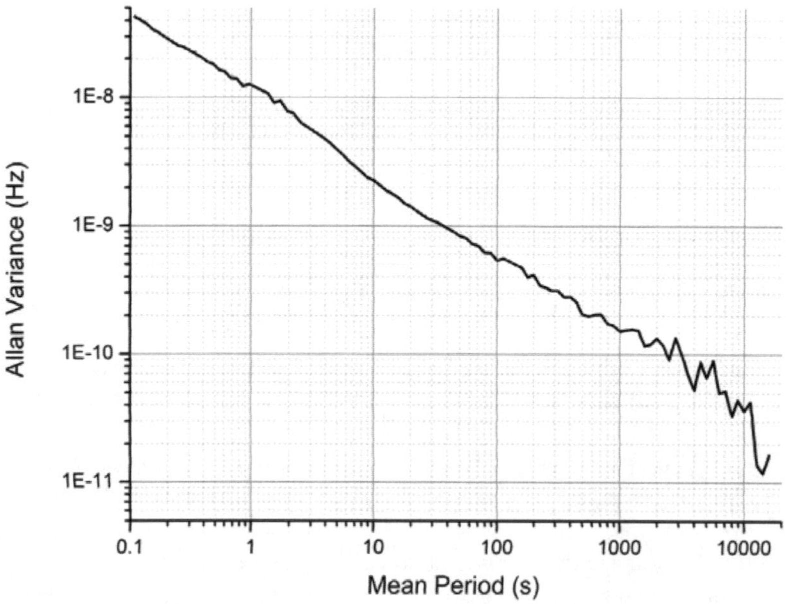

Figure 7. Allan Variance plot of the intermodal frequency of the constructed laser.

The frequency stability of the experimental laser is shown in the Figure 6 and as the Allan Variance plot in the Figure 7. The measurements were conducted for 48 hours (in the Figure 6 only first 10000s are shown). The intermodal frequency was very stable with small exception of the 0.2-2 seconds averaging range. The small "hump" in the variance plot at these averaging times was caused by either small backreflections within the testing circuit or imperfections of the laser loop control.

Registering signals from both photodetectors we compared the stability of the intermodal and absolute frequency. In the Figure 8 there is shown Allan Variance plot of the absolute frequency of the laser stabilized with "classic" and with the intermodal method. As it can be seen, the stabilization on constant value of the intermodal frequency causes worse stability of the absolute frequency. This behavior was expected as it was described in [18-20].

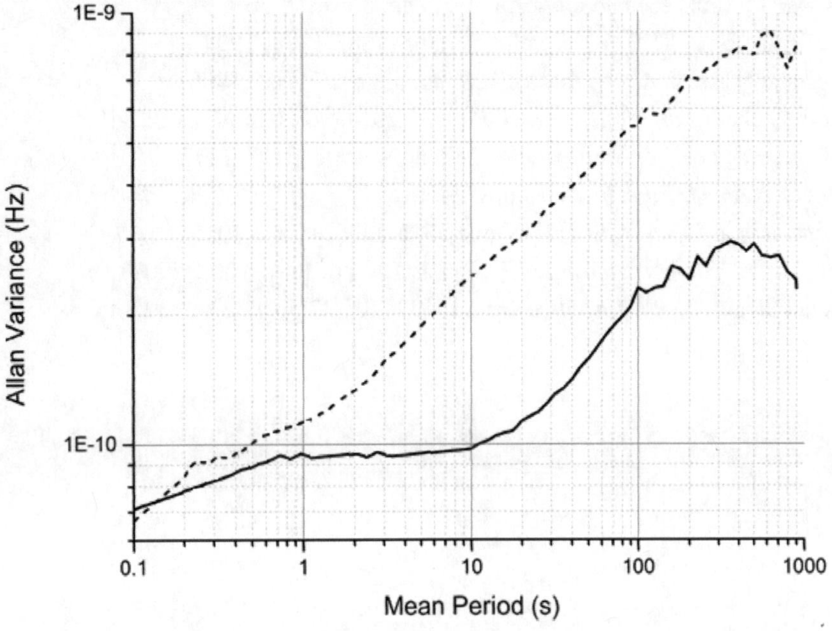

Figure 8. Allan Variance plot of the absolute frequency of the laser stabilized with signal from LF detector (solid line) and with the signal from HF detector (dashed line).

5. Conclusions

In the paper we reported a new method of intermodal frequency stabilization in two-mode HeNe lasers. The method allows obtaining excellent stability of the mode beat frequency. Laser constructed according to the proposed method can be used in the applications like laser rangefinder, where the precision of the light source is important. Required for the proposed laser, the high-frequency electronics is currently cheap and easily obtainable. The method is not optimal if the absolute frequency stability is important.

References

[1] Wallard A, *Journal of Physics E: Scientific Instruments*, vol. 6, no. 9. p. 793-807, 1973.

[2] Ciddor P and Duffy R, *Journal of Physics E: Scientific Instruments*, vol. 16, no. 12, p. 1223-1227, 1983.

[3] Niebauer T et al, *Appl. Opt.,* 27, 1285-1289, 1988

[4] Baer T et al, *Appl. Opt*., 19, no. 18, 3173-3177, 1980

[5] Budzyn G., "*Optokomutacyjna stabilizacja lasera dwumodowego*", PhD Thesis, Wrocław University of Technology, Wrocław, 2006

[6] Diao X, et all, *Applied Optics*, vol. 52, no. 3, pp 456-460, 2013

[7] Wysocki G and Abramski K.M., *Optica Applicata*, vol. XXXII, no. 1-2, pp. 207-213, 2002
[8] Sasagawa G and Zumberge M, *Appl. Opt.,* 28, no. 5, 824-825, 1989
[9] Budzyn G et al , *Appl. Opt*., 45, no. 17, pp. 3982-3984
[10] Fraczek E and Budzyn G*., Optica Applicata*. 2009, vol. 39, no 1, pp. 91-99
[11] Rowley W *Meas. Sci. Tech*., 1, no. 4, 348-351, 1990
[12] Yokoyama S et al, *Applied Optics*, vol. 33, no. 3, p. 358-363, 1994.
[13] Araki Tet al *Rev. of Sci. Instr*., vol. 65(6), pp. 1883-1888, 1994
[14] Seta K and O'Ishi T *Appl. Opt*., vol. 29, no. 3, pp. 354-359,1990
[15] Budzyn G et al *Photonics Letters of Poland*, 2014, vol. 6, no 1, pp. 17-19.
[16] Liu W et al *Optics Communications*, 282, no.8, 1602-1606, 2009
[17] Lindberg A *Am. J. Phy*., 67 (4), April 1994
[18] Budzyń, G et al *Laser Physics*, Volume 25, Issue 6, 1 June 2015
[19] Wang. L et al *Chinese Journal of Lasers 34 (9),* pp. 1198-1202, 2000
[20] Zemanek, P and Lazar, J *Applied Optics*. 1994,33, 27,pp. 6333-6339

IOP Conf. Series: Journal of Physics: Conf. Series **1149** (2018) 012032 doi:10.1088/1742-6596/1149/1/012032

Census Transform Based Optical Flow for Motion Detection during Different Sinusoidal Brightness Variations

G.Allevi [a], L.Casacanditella [b], L. Capponi [a], R. Marsili [b], G. Rossi [b]

[a] CISAS "G. Colombo", Università degli Studi di Padova - Via Venezia 15, 35131 – Padova, Italy
[b] Università degli Studi di Perugia - Via G. Duranti 1/A, 06125 - Perugia, Italy

gloria.allevi@phd.unipd.it

Abstract. This work is a first approach of the implementation of a specific Optical Flow algorithm for motion detection in cases where the brightness variation is represented by a sine wave, whose characteristics vary across the different image sectors. The final goal would be the implementation of such an algorithm in thermal films recording a component undergoing a sinusoidal load. Mapping the motion field all over the thermal video time history, and deriving it in order to obtain a strain map, would enable both the simultaneous measurement of stress (by performing Thermoelastic Stress Analysis) and strain by a unique video, and the possibility of stress calibration, thus linking the digital levels in output to real stress values. In this scenario, the authors present the implementation of Census Transform based Optical Flow on a simulated video, where the brightness variation is modeled like the superimposition of sine waves equal for all pixels and different Gaussian spatial distributions frame after frame. The latter is used for creating different sine patterns for each image sector.

1. Introduction

During the last decades, Non Destructive Testing procedures are gaining a significant role both in industrial and in research fields. In this scenario, the possibility of measuring stress concentrations during a dynamical mechanical test is very useful since it allows to highlight unpredictable stress patterns in a specific structure due, for example, to particular manufacturing processes such as 3D printing [1]. Such kind of patterns can be evaluated through Thermoelastic Stress Analysis (TSA) [2],[3],[4],[5],[6] which gives as output a thermal film where the temperature variation corresponding to the load frequency is directly linked to the stress variation during the test. Obviously, a calibration procedure aiming at finding the calibration factor allowing to convert temperature [°C] in stress [Pa] is required. Usually this procedure is performed through a strain gauge rosette, by measuring the strain distribution that will be put into relation with the stress thanks to the material constitutive equations [5],[6]. Unfortunately, this calibration methodology is invasive, thus not feasible for some applications. Another solution could be the application of Digital Image Correlation [4],[7] on thermal images by using a specific emissivity pattern as speckle. Again, this technique requires a sort of surface treatment to generate the mentioned speckle, and this is not always applicable. Therefore, in this work the authors perform a first approach that in the future will enable the evaluation of strain maps from thermal films, thus providing both a reference for calibration and the additional information about strain useful for mechanical considerations. In this work a specific Optical Flow algorithm has been chosen for reaching the fixed goal. However, most of the existing Optical Flow codes exploit gradient based methods, i.e. they assume that each point of a framed object conserves its brightness during the video sequence. On the contrary, in our case the intensity of each pixel varies both for the thermoelastic effect itself (a structure compression generates a temperature increasing revealed by the thermal camera, and vice-versa) and for the framed object is moving at the load frequency. Therefore the Horn and Schunk method [8] and Lucas-Kanade one [9] seem not to be suitable for the application. In literature some authors tried

IOP Conf. Series: Journal of Physics: Conf. Series **1149** (2018) 012032 doi:10.1088/1742-6596/1149/1/012032

to overcome the brightness variation problem, see for example [10], [11], but none of them fits well our needs, because they address to correct the typical illumination changes that can occur during a test conducted by optical systems. Therefore, they are not able to take into account the combined effect of displacement and temperature-stress variation. For these reasons, in this work it has been decided to exploit another approach for Optical Flow estimation: the Census Transform [12],[13], which in some way tolerates a certain amount of brightness variability given by the defined input parameters. In order to follow a step-by-step approach, in this work we considered a simulated video sequence where we try to reproduce conditions similar to those involved in a Thermoelastic test, where the intensity of each pixel follows a sine wave (without considering background noise) whose amplitude is different from a component region to another due to the different stress distribution across the component. Hence, we generated a video sequence by taking a thermal frame and applying both a known horizontal displacement, and a sinusoidal brightness variation for each pixel, whose amplitude varies from an image region to another, according to a Gaussian distribution, whose standard deviation changes according to a sine law in turn. At this point, the Census Transform code, which will be described in the following section, has been applied, thus providing an estimation of the displacement time histories, reconstructed thanks to the method proposed in [13].

2. Materials and Methods

In the following the authors demonstrate the feasibility of the application of Census Transform to obtain a displacement field from a thermal video sequence. The algorithm, developed in Matlab environment, has been applied to a simulated motion video starting from a thermal image. In the first section the evaluated thermal video sequence will be described.

2.1 Simulated video sequence

The considered thermal image is reported in Figure 1.
A uniform horizontal displacement field has been applied to the thermal image in Figure 1. Its expression is reported below:

$$x(t) = A * \sin(2\pi f_0 t)$$

where A was set equal to 3, while the chosen frequency was 10 Hz. The reference displacement applied to the image is reported in Figure 2. At this point the authors attempted to produce a temporal and spatial brightness variation, since during a Thermoelastic test the stored radiation changes temporally for the Thermoelastic effect itself, and spatially due to the fact that the different regions of a component can be stressed differently to each other. For this purpose, it has been stated to generate a sine wave for the temporal brightness variation, superimposed to a Gaussian light distribution (as if the component would be more stressed at the centre), whose standard deviation varies sinusoidally in turn. Figure 3 shows an example of the light distribution at a certain instant of time. In the following section the algorithm applied to the thermal video will be described.

IOP Conf. Series: Journal of Physics: Conf. Series **1149** (2018) 012032 doi:10.1088/1742-6596/1149/1/012032

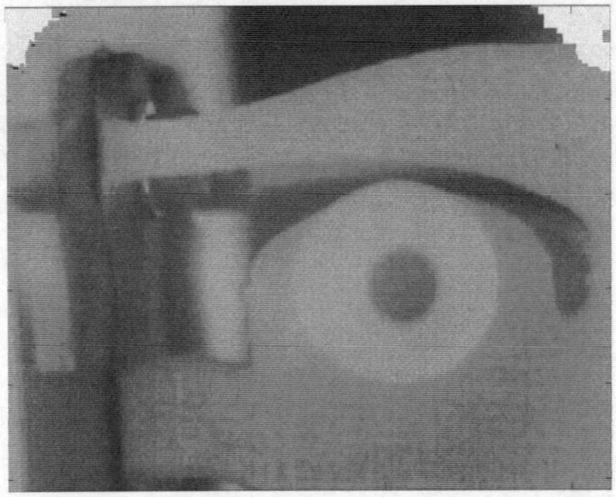

Figure 1. The considered thermal image, the video sequence has been obtained by applying a uniform displacement field to the image

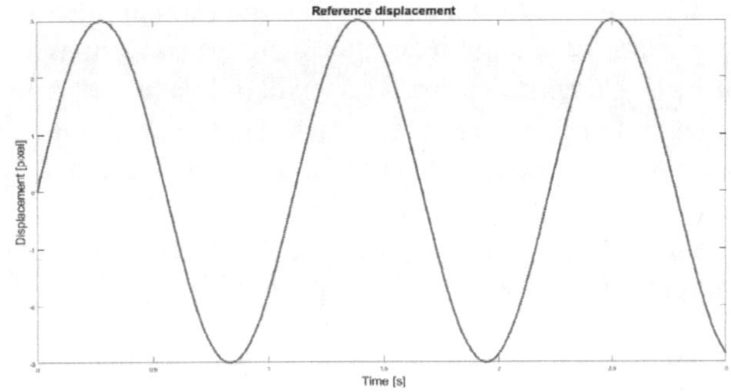

Figure 2. Sinusoidal displacement applied to the thermal image, in order to obtain the video sequence

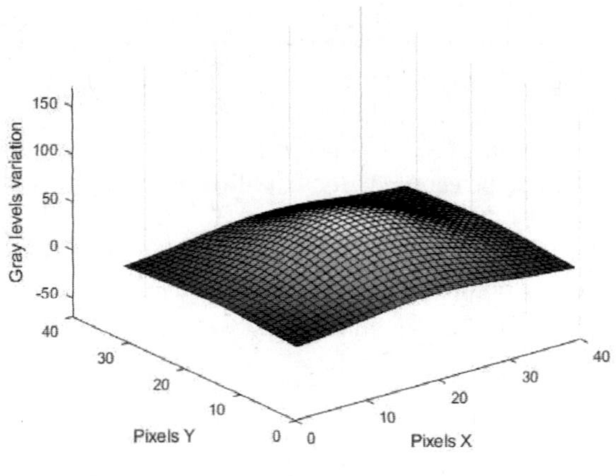

Figure 3. Variable Gaussian light distribution

2.2 The CENSUS Transform

The Census Transform is based on a non-linear transformation which associate to a pixel G a clockwise concatenation of strings representing the set of neighboring pixels whose intensity is less or more than that of G [13]. Defining a kernel that scans the overall image, the pixels surrounding the central one are numerically defined through the following digitalization:

$$\gamma(G,G') = \begin{cases} 0 & G - G' > \varphi \\ 1 & |G - G'| < \varphi \\ 2 & G' - G > \varphi \end{cases}$$

where φ is the threshold, its value depends on the format of the analysed video sequence (i.e. the number of bits of each pixel). This allows to define a signature vector for each pixel. A signature vector of length c represents n^c different patches, being n the dimension of the kernel used to construct the signature vector of each pixel. This algorithm can be used as a descriptor for image recognition or for the tracking of image features through the video sequence. In order to match the features between two consecutive images, further hypotheses are needed. The adopted algorithm is described in the flow reported in Figure 4.

- *Image filtering*: A low-pass filter with a 3x3 mean filter was applied to the video before applying the CENSUS algorithm, in order to reduce the thermal images noise.
- *Scanning of Image 1:* Computation for every pixel in Image 1 of its signature vector
- *Scanning of Image 2:* Computation for every pixel in Image 2 of its signature vector
- *Search for correspondence between Image 1 and Image 2:* search for pixels in Image 2 having the same signature vector as the pixels in Image 1.
- *Correspondence hypotheses*: each pair of pixels of two consecutive images with the same signature vector represents a correspondence hypothesis.

Therefore, in order to achieve biunivocal correspondences between the features in the two images and to correctly reconstruct the displacement of the detected features, more hypothesis are needed.

1. The first step is to filter out all the features which do not contribute to any meaningful correspondence, for example the uniform patches (which often correspond to the background) are removed.
2. All the pixels that have more than a fixed number of correspondence are not considered as significant for the displacement computation, and so they are removed. Hypotheses are only generated if there are few elements (Figure 5).
3. The third filter uses illumination and geometric constraints. In the described case study, an intensity change of 20% of the central pixel was allowed and the maximum displacement vector length was set to 10 pixels.

By means of the application of this three hypothesis, it has been possible to select the right correspondences (Figure 6) and to compute the associated displacement vectors for each pair of images in the video sequence.

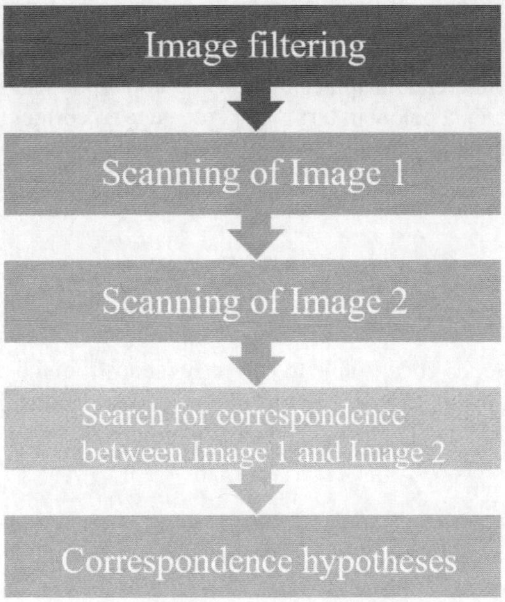

Figure 4. Flow chart of the developed code

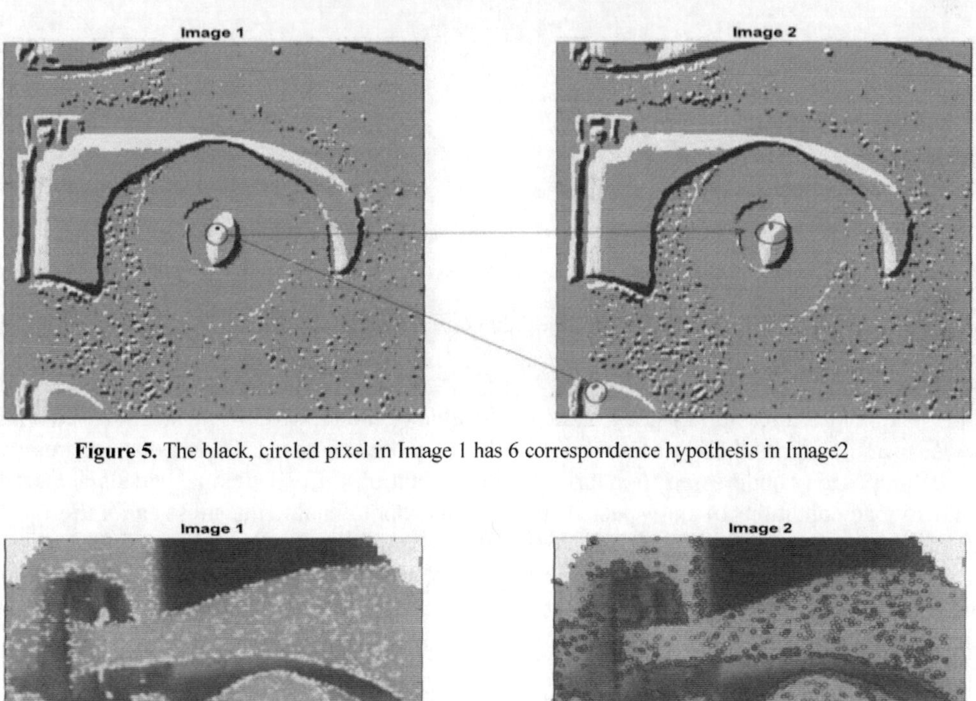

Figure 5. The black, circled pixel in Image 1 has 6 correspondence hypothesis in Image2

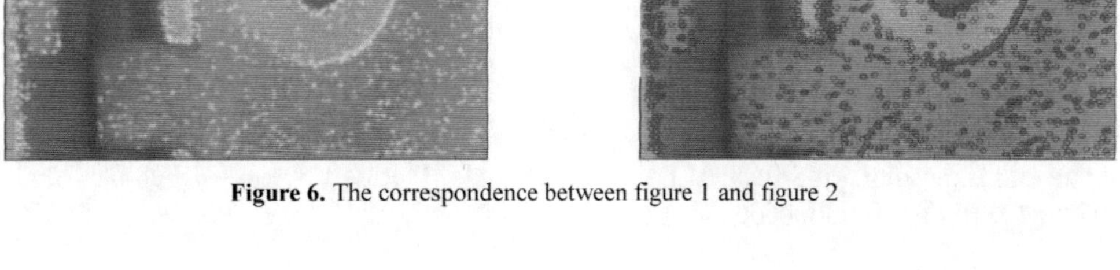

Figure 6. The correspondence between figure 1 and figure 2

IOP Conf. Series: Journal of Physics: Conf. Series **1149** (2018) 012032 doi:10.1088/1742-6596/1149/1/012032

3. Results

The displacement between each of the *n* features resulting from two consecutive images has been compared with the value of reference displacement at the correspondent instant of time. Results are reported in Figure 7 and described below in terms of percentage of correct matches on the total amount of detected features.

Max value: 99.27%
Mean value: 98.41 %
Min value: 96.74 %
Standard deviation: 0.52 %

As shown in Fig. 7, the developed code is able to correctly reconstruct the displacement time history and additionally, a sub pixel analysis could enable the very correct feature tracking.

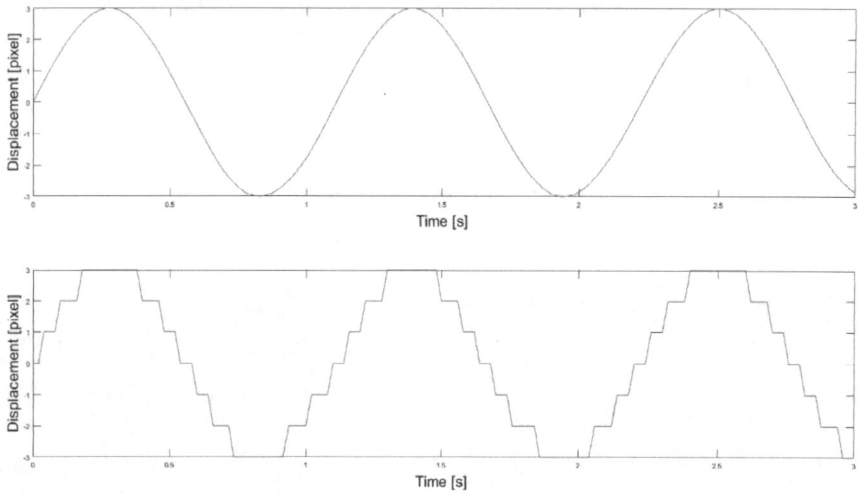

***Figure** 7. Comparison between the reference displacement and the time history of one of the detected features.*

4. Conclusions

The present work is a feasibility study of the application of the CENSUS Transform based optical flow for videos simulating ideal brightness variations that may occur during a Thermoelastic Stress Analysis. This first approach is addressed to a further implementation on real thermal videos acquired during dynamical load conditions of an inspected structure in order to enable the stress calibration without the strain gauge rosette use, and in a second step the simultaneous stress and strain-full field evaluation. The obtained results confirm the code robustness in terms of radiation changes, encouraging its implementation on a real application. As further developments, the subpixel analysis will be able to be implemented only after a robust computational optimization.

References

[1] G. Allevi, M. Cibeca, R. Fioretti, R. Marsili, R.Montanini, G. Rossi, *Qualification of additively manufactured aerospace brackets: A comparison between thermoelastic stress analysis and theoretical results*, Measurement, Volume 126, October 2018, Pages 252-258

[2] R. Marsili, M. Moretti, G.L. Rossi, *Thermoelastic Modal Stress Analysis*, IMAC XXVI Conference & Exposition on Structural Dynamic, Orlando, Florida USA, 4 – 7 February 2008; ISBN: 0912053984 ISBN 9781605600666

[3] M. Becchetti, R. Flori, R. Marsili, G.L. Rossi *Measurement of stress and strain by a thermocamera*, Proceedings of the SEM Annual Conference, 2009.

[4] M. Becchetti, R. Flori, R. Marsili, G.L. Rossi *Stress and strain measurements by image correlation and thermoelasticity*, Proceedings of the SEM Annual Conference, 2009.

[5] R. Marsili, G. Rossi, E. Speranzini, *Causes of uncertainty in thermoelasticity measurements of structural elements*, Smart Structures Systems, Vol. 20, No. 5 (2017) 539-548, DOI: https://doi.org/10.12989/sss.2017.20.5.539.

[6] A. Garinei, R. Marsili, *Thermoelastic Stress Analysis of the Contact Between a Flat Plate and a Cylinder*, Measurement: Journal of the International Measurement Confederation, 52 (1), pp. 102-110 (2014) http://dx.doi.org/10.1016/j.measurement.2014.03.005.

[7] M. Becchetti, R. Flori, R. Marsili, M. Moretti, *Comparison between digital image correlation and thermoelasticity for strain field analysis* (2010) 9th International Conference on Vibration Measurements by Laser and Noncontact Techniques and Short Course; Ancona; Italy; 22 June 2010 - 25 June 2010 AIP Conference Proceedings, Volume 1253, pp. 233-240 doi=10.1063/1.3455462 ISSN: 0094243X ISBN: 978-073540802-9.

[8] B.K. Horn and B.G. Schunck, *Determining Optical Flow,* Artificial Intelligence, vol. 17, pp. 185-204, 1981.

[9] B. Lucas and T. Kanade, *An Iterative Image Registration Technique with an Application to Stereo Vision*, Proc. DARPA Image Understanding Workshop, pp. 121±130, 1981.

[10] Horst W. Haussecker, David J. Fleet, *Computing Optical Flow with Physical Models of Brightness Variation* IEEE TRANSACTIONS ON PATTERN ANALYSIS AND MACHINE INTELLIGENCE, VOL. 23, NO. 6, JUNE 2001

[11] Oliver Demetz, Michael Stoll, Sebastian Volz, Joachim Weickert, Andrés Bruhn, *Learning Brightness Transfer Functions for the Joint Recovery of Illumination Changes and Optical Flow*, European Conference on Computer Vision ECCV 2014: Computer Vision – ECCV 2014 pp 455-471

[12] Christos Stentoumis ; Aggelos Amditis ; George Karras, *Census-Based Cost on Gradients for Matching under Illumination Differences*, 2015 International Conference on 3D Vision

[13] Fridtjof Stein, *Efficient Computation of Optical Flow Using the Census Transform*, Joint Pattern Recognition Symposium DAGM 2004: Pattern Recognition pp 79-86

Pixel-wise Amplitude Distribution Evaluation in Time Average Digital Holography

Pavel Psota[1], Vít Lédl[1], Ondřej Matoušek[2], Roman Doleček[2], Jan Kredba[2]

[1] Regional Centre for Special Optics and Optoelectronic Systems (TOPTEC), Institute of Plasma Physics, Academy of Sciences of the Czech Republic, Za Slovankou 1782/3, 182 00 Prague 8, Czech Republic

[2] Institute of New Technologies and Applied Informatics (NTI), Faculty of Mechatronics, Informatics and Interdisciplinary Studies, Technical University of Liberec, Studentská 2, 461 17 Liberec 1, Czech Republic

psota@ipp.cas.cz

Abstract. This article introduces a new holographic method for harmonic vibration measurement. The method called time average sweeping digital holography (TASDH) is based on the continuous shift of fringe pattern described by Bessel function over the object surface. The shift is done by phase modulation in one arm of holographic interferometer. Such temporal phase sweeping allows for evaluation of the vibration amplitude value independently in each pixel by application of intensity values cross-correlation of steady state and oscillating object. Main advantage of proposed technique with respect to other methods of digital holographic vibration is its ability to measure the amplitude of vibration without the risk of interference order mismatch. Thanks to above mentioned properties, the method allows measurement of amplitude modes with high slopes or discontinuous surfaces as well as partially shaded objects. The method was experimentally tested by measuring the bending beam cantilever. The standard deviation of the measured noise is below nanometer.

1. Introduction

Measurement of vibrations [1] is very important task in various technical and scientific disciplines. The most commonly used measuring instruments are laser Doppler vibrometers [2], instruments based on correlation analysis [3], ESPI (Electronic Speckle Pattern Interferometry) [4], or holographic interferometry [5]. The methods are either full-field or single-point, they provide dynamic measurement or measurement of modal shapes. Their measurement capabilities are limited in terms of maximum amplitude or frequency range. Therefore, particular attention should be paid to device parameters when selecting one for specific application.

This paper is focused on time average digital holography (TADH) [6]. Time averaging of holograms appears when the frequency of object oscillations is much higher than exposure time of camera. TADH is full-field method for harmonic vibrations measurement. One of the most significant advantages of TADH is unlimited frequency of measured vibration. The intensity map reconstructed from the time averaged digital hologram is modulated by the Bessel function where argument of the function is value of vibration amplitude. Maxima of the function creating Bessel fringe pattern play role of amplitude distribution contours. One of the most important tasks in TADH is to quantitatively determine the amplitude distribution from the Bessel function modulated fringe pattern. Commonly used evaluation methods are based on fringe counting [7], inversion Bessel function [8] or phase shifting (PS). PS approach allows for accurate determination of interference phase where phase modulation by either a mirror mounted on a piezoelectric transducer (PZT) [9] or acoustic-optical modulator [10] is employed. Modulation depth introduced by the phase modulation leads to shift of the Bessel fringe pattern similarly like phase shifts in the case of cosine fringe pattern well known from PS interferometry. The result of PS evaluation method is very accurate phase value but bounded within 2π interval. In order to remove the 2π jumps, the spatial phase unwrapping must follow. Such spatial unwrapping can be very problematic or even impossible in some cases.

Easy use of PS approach is derived from the harmonic cosine nature of interference phenomenon. In TADH the intensity is modulated by Bessel functions and therefore the straightforward phase recovery in the very same way as for cosine fringes is not possible. Possible solution is to exploit nearly periodic nature of the Bessel functions and consider the modulation to be cosine. Such simplification must be compensated by application of correction function derived from the difference between cosine and Bessel approach.

Although PS methods are considered to be the most convenient, there are definitely limitations. On the one hand, the result needs to be corrected due to the difference between cosine and Bessel functions. Further, a spatial unwrapping algorithm is required to demodulate the wrapped phase field and "zero point" (oscillation node) must be determined - to which other pixels are relatively measured. PS method naturally fails in vibrational modes with high slopes that lead to very dense fringe patterns. In addition, the field of view mustn't contain discontinuities and has to be free of shaded places, etc.

In order to address the above mentioned drawbacks, we propose time average sweeping digital holography (TASDH). The proposed method is also based on phase modulation of reference or object waves; however, the modulation depth is now continuously swept over certain range. Using the non-harmonic properties of a zero-order first kind Bessel function, vibration amplitude is determined absolutely in each pixel as a lag of the Bessel function at oscillating state behind the steady state. Therefore, no spatial phase unwrapping or Cosine-Bessel correction is necessary. Compared to PS methods, TASDH needs more acquired data, which is not a problem for today's computers. This paper presents the principle of TASDH, experimental evaluation done on bending beam cantilever and simple noise estimation is performed.

2. Recording and reconstruction of digital holograms

Digital holography (DH) includes the recording and reconstruction of optical waves. Capturing of hologram is done by recording of the superposition of the wave scattered from the surface of the object (object wave) O with the known reference wave R. This captured interference pattern is called a digital hologram and can be described by an interference equation:

$$H = |O+R|^2 = |R|^2 + |O|^2 + O^* R + OR^* . \tag{1}$$

Hologram H contains information about the intensity and phase of the waves and is stored in computer as a matrix of numbers.

The reconstruction process restores information about the amplitude and phase of the digital hologram. The digital hologram H is at first multiplied by a numerical representation of the conjugated reference wave R^*. It results in a complex wave field in the holographic plane (with a coordinate notation ξ , η). This wave field is then numerically propagated in free space according to the laws of diffraction, and the resulting complex array is calculated at a reconstruction distance d - called image plane (coordinates notations x, y). Propagation in the free space can be calculated by Fresnel transform:

$$U(x,y) = \Im^{-1}\left\{ H(\xi,\eta) R^*(\xi,\eta) \exp\left[-\frac{j\pi}{\lambda d}\left(\xi^2+\eta^2\right)\right]\right\}, \tag{2}$$

where $\Im^{-1}$ denotes the inverse discrete Fourier transform and λ is the wavelength of the laser. The both hologram and image plane are sampled by $N \times M$ pixels. Pixel extension in the hologram plane $\Delta\xi \times \Delta\eta$ is naturally defined by the digital sensor parameters, while the image plane pixel dimensions Δx, Δy are given by parameters of the reconstruction:

$$\Delta x = \frac{\lambda d}{N\Delta\xi} \quad and \quad \Delta y = \frac{\lambda d}{M\Delta\eta}. \tag{3}$$

The output of the reconstruction algorithm is a complex wave field with imaginary (*Im*) and real (*Re*) parts in the image plane from which the intensity and phase distributions can be calculated as:

$$I(x,y) = |U(x,y)| \quad and \quad \varphi(x,y) = \tan^{-1}\left(\mathrm{Im}(U(x,y)) / \mathrm{Re}(U(x,y))\right). \tag{4}$$

3. Principle of phase sweeping time average digital holography

Holographic recording of a harmonically oscillating object with an exposure time much longer then the period of measured object vibrations results in time average holography. Let us consider out-of-plane amplitude distribution $d(x,y,t) = D(x,y)\sin(\omega t)$ oscillating in time *t*, where *D* is the amplitude of vibration and ω stands for the angular frequency. Such phase modulation of object wave results in $O(x,y,t) = O(x,y)\exp(j\varphi(x,y)\sin(\omega t))$ and the intensity of the reconstructed image is proportional to magnitude of the first kind zero-order Bessel function J_0:

$$I(x,y) = |J_0(\varphi(x,y))|. \tag{5}$$

The argument of the Bessel function (further called interference phase) in equation (5) is proportional to the out-of-plane amplitude distribution:

$$\varphi(x,y) = D(x,y)4\pi/\lambda. \tag{6}$$

The TASDH requires secondary modulation of phase within the reference *R* or the object *O* beam. The frequency of the secondary modulation has the same frequency ω as the oscillating object and a modulation depth Ω. Considering the object wave including the secondary phase modulation:

$$O(x,y,t) = O(x,y)\exp(j\varphi(x,y)\sin(\omega t))\exp(j\Omega\sin(\omega t)). \tag{7}$$

and substituting (7) into interference equation (1) modifies the equation (5) to be:

$$I(x,y,\Omega) = |J_0(\varphi(x,y) - \Omega)|. \tag{8}$$

Naturally, no secondary modulation $\Omega = 0$ leads to equation (5). As follows from (8) the unambiguously loci of bright zero fringe appears where modulation of the secondary modulation and amplitude of vibration at point *(x,y)* are in agreement:

$$\varphi(x,y) = \Omega. \tag{9}$$

Sweeping of the modulation depth Ω gives an intensity signal for each pixel *(x,y)* of the reconstructed intensity map as a function of the modulation depth (8). The modulation depth generated by the secondary phase modulator is not exactly known since it is defined by transfer function between user controllable electronic device and the real phase modulation of the light beam. Therefore the interference phase φ is computed as the shift between steady-state ($\varphi = 0, \Omega = 0$) with intensity:

$$I(x,y,0) = |J_0(0)| \tag{10}$$

and the oscillating state defined in equation (8). Although there are several strategies how to search for the shift, due to noise resistance cross-correlation formula (operator denoted by $*$):

$$C(\hat{\Omega}) = I_0 * I = \int I_0(\Omega) I(\Omega + \hat{\Omega}) d\Omega \qquad (11)$$

was used. In (11) the lag of the cross-correlation function $\hat{\Omega}$ has the physical meaning of the modulation depth Ω. Lag of the cross-correlation function $C(\hat{\Omega})$ for its maximum indicates where the signals are best aligned and $\hat{\Omega} \approx \varphi$. The interference phase therefore can be computed as:

$$\varphi = \arg\max(I_0 * I). \qquad (12)$$

Hence, the out-of-plane amplitude in each pixel $D(x, y)$ is computed as simple inversion of (6):

$$D(x, y) = \varphi(x, y)\, \lambda/4\pi. \qquad (13)$$

4. Measurement of bending beam cantilever

The schematics of experimental arrangement based on off-axis digital holography is provided in Figure 1. The object under investigation is a bending beam cantilever that is *d=0.7 m* apart from the sensor of AVT Stingray digital camera. The sensor has *N=M=2048* pixels with extensions $\Delta\xi=\Delta\eta=3.45$µm. Light of wavelength $\lambda=532nm$ was emitted by Nd:YAG laser. Superposition of the reference and the object wave (digital hologram) is captured by the digital camera.

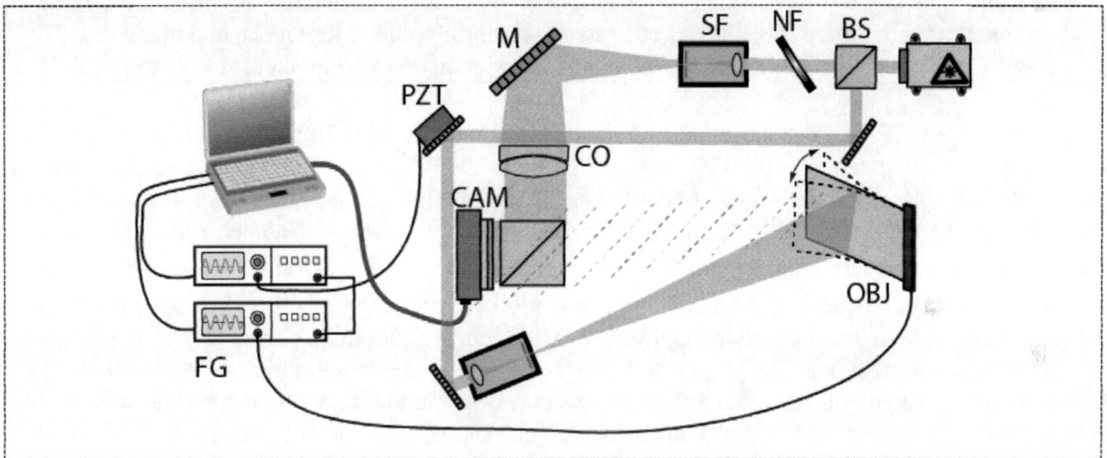

Figure 1. Experimental arrangements for testing of TASDH: BS - beam splitter, NF - neutral density filter, SF - spatial filter, CO - collimating objective, OBJ - object, FG - arbitrary waveform generator, CAM - digital camera, M - mirror, PZT – piezoelectric transducer.

The phase of the object wave is modulated by the object vibrations and also by oscillating mirror mounted on an electronically controlled piezoelectric transducer (PZT). The PZT mirror can also be placed in the reference arm. The object under investigation - bending beam cantilever – and the PZT are driven by a two channel waveform generator. One channel of the waveform generator sets the frequency and amplitude of the harmonic object vibrations while the second channel controls the PZT mirror.

In the first step, the reference state is measured. The object is in steady-state and the PZT is oscillating with frequency of 1000 Hz. The driving amplitude of PZT was swept in the range from -20 V_{PP} (peak-to-peak) to 20 V_{PP} with the step of 1 V_{PP}. The reconstructed intensity at any object point follows equation (8) with $\varphi = 0$. Results of the Bessel function fitting provides the transfer function between the electronic output signal (PZT driven amplitude) from the waveform generator and the phase modulation depth of the light wave. The measured or fitted result can be used as reference for the measurement.

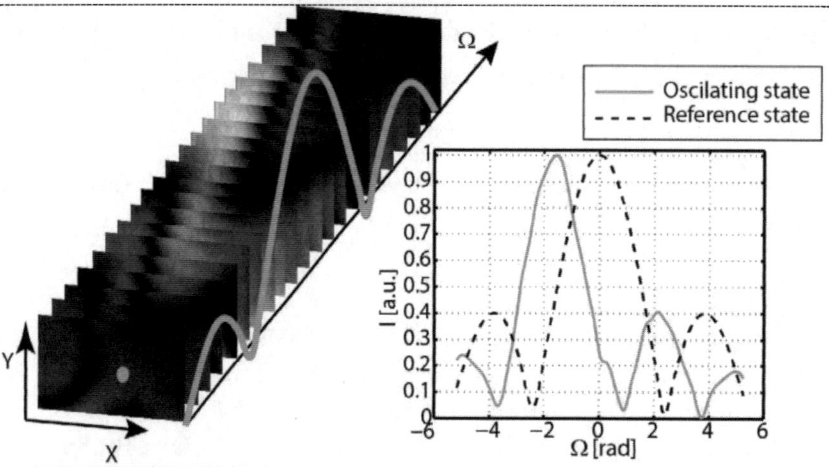

Figure 2. Reconstructed intensity maps with different phase modulation depth Ω creating a stack of images. Intensity profile in each pixel is described by Bessel function that is shifted according to amplitude of vibration with respect to steady state (reference). Reference and measured intensity profiles in pixel denoted by green mark are plotted on the right hand side of the figure.

In the second step we set the object frequency to 1000 Hz with 100 mV_{PP} driving amplitude. Frequency of the PZT remains 1000 Hz and start phase of the PZT is aligned with the object vibration phase. The driving amplitude is again swept from -20 V_{PP} to +20 V_{PP} with the step of 1 V_{PP} leading to sweeping of modulation depth. Maximum of cross-correlation function between the reference and oscillating states indicates the lag corresponding to the interference phase $\hat{\Omega} \approx \varphi$ as discussed in equation (12). This procedure is realized independently in every single pixel and using formula (13) the interference phase is converted to amplitude distribution, see Figure 3. Values along the white line in Figure3a were plotted in Figure 3b and fitted with 8th order polynomial function. Difference between the measured and the fitted values is introduced in Figure 3c. Values represent noise of the measurement since such high spatial frequencies are not supposed to be real movement of the object. Standard deviation of the noise is 0.8 nm.

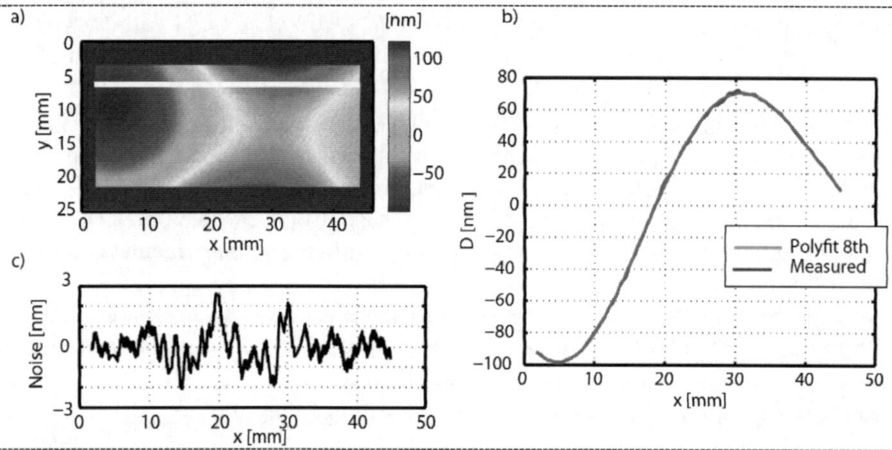

Figure 3. a) out-of-plane amplitude distribution measured by TASDH represented as false colour

IOP Conf. Series: Journal of Physics: Conf. Series **1149** (2018) 012033 doi:10.1088/1742-6596/1149/1/012033

image; b) measured values (blue) of vibration amplitude values along white line and fitted curve using 8th order polynomial (green); c) measured noise as a difference between measured and fitted values in b). Standard deviation of the noise is 0.8 nm.

Conclusion

Time average sweeping digital holography (TASDH) introduced in the paper is a powerful technique for evaluation of time averaged holograms. TASDH is used to quantitative measurement of harmonic vibration amplitudes. The reconstructed intensity image from the time average digital hologram is modulated by the Bessel function that is unambiguous function. For phase shifting techniques such disadvantage must be corrected while it is a key property for TASDH. The principle of TASDH is based on phase modulation of object or reference wave of a holographic interferometer. As the depth of phase modulation changes, the interference pattern continuously shifts over the surface of the object and due to the unambiguous behavior of the Bessel function, the lag of steady and oscillating state can be found in each pixel. The interference phase value is uniquely known and can therefore be easily converted into amplitude value. This is done independently in each pixel, so there is no need for spatial unwrapping. A beneficial feature of TASDH over other quantification methods based on TADH is the ability to measure vibration amplitudes without the risk of interference order errors. For this reason, the method can provide amplitude distribution of amplitudes with high slopes, discontinuities or in case of partially shaded objects. The correctness of the method was experimentally verified by measurement of bending beam cantilever where standard deviation of the measurement noise reached only 0.8 nm.

Acknowledgement

This work was supported by Czech Science Foundation Project No.: GACR GA 16-11965S.

References

[1] McConnell K G and Varoto P S 1995 *Vibration testing: theory and practice*. John Wiley & Sons
[2] Lutzmann P, Göhler B, Hill Ch A and van Putten F D N 2016 *Opt. Eng.* **56** 031215
[3] Butters J N and Leendertz J A 1971*Measurement and Control* **4** 12.
[4] Macovski A Ramsey S D and Schaefer L F 1971 *Applied Optics* **10** 12
[5] Kreis T 2005 *Handbook of holographic interferometry: optical and digital methods* Wiley-VCH. xii.
[6] Powell R L and Stetson K A 1965 *Journal of the Optical Society of America* **55** 12
[7] Joud F, Laloë F, Atlan M, Hare J and Gross M 2009 *Optics Express* **17** 4
[8] Borza D N 2005 *Measurement Science and Technology* **16** 9.
[9] Stetson K A and Brohinsky W R 1988 *J. Opt. Soc. Am. A* **5** *9*
[10] Psota P, Lédl V, Doleček R, Mokrý P, Vojtíšek P, Václavík J 2016 Opt. Eng. **55** 121726

IOP Conf. Series: Journal of Physics: Conf. Series **1149** (2018) 012034 doi:10.1088/1742-6596/1149/1/012034

A high-accuracy image-based crack displacement sensor

Xinxing Chen[1,*], Jiannan Xiang[2], Chaobo Zhang[1], Chih-Chen Chang[1], Ming Liu[2]

[1]Department of Civil and Environmental Engineering, Hong Kong University of Science and Technology, Clear Water Bay, Kowloon, Hong Kong
[2]Department of Electronic and Computer Engineering, Hong Kong University of Science and Technology, Clear Water Bay, Kowloon, Hong Kong

Corresponding author: xchenak@connect.ust.hk

Abstract. In this study, a new image-based sensor with high accuracy was developed. This wireless sensor is composed by optical sensor, micro-processor, wireless modules and rechargeable battery. Highly sensitive digital sampling moiré (DSM) method is used to calculate displacements. Its high accuracy was demonstrated by a numerical simulation, and a laboratory test was conducted to verify its applicability in crack monitoring.

1. Introduction

Cracks are commonly observed on concrete structures and the measurement of crack propagation is crucial for the health monitoring of structures [1]. A variety of sensors have been developed to monitor the propagations of cracks, such as the crack-meter, linear variable differential transformer (LVDT) and optical fiber [2, 3]. Crack-meters with resolution of 10 μm were installed on the oval drum-dome system for the crack monitoring for 10 years [2]. LVDTs and optical fibers were used to monitor the crack's propagation on a masonry walls during the settlement [3]. However, these sensors can only measure one-dimensional displacement, while the cracks may propagate in multiple directions [4]. To achieve multi-dimensional monitoring of cracks, an image-based crack propagation sensor was developed by Man and Chang in 2016 [5]. This sensor was installed across the crack and captured images of the pattern attached on the other side of crack. Based on the images of the pattern captured between motion, this sensor can calculate the 2D displacement of cracks. The algorithm adopted by the sensor for the displacement calculation is the normalized cross correlation (NCC) method, which can achieve the accuracy of 13.2 μm in the laboratory test [5].

Moiré method is widely used for rotational and translational displacement measurement [6]. The traditional moiré technique requires the overlapping of two gratings to generate the moiré pattern for motion amplification. In 2010, Ri. et al developed the digital sampling moiré (DSM) method [7], which can generate moiré patterns from the images of single grating. It can achieve the accuracy of 1/500 of grating pitch length in the bending test. In this study, a new crack propagation sensor is proposed to use the DSM method to achieve high-accuracy 2D displacement measurement. Specifically, the DSM method is compared with the NCC method at similar computational cost via the numerical simulation. The result shows that the DSM method can achieve higher accuracy than the NCC method. Moreover, the proposed DSM-based sensor is applied for crack monitoring in a corrosion test to verify its applicability for real crack measurement.

2. The proposed sensor

Fig. 1(a) shows the schematics of the proposed sensor. As can be seen, the sensor is fixed at one side of the crack and taking images of the pattern attached on the other side. In this way, the motion of the crack is identical with the relative motion between the fixed support of the sensor and the pattern. Therefore, the crack's motion can be calculated from the captured image of the pattern. Fig. 1(b) illustrates the functionality of four principal components inside the sensor. The optical sensor (ADNS 3080) captures the raw images and sends them to the processor (Arduino UNO) for displacement calculation. The algorithm built inside the processor calculates the translations along *X*- and *Y*-directions, which are presented as T_X and T_Y, respectively. Then, the obtained T_X and T_Y are sent to XBee for wireless transmission. All these components are powered by a battery connected to Arduino UNO.

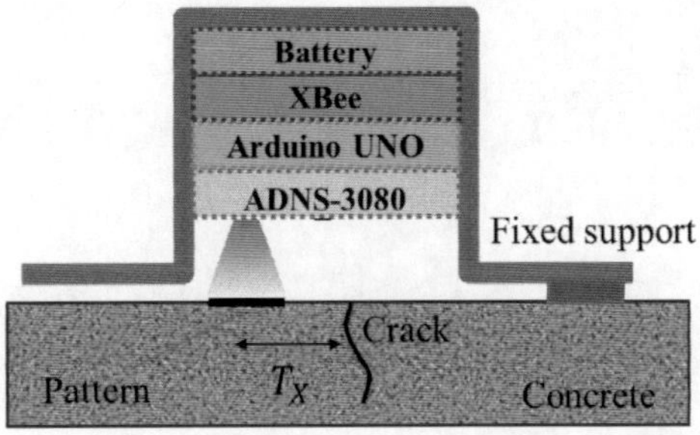

(a)

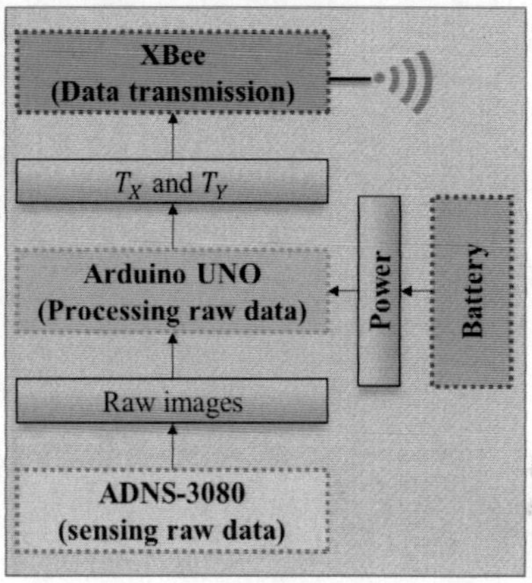

(b)

Fig. 1 (a) The alignment of the sensor; (b) the functions of principal components

The proposed sensor adopts the DSM method [7] as the algorithm for displacement measurement. Fig. 2 shows the enlarged image of the 2D grating pattern used for the DSM method. The sampling pitch is the integer value of pitch length on the image plane. The sampling pitch of the *X*- and *Y*-directional gratings are s_x and s_y, respectively.

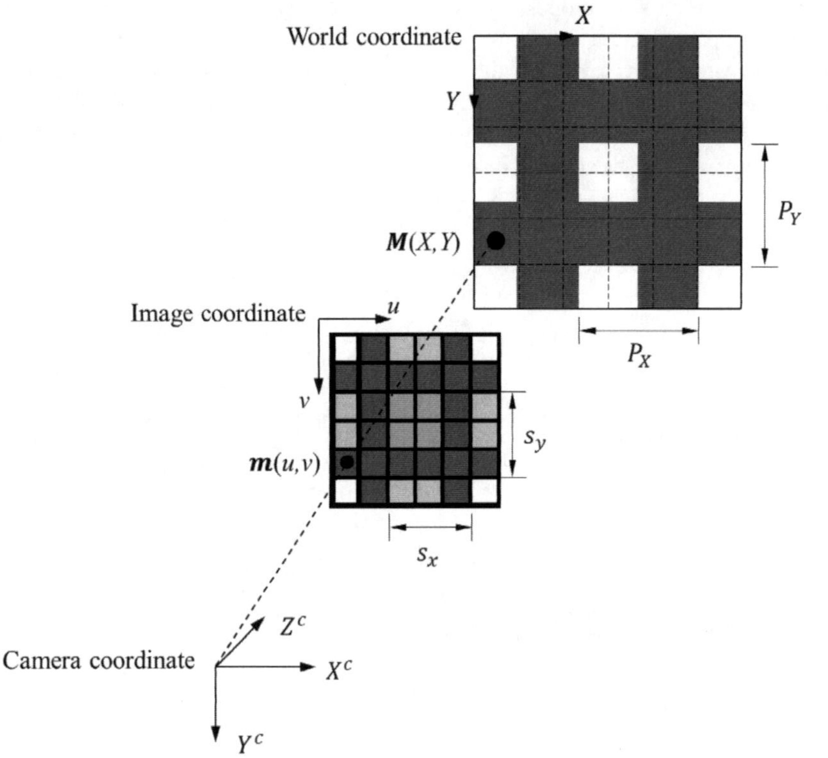

Fig. 2 The projection of 2D grating pattern on the image sensor

After the recording of 2D grating pattern, the grating image should be separated into individual dimensions for further processing with the DSM method. As shown in Fig. 3, the 2D grating pattern is decomposed into *X*- and *Y*-directional gratings using image averaging filters. The filtered *X*- and *Y*-directional gratings were shown in Fig. 3(b) and (c), respectively. To demonstrate the procedures of the DSM method, one line of the filtered *X*-directional grating was extracted and processed by the DSM method to calculate its moiré fringes and phase distribution.

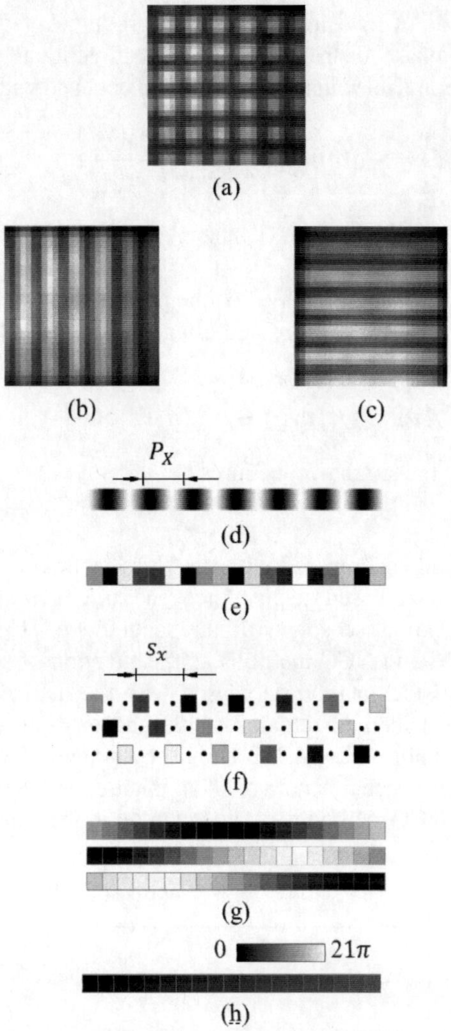

Fig. 3 Separation of 2D grating and the calculation processes of the DSM method

Fig. 3(d) shows a line of filtered grating in physical domain and its intensity distribution in image domain is shown in Fig. 3(e). Totally s_x moiré patterns are generated through down-sampling and up-sampling (s_x = 3). Firstly, down-sampling was performed by picking up one pixel in every s_x pixels. As shown in Fig. 3(f), the 1st down-sampled frame pick the 1st point from the *x*-directional grating, and then it picks one pixel in every s_x pixels. Similarly, the 2nd and 3rd down-sampled frames pick the 2nd and 3rd points and generate corresponding down-sampled grating lines. Interpolation was then performed to calculate the light intensity of pixels between the down-sampled points. After interpolation, three phase shifted moiré fringes are shown in Fig. 3(g). Finally, the phase distribution of the moiré fringe can be obtained, as shown in Fig. 3(h). Eq. (1) and (2) are the light intensity distributions of the *x*-directional grating and the *k*th moiré fringes, respectively.

$$I_x(u,v) = L_a(u,v)\cos[\varphi_x(u,v)] + L_b(u,v) \quad (1)$$

$$I_m(u,v;k) = L_a(u,v)\cos\left[\theta_x(u,v) + \frac{2\pi(k-1)}{s_x}\right] + L_b(u,v) \quad (k=1,\cdots,s_x) \quad (2)$$

where $L_a(u,v)$ is the amplitude of light intensity of the grating; $L_b(u,v)$ is the light intensity of the background; $\varphi_x(u,v)$ is the phase distribution of the x-directional grating; $\theta_x(u,v)$ is the phase distribution of the moiré fringe, and its wrapped value can be obtained with Eq. (3):

$$\theta_{x\_w}(u,v) = -atan\frac{\sum_{k=1}^{s_x} I_m(u,v;k)\times\sin\left(\frac{2\pi k}{s_x}\right)}{\sum_{k=1}^{s_x} I_m(u,v;k)\times\cos\left(\frac{2\pi k}{s_x}\right)} \tag{3}$$

The wrapped phase value θ_{x_w} is discontinuous and it ranges between -π and π. Unwrapping should be performed to get θ_x.

The DSM method calculates the translation based on the phase difference. $\theta_x(u,v)$ and $\theta'_x(u,v)$ are the phase distributions of the moiré fringes before and after motion. Translation along X-direction T_X can be obtained from Eq. (4):

$$T_X(u,v) = [\theta_x(u,v) - \theta'_x(u,v)] \times \frac{P_X}{2\pi} \tag{4}$$

T_Y can be calculated following the similar procedures.

3. Simulation

In this section, the DSM method is compared with the NCC method at similar computational cost via numerical simulation. As an image-based displacement measurement algorithm, the NCC method has been used in a crack propagation sensor with an accuracy of 13.2 μm [5]. Fig. 4 shows the computational cost of the DSM and NCC methods as the function of searching step. As can be seen, the computational cost of the NCC method decreased with the increase of searching step, while the computational cost of the DSM method remains stable and is slightly larger than that of the NCC method with searching step equal to 0.2 pixel. It should be noticed that the NCC method with higher computational cost could achieve higher accuracy. To compare the DSM and NCC method in a critical way, the searching step of the NCC method is set as 0.1 pixel in the simulation.

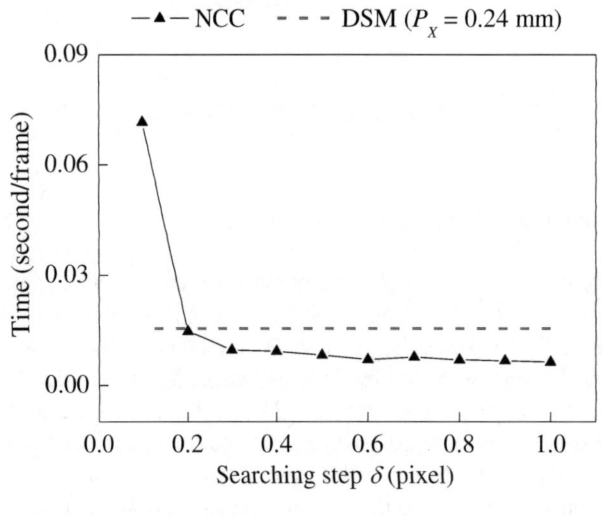

Fig. 4 Computational cost of the NCC and DSM methods

The NCC method was firstly calibrated to obtain the scaling factor using the method stated by Man and Chang (2016). As shown in Fig. 5, the pattern was simulated to do step motion and its displacements were calculated by the NCC method. A line was used to fit the step motion results, and the obtained scale factor is 16.5 pixel/mm when P_X equal to 0.24mm. Then, the pattern was simulated to do harmonic motion, and the displacements measured by the NCC and DSM methods are shown in

IOP Conf. Series: Journal of Physics: Conf. Series **1149** (2018) 012034 doi:10.1088/1742-6596/1149/1/012034

Fig. 6. It demonstrates that both two methods track 2D motion successfully. Fig. 7 shows the corresponding errors of two methods in the harmonic motion detection. As can be seen, the errors of the NCC method are significantly larger than those of the DSM method. Therefore, it can be concluded that the DSM method is of higher accuracy comparing with the NCC method.

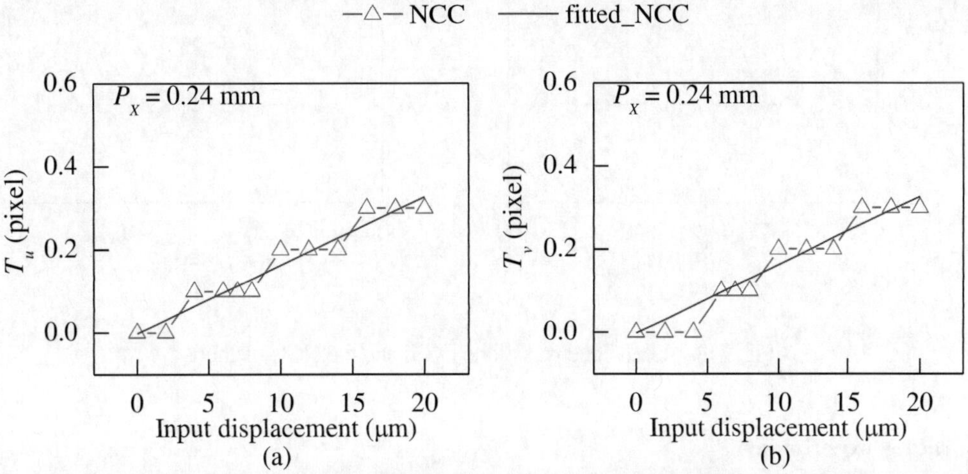

Fig. 5 Calibration for the NCC method

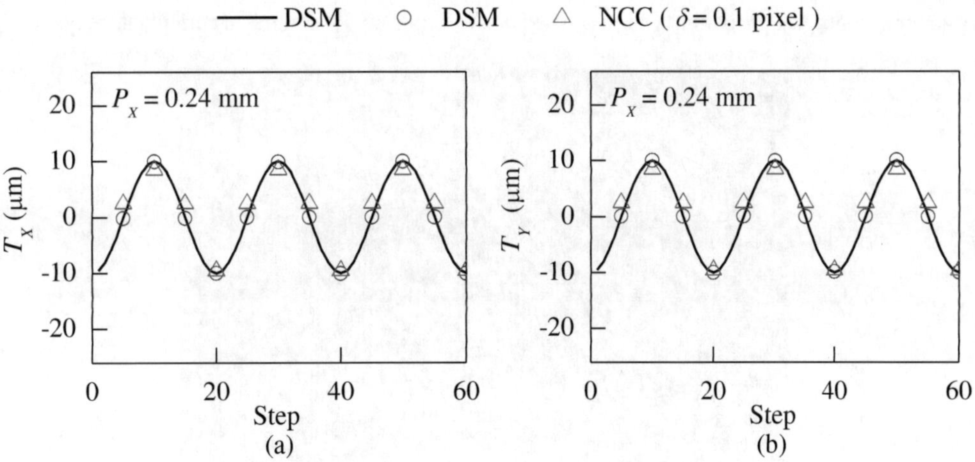

Fig. 6 Displacement results of the DSM and the NCC methods

IOP Conf. Series: Journal of Physics: Conf. Series **1149** (2018) 012034 doi:10.1088/1742-6596/1149/1/012034

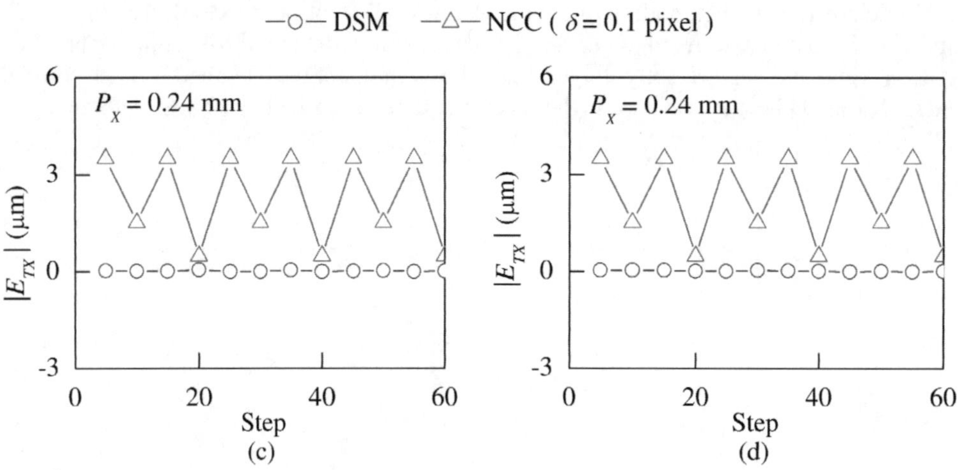

Fig. 7 Displacement errors of the DSM and the NCC methods

4. Laboratory experiment

Corrosion of steel bars is the major cause of concrete structures' failure. After the initiation of corrosion in steel, products of corrosion expand usually result in the formation of crack on the concrete [8].

As shown in Fig. 8, an accelerated corrosion test was carried out on a rectangular concreter sample with an embedded steel bar. To accelerate corrosion, specimens were partially immersed in sodium-chloride solution in tanks. Meanwhile, direct electric current was impressed on the steel bar with controlled current intensity.

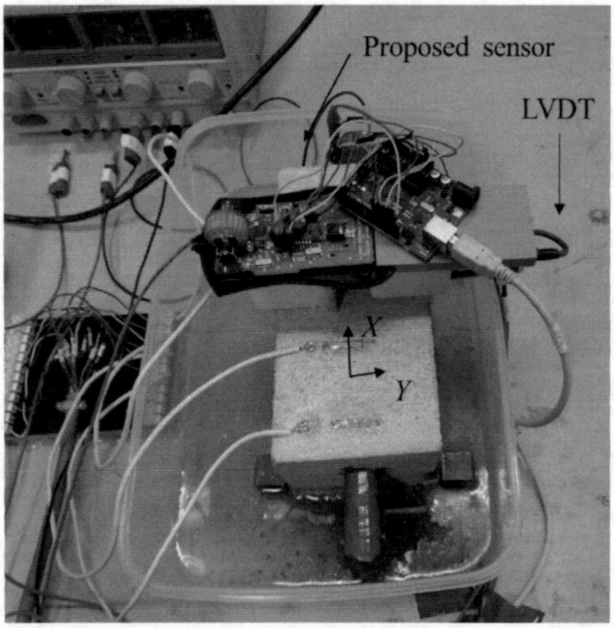

Fig. 8 The proposed sensor on the concrete cubic

The test was performed for 14 days. With the expanded corrosion of the embedded bar, crack was observed on the concrete specimen. The proposed sensor was installed across the crack to monitor its displacement. A LVDT with a resolution of 5 μm was installed under the proposed sensor to provide ground truth.

The measured displacements of the crack were shown in Fig. 9. The proposed sensor can track 2D displacements of the crack simultaneously, while the LVDT provided the single dimensional measurement along Y direction. The results show that the Y-directional displacement measured by the LVDT matches well with that measured by the proposed sensor. This test demonstrates the applicability of the proposed sensor in the long-term monitoring of the 2D propagation of cracks.

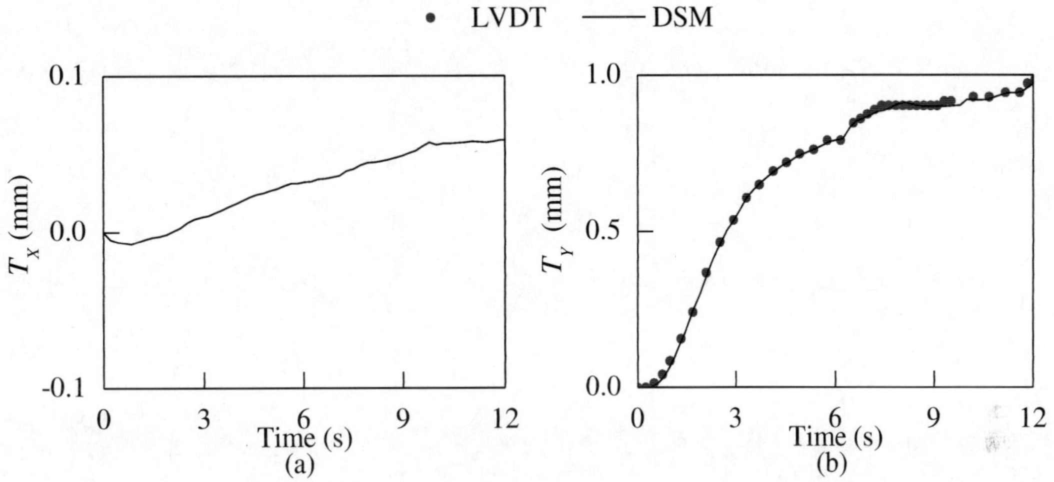

Fig. 9 Crack displacement monitoring result

5. Conclusion

In this study, a new wireless sensor has been proposed for 2D displacement monitoring of cracks. Comparing with the previous sensor using the NCC method, this proposed sensor adopts the DSM method as algorithm and can achieve higher accuracy. Numerical simulation and laboratory test have been performed to show its accuracy and applicability. The results demonstrate that this sensor can be used as a powerful tool for long-term monitoring of crack propagations.

References

[1] Y. Liu, "Modeling the Time-to-Corrosion Cracking of the Cover Concrete in Chloride Contaminated Reinforced Concrete Structures," Modeling the time-to corrosion cracking of the cover concrete in chloride contaminated reinforced concrete structures (1996).

[2] R. Ceravolo, A. De Marinis, M. L. Pecorelli, and L. Zanotti Fragonara, "Monitoring of masonry historical constructions: 10 years of static monitoring of the world's largest oval dome," Structural Control and Health Monitoring 24, e1988 (2017).

[3] E. Verstrynge, K. De Wilder, A. Drougkas, E. Voet, K. Van Balen, and M. Wevers, "Crack monitoring in historical masonry with distributed strain and acoustic emission sensing techniques," Constr. Build. Mater. 162, 898-907 (2018).

[4] G. Sih and B. Barthelemy, "Mixed mode fatigue crack growth predictions," Eng. Fract. Mech. 13, 439-451 (1980).

[5] S. H. Man and C. Chang, "Design and performance tests of a LED-based two-dimensional wireless crack propagation sensor," Structural Control and Health Monitoring 23, 668-683 (2016).

[6] D. Post, B. Han, and P. Ifju, *High sensitivity moiré: experimental analysis for mechanics and*

materials (Springer Science & Business Media, 2012).
[7] S. Ri, M. Fujigaki, and Y. Morimoto, "Sampling moiré method for accurate small deformation distribution measurement," Exp. Mech. 50, 501-508 (2010).
[8] S. K. Verma, S. S. Bhadauria, and S. Akhtar, "Monitoring corrosion of steel bars in reinforced concrete structures," ScientificWorldJournal 2014, 957904 (2014).